Superlattice to Nanoelectronics

To dear Mohamed

I am grateful to you for your participation to our workshop May 5-7, 2013

Enclosed is my book as a token of thanks

Ray

This page intentionally left blank

Superlattice to Nanoelectronics

Second Edition

Raphael Tsu

ELSEVIER

AMSTERDAM • BOSTON • HEIDELBERG • LONDON • NEW YORK • OXFORD
PARIS • SAN DIEGO • SAN FRANCISCO • SINGAPORE • SYDNEY • TOKYO

Elsevier
32 Jamestown Road London NW1 7BY
30 Corporate Drive, Suite 400, Burlington, MA 01803, USA

Second edition 2011

Notices
Knowledge and best practice in this field are constantly changing. As new research and experience broaden our understanding, changes in research methods, professional practices, or medical treatment may become necessary.

Practitioners and researchers must always rely on their own experience and knowledge in evaluating and using any information, methods, compounds, or experiments described herein. In using such information or methods they should be mindful of their own safety and the safety of others, including parties for whom they have a professional responsibility.

To the fullest extent of the law, neither the Publisher nor the authors, contributors, or editors, assume any liability for any injury and/or damage to persons or property as a matter of products liability, negligence or otherwise, or from any use or operation of any methods, products, instructions, or ideas contained in the material herein.

British Library Cataloguing-in-Publication Data
A catalogue record for this book is available from the British Library

Library of Congress Cataloging-in-Publication Data
A catalog record for this book is available from the Library of Congress

ISBN: 978-0-08-096813-1

For information on all Elsevier publications
visit our website at www.elsevierdirect.com

This book has been manufactured using Print On Demand technology. Each copy is produced to order and is limited to black ink. The online version of this book will show color figures where appropriate.

Working together to grow
libraries in developing countries

www.elsevier.com | www.bookaid.org | www.sabre.org

ELSEVIER   BOOK AID International   Sabre Foundation

Table of Contents

This page intentionally left blank

Preface

Survival of the Fittest Applied to Biological Species Is Overshadowed by Survival of the Fittest Applied to Technology

When I was leaving China, I was at the deathbed of my great-uncle who was a pioneer in heavy industry in Shanghai. He told me that I should remember and be guided by the old Chinese proverb, "To succeed in a given task, must take advantage of the tools." The success story of the man-made superlattices is a good example of this old saying. The basic tools for the development were in place, including vacuum technology, growth facilities, and above all, diagnostic tools such as XRD, HRTEM, Raman, C-V and DLTS. Something new entered the list of new tools, STM and AFM. These new capabilities of looking into the atomic arrangements help to expand the role of superlattices and quantum wells into nanoelectronics, and eventually into nanoscience in general.

When I was young, I looked up at the sky and marveled at the mystery of the universe, thinking that I would understand more if I had taken more courses in advanced physics. At The Ohio State University, I got together five others and convinced Roger Mills to give a course on advanced quantum field theory. One day I was early for the class and found him eating his lunch. I said, "Professor Mills, I thought if I had taken the most advanced course in physics, I would have a good idea on the mystery of the universe. Now I found out that nobody knows it, including Einstein." He replied, "Now you know!"

I left BTL to teach physics at Trinity University for a couple of years, including a course on general relativity. At least I know now that basic understanding comes and goes, but technology and human achievements always advance. I would modify Darwin's survival of the fittest for biological species to include "survival of the fittest in technology." *Today, I do not think it is too meaningful to ask the origin of the universe, rather, the origin of living things.*

The man-made superlattice was launched to utilize the phenomena belonging to the world measured in nanometers arranged in such a way that the entity serves the macroscopic world in which we live. With structures smaller than the phase-coherent length of electrons in solids, we need instruments capable of characterizing in the nanometer regime. To be sufficiently large for the purpose of interfacing with human activities and power handling, we come up with repeating the structure in a periodic way similar to nature's formation of solids to mimic a continuum with elementary particles to molecules. Today, nanocomposites have joined the rank of nanoelectronics and photonics that Esaki and I envisioned 40 years ago.

At first, IBM management gave us token backing, but the U.S. Army Research Office soon furnished us with partial support. After the first experimental verification of NDC in a GaAs/GaAlAs superlattice, Gunn pointed out that the observed NDC may be due to domain oscillation, a non-quantum mechanical effect. To eliminate this possibility, 2 years later we introduced the resonant tunneling diode (RTD), a "superlattice" consisting of a single quantum well. Ten years after terahertz RTDs were realized and the age of man-made quantum structures had been universally recognized. In 1996, Robert Lontz, who directed the Physics-Electronic Division of the U.S. Army Research Office during the initial 5 years of the superlattice development, was asked by Mikael Ciften, head of the physics division at ARO during 1996—1997, to put together a document, "The Superlattice Story." This document was distributed to many agencies, including the White House, to promote awareness and to stimulate the direction of research in electronics in the general areas known as nanoscience.

Last, I want to give some indication of what sort of technical background is required of the reader. Naturally, one needs some working knowledge of basic mathematics such as complex variables and partial differential equations, as well as some skill in computer programming. Intermediate to advanced courses in electromagnetics, quantum mechanics, solid-state, and semiconductor physics are necessary to understand and appreciate many concepts presented in this book. However, this book requires the reader to have a generally rigorous mindset rather than being overly concerned with detail. I used parts of this book in my course on semiconductors.

Apart from a few corrections to typos and misprints in the first edition, a few sections are repackaged in this second edition e-book to include some new materials.

Semiconductor atomic superlattice represented by the silicon/oxygen is really a new type of superlattice somewhat similar to polymers, which are represented by two heterostructures forming a unit consisting of a well and a barrier.

The section dealing with the capacitance of a quantum dot includes new results dealing with the role of symmetry applying to systems of few electrons.

A new section on the phonon laser is added.

The concept of conductance and resistance defined in lump networks applies to waves only at the junctions of waveguides, terminations of antennas, etc. For this reason, the section on electron wave impedance has been expanded.

A concluding section was added to emphasize the role of symmetry. For example, chemistry deals with symmetry between atoms and molecules, while nanochemistry must include the symmetry relationship of a boundary between a group of atoms and molecules.

In a way, nanoscience and nanotechnology are common to many disciplines, breaking down the barriers between physics, chemistry, engineering, and biology.

Recently I was shown the FOXCONN Laboratory of Research in Tsinghua University, Beijing, and it reminded me of the great BTL when I started my career. I was sad because such a great institution was basically dissolved by the action of the U.S. Supreme Court, as part of the divestiture of AT&T.

I thank Jennifer and Jamie Stewart for helping me in the preparation of the first edition of the book. Leo Esaki has always been my friend and my teacher. He believed in me. When he took me on, he told me that he had checked with a few people at Bell Laboratories about me and what prompted him to hire me was a remark that "Ray likes to do theory, but he also likes experiments. But do not get him involved with too complicated experiments." He confided in me that he himself did not like very complicated experiments.

I dedicate this book to my wife Danusia and to Leo Esaki.

This page intentionally left blank

Introduction

This book follows the development of superlattices and quantum wells from their inception in 1969. My role paused after 1977, and I re-entered the field in 1984 after devoting 7 years to laser annealing and amorphous silicon. During the initial period, I was involved with most phases, beginning with the creation of man-made superlattices with Esaki, later joined by Chang and followed by efforts to characterize the materials, structures, and properties, and branching into resonant tunneling, optical response, Raman scattering, and type-III superlattices. The theoretical results that helped to launch the endeavor came from a few things we learned from a published book by Pippard (1965).

The reason we became involved in Raman scattering was that we used it to characterize alloy compositions and structures. Besides, I have always had a fascination with Raman scattering, partially because it was a personal challenge to master a subject which I considered to be very complex, one that needed a working knowledge of group theory. It was not apparent when we were tackling the theory of resonant tunneling that we needed to introduce a new method of tunneling calculation. This is mainly because methods of calculating tunneling—such as the Bardeen's transfer Hamiltonian and the Wentzel−Kramers−Brillouin (WKB) approximation, subjects covered in detail by Duke—simply could not work. In retrospect, it was the use of a computer that provided us with a handle on resonant tunneling through a double-barrier structure.

In 1984, I went to Brazil for an extended stay. Ioriatti and I started calculating the dielectric constant and doping of a superlattice. During this period, I reopened some old data and computations using a complex k-vector, which were replaced by a formulation with Green's function with a finite self-energy function, usually denoted by $\sum$, with $\sum \neq 0$. The usual way of summing the diagrams is nothing more than a perturbation calculation, only that perturbation may be carried to infinite order. My view was that let us tackle the "holy cow" instead, by assuming the Hamiltonian of an open system to be non-Hermitian, allowing a simple way of accounting for losses and their effects on the resonant state. Basically, we gave up pretending that there is an eigenstate. The Im G gives us the local spectral density and that is all we need. In fact, Sir Mott took a fairly receptive view toward my contention, after I pointed out to him that some parameters are directly measurable, such as the mobility or the scattering time. This simple theory does not single out the details of scattering such as those from impurities, phonons, and deformations. Nevertheless, they are not parameters that fit; rather, they represent a quantity such as the mobility of electrons in the quantum well.

After I joined the University of North Carolina at Charlotte in 1988, Zypman became my first postdoctoral fellow. We calculated the surface states of GaP on GaAs without fitting parameters, and the results we obtained are applicable to the Si–O superlattice. Babic and I started a series of calculations on capacitance, doping, and exciton of a quantum dot. My interest clearly moved into quantum dots, a subject very much in fashion today. The main finding was rather surprising. It is obvious that the dielectric screening should go down because as the electron becomes more and more confined, it cannot even move, so how can it screen? During this initial phase of the theory, I called Ioriatti on the phone and asked him point-blank whether the Kramers–Kroenig relation (the "holy cow") should be taken as gospel. He told me that the K–K relation applies when $\varepsilon(r,\ r';\ t,\ t') = \varepsilon(r - r';\ t - t')$; quantum dots do not satisfy this assumption. I invited him to spend 3 months with me because I needed someone who was not bound by tradition. My original paper on the size-dependent dielectric constant was delayed for 3 years before it finally appeared in print. This story has a happy ending because today size-dependent dielectric function, perhaps more appropriately called size-dependent response function, is well accepted.

A new type of superlattice, semiconductor atomic superlattice, SAS, presented in Chapter 6, was introduced in mid-1990s as an attempt to breakdown the indirect nature of the silicon band structure. By replacing an entire plane of Si with atomic oxygen, photoluminescence and electroluminescence at ~2.25 eV were experimentally measured. The origin of light emission was found to be local, rather than quantum size effects. This work was discontinued after losing ARO support. In retrospect, there is no reason to expect that all superlattice structures should exhibit size effects, and the discontinuation of this work was due to the lack of appreciation for something new.

Chapter 7 deals with silicon quantum dots fabricated from annealing amorphous silicon. I want to mention something about the work done by the postdoctoral fellow Quiyi Ye who came to us from F. Koch in Germany. It was because of her willingness "secretly" to disagree with us that progress was made at the beginning. Very strange $I–V$ led to a few years of additional effort by Nicollian and me. After he passed away, I lost the staying power to continue the study on such complicated data. In retrospect, we would have been better off focusing our energy onto the study of quantum dots with InAs on GaAs. There is something to be learned from this experience. If one considers himself as alert and hard working, generally speaking, it is time to make a change after 2 years.

Chapter 8 contains additional discussions regarding capacitance and doping. The traditional dielectric function is global; however, the size-dependent dielectric function, like any response function, is local. We learn in elementary electricity and magnetism that the capacitance of a sphere may be simply calculated with two lines using the Gauss law. This is so if we assume electronic charges may be infinitesimally divisible. A purely electrostatic calculation with discrete electronic charges is quite involved. A new section is added in this introduction dealing with a model where N electrons are confined inside a dielectric sphere, with all the

Coulomb interaction between the electrons as well as all the induced interaction energy of the multipoles. Results are striking, basically resembling the periodic table of the chemical elements because we chose a sphere for the N electrons. The spherical harmonics for the spherical geometry led to results similar to the atomic systems having the s, p, d states. If we were to choose a tetrahedron, results should be quite different. We were trying to extend out previous calculation for two electrons to three electrons using the difference in energy for the He- and H-like cases. T.J. LaFave graduated under me with the classical computation without completing the quantum mechanical computations partially because of the complexity of using the Schrodinger equation instead of the La place equation. However, the main finding is clear: capacitance can only be defined by mono-phasic system. Each time an extra electron is introduced, the change of symmetry leads to a new phase.

I want to mention something about the work done on silicon quantum dots with Nicollian and our postdoctoral fellow Quiyi Ye. Ye who came to us from F. Koch in Germany. It was because of her willingness to "secretly" disagree with us that progress was made at the beginning. When we first obtained results in 1990 on conductance steps, they were numerous and had variable steps instead of equally spaced jumps in conductance. We were having a hard time in getting the work published. Today, I can state with some certainty that most of our (or my) work that had problems being accepted for publication at that time turned out to be well cited later. I have been praised by the reviewer on several occasions upon acceptance, but the work turned out to be poorly cited by others. So again, I take this opportunity to urge you not to despair when your manuscript is brutally rejected.

I was one of the first to get involved when visible light emission in porous silicon was reported. I was able to get started quickly while working mainly with J. Harvey at the U.S. Army Research Laboratory (called ETDL in 1990), because there were plenty of human and material resources. We have determined experimentally that quantum confinement at least contributed to pushing up the emission, and the refractive index reduction is greater than could be explained by porosity alone. In this book, I have clearly accepted the dual role of quantum confinement and surface complexes. It is a good example of why well-qualified groups disagree and usually both are right.

In this second edition, Chapter 10 on novel devices has been enlarged to include recent work on field emission using electrons stored in the surface region, created by surface field, or an additional DBQW on the surface. Actual measured field emission as a cold cathode shows more than an order of magnitude lower in the electric field required. A new section on THz phonon laser as well as quantum cascade lasers has been added together with discussion dealing with the recent feverish activity with graphene and graphene-related activities.

Quantum conductance, explained in terms of an input conductance to a quantum system such as a quantum wire with a wave conductance, in my view, has a more general role than contact conductance. The discussion centered on the difference between an open and closed system has deeper conceptual meaning. It may become quite important if phase-coherent electronic devices become more widely used in the future.

The title of Chapter 12 has been changed to "Why Super and Why Nano?" As size of devices approaches nanometer regime, even superlattices are defined within a new domain, global features are replaced by local features. The most fundamental consequence is the loss of traditional translational symmetry in solids and simple point symmetry for atoms and molecules. Superlattice is giving way to lumped entities similar to the QCL with three components, the injector, the optical transition, and the collector. What is obvious now is the fact that we need to re-orient ourselves in the traditional thinking that multiple interactions can always be treated by some statistical average. What happens in each component having only a few interactions due to defects? Obviously, statistical averaging can only be defined in terms of an assemble average. However, the main question is that how do we ensure redundancy and robustness?

Forty years ago I was able to read a book quickly and retain a good part of what I read. Unfortunately I cannot quite do the same now. I do urge younger researchers to take every advantage of being young, energetic, and having an inquisitive mind. The old song, "Anything You Can Do I Can Do Better," should be a motto for us all. Last, every book has some good passages, but we must realize that we cannot find all the answers in one book. The thing to do is to read one and then another, until you find one that has some of the details you are looking for. Then, spend all the time necessary to absorb it all. Even though I have taken care to eliminate any mistakes in this book, I apologize for any that do remain. Below is a partial list of the works that I found useful in preparing this book.

Good luck and best wishes.

Raphael Tsu, Charlotte NC 2010

(1) Textbooks

Bottcher, C.J.F., 1973. second ed. Theory of Electric Polarization, vols. 1 and 2. Elsevier, Amsterdam.

Harrison, W.A., 1970. Solid State Theory, McGraw-Hill, New York.

Heitler, W., 1954. Quantum Theory of Radiation, Oxford University Press, Oxford.

Herzberg, G., 1944. In: Spinks, J.W.T. (Ed.), Atomic Spectra and Atomic Structure. Dover Publications, New York.

Kittel, C., 1958. Elementary Statistical Physics, Wiley, New York.

Kittel, C., 1973. Introduction to Solid State Physics, third ed. Wiley, New York.

Mahan, G.D., 2000. Many-Particle Physics, third ed. Kluwer Academic/Plenum Publications, New York.

Merzbacher, E., 1961. Quantum Mechanics, Wiley, New York.

Morse, P., Feshbach, H., 1953. Methods of Theoretical Physics, McGraw-Hill, New York.

Moss, T.S. (Ed.), 1980. Handbook on Semiconductors, vol. 2. North-Holland, Amsterdam.

Schiff, L., 1955. Quantum Mechanics, McGraw-Hill, New York.

Smith, R.A., 1961. Wave Mechanics of Crystalline Solids, Wiley, New York.

Ziman, J.M., 1988. Principle of Solids, second ed. Cambridge University Press, Cambridge.

(2) Topical books

Allan, G., Bastard, G., Boccara, N., Lannoo, M., Voos, M. (Eds.), 1986. Heterojunctions and Semiconductor Superlattices. Springer, Berlin.

Capasso, F. (Ed.), 1989. Physics of Semiconductor Devices. Springer, Berlin.

Carslaw, H.S., Jaeger, J.C., 1959. Conduction of Heat in Solids, Oxford University Press, Oxford.

Cathay, M., 1997. In Quantum Confinement: Nanoscale Materials, Devices and System, ECS Proceedings, Montreal.

Chang, L.L., Mendez, E.E., Tejedor, C. (Eds.), 1991. Resonant Tunneling in Semiconductors. Plenum Press, New York.

Datta, S., 1995. Electronic Transport in Mesoscopic Systems, Cambridge University Press, Cambridge.

Davies, J.H., 1998. The Physics of Low Dimensional Semiconductors, Cambridge University Press, Cambridge.

Duke, C.B., 1969. Tunneling in Solids, Academic Press, New York.

Dutta, M., Stroscio, M.A. (Eds.), 2003. Advanced Semiconductor Heterostructures. World Scientific, Singapore.

Feng, Z.C., Tsu, R. (Eds.), 1994. Porous Silicon. World Scientific, Singapore.

Hayes, H., Loudon, R., 1978. Scattering of Light by Crystals, Wiley, New York.

Neuberger, M., 1971. Handbook of Electronic Materials, vol. II, III−V Semiconducting Compounds, IFI/Plenum Press, New York.

Ng, K.K., 2002. Complete Guide to Semiconductor Devices, Wiley-IEEE Press, New York.

Kelly, M.J., 1995. Low Dimensional Semiconductors—Materials, Physics, Technology, Devices, Clarendon Press, Oxford.

Koch, C.C. (Ed.), 2002. Nanostructured Materials. Noyes Publications, Norwich.

Mitin, V.V., Kochelap, V.A., Stroscio, M.A. (Eds.), 1999. Quantum Heterostructures. Cambridge University Press, Cambridge.

Moss, T.S. (Ed.), 1980. Handbook on Semiconductors, vol. 2. North-Holland, Amsterdam.

Nicollian, E.H., Brews, J.R., 1982. MOS Physics and Technology, Wiley, New York.

Noteborn, H., 1993. Quantum Tunneling Transport of Electrons in DB Heterostructures, Physics, Eindhoven University of Technology. Eindhoven, Netherlands.

Pippard, A.B., 1965. Dynamics of Conduction Electrons, Gordon & Breach, New York.

Shah, J. (Ed.), 1992. Hot Carriers in Semiconductor Nanostructures: Physics and Applications. Academic Press, Boston.

1993. Special Issue of NATO ASI Series for NATO Workshop on Optical Properties of Low Dimensional Silicon Structures, CNET, Grenoble, France, Kluwer, Dordrecht.

Vial, J.C., Derrien, J. (Eds.), 1994. Porous Silicon Science and Technology, Winter School Les Houches, February 1994. Springer, Berlin.

Wallis, R.F. (Ed.), 1963. Lattice Dynamics. Pergamon Press, New York.

This page intentionally left blank

1 Superlattice

1.1 The Birth of the Man-Made Superlattice

To appreciate why man-made superlattices were conceived, we need to understand why of all the elements in the periodic table, only a handful of elements are suitable for electronic devices. Before the advent of semiconductor devices, vacuum tubes started the electronic revolution, whereby electrons are emitted into the vacuum from a heated filament, intercepted by a controlled grid, and collected by an anode. Even with all the wonderful functions that semiconductor devices can offer, particularly in integrated circuits (ICs), even today many high-power devices—like magnetrons used as electromagnetic generators for high frequency and traveling wave tubes used as high-power and high-frequency amplifiers—are still in use. However, hot filament is replaced by the search for a cold cathode. Other endeavors consist of getter for adsorption of the residual gas inside a vacuum tube and metals capable of stable operation at higher temperatures. After the introduction of solid-state devices such as the transistor, a search for appropriate materials launched a new discipline, material science, a term introduced by Gerald Pearson. He joined Stanford University from Bell Telephone Laboratories (BTL) after contributing to the first transistor. Even though incredibly rapid development has led to the present IC, a planar structure consisting of literally millions of circuits known as the chip, only a handful of materials are used, and of these mainly silicon.

Let us briefly summarize what types of material are involved in the modern-day electronic revolution: metals for contacts, semiconductors for active components of a device, and high band-gap materials for insulation. When I first started my career at BTL, I was told that we should all look for new materials rather than inventing new schemes for devices. Group IV covalent materials like Si and Ge are good for basic transistors; Group III−V semiconductors, GaAs for example, are used for detectors and photonic devices; Perovskite structures such as barium titanate provide high dielectric constants; lithium niobates are good for nonlinear optical devices; and rare-earth-doped materials such as Nd-doped YAG garnets are used for high-power lasers, and so on. Specifically, injection lasers were limited to GaAs, the best LED was GaP utilizing nitrogen-doped bond-excitons, high-frequency transducers used quartz, and the best photoconductor used CdS, a highly defective Group II−VI compound semiconductor. Some crystals, such as hexagonal SiC, have a number of different structural forms known as polytypes (Choyke et al., 1964), forming a natural superlattice structure with

Superlattice to Nanoelectronics. DOI: 10.1016/B978-0-08-096813-1.00001-1

a period ranging from 1.5 to 5.3 nm; however, these polytype structures cannot be controlled and the resulting energy gaps are too small to provide any useful electronic novelty. These considerations were the leading reasons that drove me and Esaki to contemplate man-made solids.

In attempting to create a man-made solid, at the onset we recognized that mimicking the translational symmetry can best be physically realized using a planar structure that has a modulation of potential energy only in one direction, the direction normal to the planar layers. This modulation can be achieved either by a periodic pn-junctions or layers of A/B, two different materials arranged in a periodic way. To form an artificial energy band structure, the distances involved must be smaller or at least no greater than the junction width of a tunnel diode (Esaki, 1958); more precisely, the mean free path of the electron must be at least greater than the period of modulation in order to preserve phase coherence. Let us take a closer look at the comparison between the pn-junction and the heterostructures in forming a modulated potential variation. Almost all semiconductor devices, transistors, detectors, transducers, and switches were based on pn-junctions, a junction separating two different forms of doping, n for negative owing to the surplus electrons in the conduction band, and p for positive, owing to the deficiency of electrons in the valence band. Doping, such as phosphorus on silicon sites, results in ionization, at normal operating temperatures, of the extra electrons (not needed for fourfold coordination) into the conduction band, contributing to free carrier transport. A junction is formed when dopings on both sides develop a junction voltage caused by the alignment of the Fermi levels. Extra electrons ionized into the conduction band from the phosphorus dopant sites in n-doped Si fall into the boron dopant sites in p-doped Si, leaving positive charges in the vicinity of the n-side and negative charges in the vicinity of the p-side, producing an electric field across the junction. At equilibrium, the potential developed across the junction prevents further transfer of charges from the n-doped side to the p-doped side. We must recognize that the potential difference is spread over the depletion width of the pn-junction, because transferring charges involves a monopole with a range extending over a fairly long distance. Heavier doping reduces the distance needed to acquire sufficient charges to move the potential. However, the solid solubility limit prevents the excessive doping needed to reduce the depletion width below the mean free path, or in general, the coherence length of the electrons. Simply put, the solid solubility limit arises because whenever two P atoms are next to each other, they would rather be bonded metallically than covalently, resulting in no contribution to the extra electron in silicon in conventional n-doped silicon. The same reason applies to B-doping in Si, which results in extra holes, a deficiency in electrons for p-doped silicon. Before we touch on the heterostructures used in superlattices and quantum wells, we need to develop an appreciation of why heterojunctions that have a potential difference on each side maybe well defined or "sharp" to within a few tenths of a nanometer. We know that the band-edge offset between two materials cannot be predicted using only an argument involving the alignment with respect to the vacuum level, which is based on the work functions of the two materials. Usually, one is left with measurements for the band-edge offset, or some complicated *ab-initio* calculation such as the use of the density functional theory. What I want to convey is a simple rule as to why the heterojunction maybe very sharp. We know that in a multipole expansion of a potential functions, the monopole term falls off more slowly than the dipole term and the dipole term falls off more slowly than the quadrupole term, and so on. Whenever positive and negative pairs are neutralized in a given region, more precisely, in a unit cell of the solid, it is the multipole potential that an electron sees, thus, a much sharper potential profile is experienced by an electron. Heterostructures are neutral and therefore are quite similar to multipoles. In a man-made solid, the period must be less than the mean free path, which is no more than few tens of nanometers at room temperature. Except at temperatures close to 0 K, doping the superlattice can barely make the grade. The need to make contact to a given active region of a device calls for a planar structure with alternating layers of two materials that have a sufficient band-edge offset, mimicking a periodic variation along the direction perpendicular to the layers. In short, the criterion of a man-made superlattice introduced by Esaki and Tsu (1969, 1970), is to develop alternating layers of A/B, having sufficient band-edge offset at a distance well below the period of the layers. Being a planar structure, input/output, has as usual, contacts capable of handling large current density.

Figure 1.1 shows the original drawings depicting the two cases of modulation. Because of the reasons elaborated herein, only the heterojunction scheme was pursued at IBM Research.

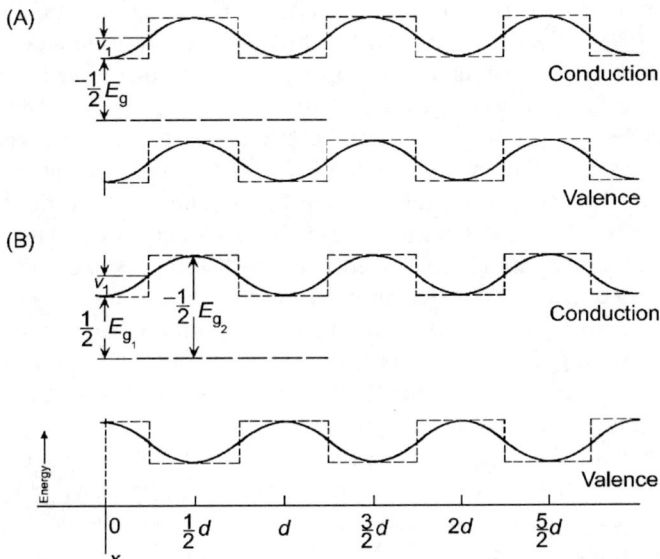

Figure 1.1 Electron energy profile in the conduction- and valence-band for pn-junctions (A) and periodic variation of the heterojunction (B).
Source: After Esaki and Tsu (1969, 1970), with permission.

1.2   A Model for the Creation of Man-Made Energy Bands

We represented the periodic potential $V(x) = V(x + nd)$ with a period d typically $10-20$ times greater than the lattice constant a in the host solid. We assumed that the wave equation is that of a free particle Schrödinger equation with an effective mass m and a potential term $V(x)$. With variable separable, we only deal with the x-direction,

$$-\frac{\hbar^2}{2m}\frac{d^2}{dx^2}\psi_x - V(x)\psi_x = E_x\psi_x. \tag{1.1}$$

A general solution for a periodic $V(x)$ results in the Bloch state,

$$\psi_x = e^{ikx}\sum_{n=-\infty}^{+\infty} u_n(k,x). \tag{1.2}$$

For the sinusoidal case, $V(x) = V_1(\cos 2k_d x - 1)$, Eq. (1.1) becomes the Mathieu's equation, well described by McLachlan (1947):

$$\frac{d^2}{d\vartheta^2}\psi_\vartheta + (\eta + \gamma \cos 2\vartheta)\psi_\vartheta = 0, \tag{1.3}$$

where the reduced energy $\eta = (E_x - V_1)/E_0$, $\gamma = V_1/E_0$, and the dimensionless momentum $\theta = k_d x$, and $E_0 = \hbar^2 k_d^2/2m$, with $k_d = \pi/d$. We used a sinusoidal variation of the modulation potential instead of a periodic square-well potential at the very beginning because we thought that the potential variation even with hetero-junctions maybe quite "soft" owing to interdiffusions. The comparison of the sinusoidal potential modulation and the periodic square-well potential is shown in Figure 1.2, for the parameters shown in the figure. Note that the difference for the dispersion relation was not significant leading us, exclusively in later work, to use the periodic square potential, calculated with the Krönig–Penney potential (Smith, 1961), for subsequent model calculations. Note that the Brillouin zone (BZ) boundary is reduced from π/a to π/d, resulting in the formation of the minizones. For $d \sim 20a$ with a being the lattice constant of the host solid, the minizone is 20 times smaller allowing electrons reaching the minizone boundary to create some intriguing transport under tolerable values for the electric field.

Before we proceed to show the transport property, current versus applied voltage in the next section, I want to discuss several points that led to the resolution of some initial concerns. We pointed out that we were not pursuing the doping superlattice, instead concentrating on the A/B alloys. The Group III–V compounds of GaAs for the well and GaAlAs for the barriers were selected because these materials have been used for the double heterojunction (DH) lasers developed at BTL where GaAlAs was used to confine the charge carriers in the so-called DH lasers. We must distinguish between quantum confinement, which Esaki and I were seeking, and charge

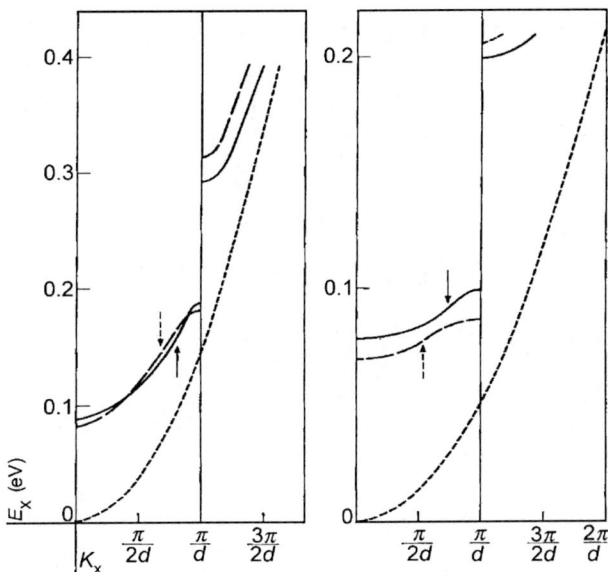

Figure 1.2 Electron energy versus wave vector of superlattices:−, sinusoidal; − − −, square-well; ..., free electron. Arrows denote the points of inflection; the second derivatives of $E-k$ become negative.
Source: Esaki and Tsu (1969, 1970), with permission.

confinement, used in the DH lasers. The first problem developed because we were told that GaAs and GaAlAs form an ohmic junction. If this were true, it would not have been possible to serve as a superlattice potential modulation. I was very concerned, but Esaki set my concerns aside. He said to me, "Experts are not always right." The next big step was asking, short of measurements, what values can one use for the modulation V_1 in Eq. (1.1)? Esaki called on Frank Herman, who was in charge of the Department of Large Scale Computation at IBM Research at San Jose. He used an LCAO calculation and presented us with his result that 80% of the band gap aligned with the conduction band and 20% with the valence band. In other words, we could take 80% of the difference of the band gap of GaAs and GaAlAs as the modulation parameter for the square-well potential in the conduction band. However, he warned us that his calculation was based on alignment of the 1S state of the As atom in GaAs and the 1S state of the As atom in GaAlAs, so that his error bar could be as high as 1 eV. Again I was shocked, but Esaki quickly reassured me that his only concern was to ask Herman to wait before submitting his calculation for publication until we had had a chance to demonstrate our superlattice. As I pointed out in the Introduction, one of my goals in writing this book is to provide a fairly detailed discussion of the difficulties and stages we passed before this venture succeeded. The preceding concerns can provide some insight as to the extent that one needs to stand firm. Frankly, without Esaki's experience and forcefulness, I would have given up.

1.3   Transport Properties of a Superlattice

A simplification by Pippard (1965) of the path integration method of Chambers (1952) was used to obtain current versus applied field F. The equations of motion are

$$\hbar \frac{dk_x}{dt} = eF, \quad v_x = \hbar^{-1} \frac{\partial E_x}{\partial E_x} \tag{1.4}$$

$$dv_x = eF\hbar^{-2}(\partial^2 E_x / \partial k_x^2)dt, \tag{1.5}$$

and the drift velocity is

$$v_d = eF\hbar^{-2} \int (\partial^2 E_x / \partial k_x^2) \exp(-t/\tau) dt. \tag{1.6}$$

Taking a sinusoidal $E-k$ relationship, the so-called tight-binding dispersion relation,

$$v_d = g(\xi)[\hbar k_d / m(0)], \quad \text{where } g(\xi) = \xi/(1 + \pi^2 \xi^2), \tag{1.7}$$

in which $\xi \equiv eF\tau/\hbar k_d$, $m(0) = 2\hbar^2/E_1 d^2$, and $k_d = \pi/d$. We see in Figure 1.3 that the drift velocity versus the applied field F reaches a maximum value and beyond this point the slope is negative and the so-called negative differential conductance (NDC) appears. What happens is that electrons driven to the BZ boundary turn around because of the Bragg reflection. Without scattering, $\tau \to \infty$, oscillation results in zero constant current. With scattering, a constant current appears, but decreases with increase of ξ. Therefore, the source of the NDC is precisely what leads to oscillation, the so-called Bloch oscillation. It is probably unknown to most researchers today that Krömer (1958) proposed using the heavy-hole band in semiconductors to create a negative effective mass amplifier. However, the scheme was not developed owing to difficulty in controlling the presence of transverse negative masses. In any case, this book deals with the man-made superlattice that has a negative effective mass only along the direction of the superlattice.

1.4   More Rigorous Derivation of the NDC

A more rigorous derivation involving the Boltzmann equation with a tight-binding $E-k$ relationship was obtained 1 year after the simple derivation using the impulse method of Pippard (1965), presented in the last section as suitable for low carrier concentration. This was the version that formed the basis for launching the

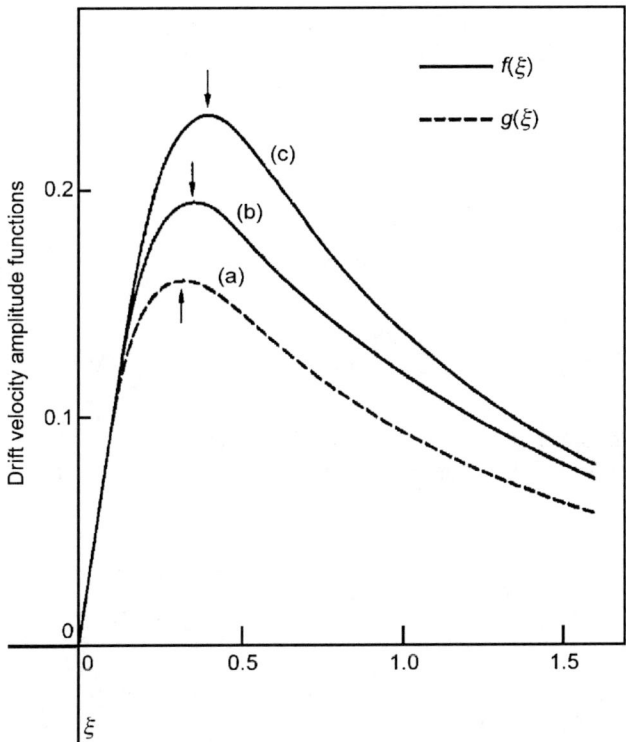

Figure 1.3 Drift velocity versus the reduced field parameter ξ: (a) for sinusoidal potential, (b) periodic square-well potential with $k_i/k_d = 0.5$, and (c) with $k_i/k_d = 0.82$, where k_i is the point of inflection. Arrows show the peaks of the drift velocities.
Source: Esaki and Tsu (1969, 1970), with permission.

man-made superlattice program at IBM Research. Esaki and I made a survey of what properties maybe exclusively attributed to the formation of a man-made superlattice. We decided that the appearance of an NDC serves as the key criterion. The basic assumption involves essentially the use of two-dimensional electron gas (2DEG), where the $E-k$ in the transverse direction is parabolic, free electron energy momentum, and scattering time is assumed to be energy independent. The latter assumption is in fact quite acceptable owing to the cancellation, to some degree, of the primary scattering from acoustic phonons and impurities. In most elementary treatments of the Boltzmann transport equation, a simple shifted distribution results in Ohm's law. Lebwohl and Tsu (1970) found that for a one-dimensional nonparabolic $E-k$ relationship in the Boltzmann equation, an exact solution is possible for constant relaxation time. They showed that their result for Fermi–Dirac distribution is identical to the proof by Budd (1963) of Chambers's (1952) path integral method for Maxwellian distribution.

Taking the electric field $\mathbf{F}$ in the x-direction, the Boltzmann transport equation without the time and spatial variation becomes

$$k_0 \frac{\partial f}{\partial k_x} + f = f_0, \tag{1.8}$$

in which $k_0 = eF\tau/\hbar$ and $f_0(k)$ is the equilibrium distribution function. The periodicity of the crystal along $\mathbf{F}$ is described by the reciprocal lattice vector $\mathbf{K}$. The energy bands are assumed to be:

$$E = (\hbar^2 k_\perp^2 / 2m) + E_x(k_x), \tag{1.9}$$

where $k_\perp$ is the component of the wave vector perpendicular to the direction of the superlattice and $E_x(k_x)$ is periodic in k_x with period K. Eq. (1.8) has the general solution

$$f(k_\perp, k_x) = \int_{K/2}^{K/2} G(k_x, k_x') f_0(k_\perp, k_x') dk_x', \tag{1.10}$$

in which the Green's function G satisfies boundary conditions with period K. The adjoined equation that G satisfies is

$$-k_0 \frac{\partial G}{\partial k_x} + G = \sum_{-\infty}^{\infty} \delta(k_x - k_x' - mK). \tag{1.11}$$

Figure 1.4 shows the deformation of contour from C to C' in the complex Z-plane.

$$
\begin{aligned}
G &= \frac{1}{K} \sum_{-\infty}^{\infty} \frac{e^{i2\pi m(k_x - k_x')/K}}{1 - i2\pi m k_0/K} = \pm \frac{1}{2\pi i K} \int_c \frac{e^{(k_x - k_x')Z/K}}{(1 + k_0 Z/K)(e^{\pm Z} - 1)} \\
&= \frac{e^{(k_x - k_x')/k_0}}{k_0} \frac{(1 - e^{-K/k_0})^{-1}}{(e^{K/k_0} - 1)^{-1}} \quad \begin{matrix} k_x < k_x' \\ k_x < k_x' \end{matrix}.
\end{aligned} \tag{1.12}
$$

Taking

$$f_0(k_\perp, k_x) = \theta(k_\perp - k_{\perp m})\{\theta[k_x + k_{xm}(k_\perp)] - \theta[k_x - k_{xm}(k_\perp)]\}$$

with θ being the heaviside unit-step functions and k_{xm} the maximum k_x, defined by

$$E_x(k_x) = E_f \quad \text{and} \quad E(k_m) = E_f.$$

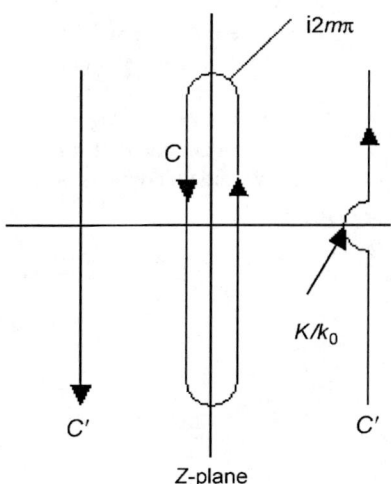

Figure 1.4 Deformation of contour from C to C' in the complex Z-plane.

At $k_\perp = 0$, and $k_x = 0$, respectively, $f(k_x)$ is given by

$$f(k_x) = \begin{cases} \exp\left[-(k_x + K/2)/k_0\right]\dfrac{\sinh(k_{xm}/k_0)}{\sinh(K/2k_0)} & k_x < -k_{xm} \\[2ex] 1 + \exp(-k_x/k_0)\dfrac{\sinh[(k_{xm} - K/2)/k_0]}{\sinh(K/2k_0)} & |k_x| < k_{xm} \\[2ex] \exp\left[-(k_x - K/2)/k_0\right]\dfrac{\sinh(k_{xm}/k_0)}{\sinh(K/2k_0)} & k_x > k_{xm} \end{cases} \tag{1.13}$$

and $f(k) = \theta(k_\perp - k_{\perp m})f(k_x)$. The average current, as usual, is given by

$$j_x = \frac{1}{4\pi^3 \hbar}\int f(\mathbf{k})\frac{\partial E}{\partial k_x}\, d\mathbf{k}.$$

This procedure leads to the same result as Chambers's path integral method for the distribution function f with force $\mathbf{F}$ and velocity $\mathbf{v}$,

$$f = \frac{1}{\tau}\int_{-\infty}^{t'} f_0(E - \Delta E)\exp(-(t' - t)/\tau)\, dt', \quad \text{with} \quad \Delta E = \int_t^{t'} \mathbf{F}\cdot\mathbf{v}\, dt''. \tag{1.14}$$

With the tight binding approximation for the superlattice $E\text{--}k$, $E_x = E_0 - E_1 \cos(2\pi k_x/K)$, the average current along the superlattice direction becomes

$$j_x = \frac{neE_1 d}{\hbar}\, \frac{\omega_B \tau}{(\omega_B \tau)^2 + 1}\, H \tag{1.15}$$

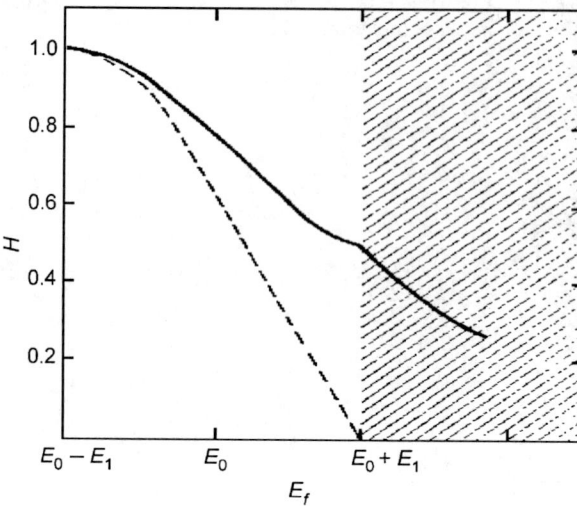

Figure 1.5 H versus E_f for the sinusoidal $E-k$ relation. The solid curve shows the case with a transverse degree of freedom and the dotted curve shows the one-dimensional case.

in which $\omega_B \equiv eFd/\hbar$, and

$$H \equiv \frac{1}{2}\left[(Q - \sin Q \cos Q)/(\sin Q - Q \cos Q)\right], \quad E_0 - E \le E_f \le E_0 + E_1, \quad (1.16a)$$

$$\frac{1}{2}\left[E_1/(E_f - E_0)\right], \quad E_f > E_0 + E_1, \quad (1.16b)$$

where the superlattice period d is defined by $K = 2\pi/d$, and $Q = \cos^{-1}((E_0 - E_f)/E_1)$. Note that Eq. (1.15) is essentially same as Eq. (1.7) except the factor H defined in Eq. (1.15). Figure 1.5 shows H versus E_f for the sinusoidal $E-k$ relation. Note that H in Eq. (1.15) depends only on the Fermi energy and the parameters specifying the $E-k$ relation, but does not involve the applied electric field **F**. Including $k_\perp$ shows that H is not zero even inside the minienergy gap. This is correct because electrons have transverse motion even within the minigap. It is noted that if one simply includes the transverse degree of freedom in the original derivation of Eq. (1.7), integration over the density of states (DOS) results in Eq. (1.15). Again we show that the Boltzmann equation for constant relaxation time offers the same result the same as using Pippard's method.

1.5 Response of a Time-Dependent Electric Field and Bloch Oscillation

The condition for NDC whenever $\xi \sim 0.4$ or $eFd\tau/\hbar \ge 1$ in Eq. (1.7) maybe grouped differently by defining the Bloch frequency $\omega_B \equiv eFd/\hbar$, then the condition for

NDC becomes $\omega_B \tau \geq 1$. Therefore, the physics of NDC in a man-made superlattice is intimately related to the subject of the Wannier Stark ladder (SL), introduced by Wannier (1960, 1961). Actually the concept, if $\Psi(x)$ is a solution of the wave equation for energy $E - neFa$, with a being the lattice constant and n being any integer, was first discussed by James (1949). The explicit localization induced by the application of a constant electric field was given by Kane (1959). Since these earlier treatments, many arguments have appeared against the existence of the SL, principally raised by Rabinovitch and Zak (1971), with arguments in favor of SL and localization by Shockley (1972). Basically, these objections arise because of the finite dimensions of the solid, the failure of representation of the wave function with an operator x in the Hamiltonian by a superposition of Bloch functions and tunneling into other bands. Using a vector potential to preserve symmetry, Krieger and Iafrate (1986) were able to remove these objections. Unlike the case in which a scalar potential is used to represent the effect of the field, the field-dependent terms are continuous functions of time, permitting the eigenfunctions of the unperturbed Hamiltonian to be the eigenfunction of the Hamiltonian at the instant the field is turned on. The time development of the system for times small compared to the inverse tunneling rate is represented by a linear combination of the eigenfunctions of the instantaneous Hamiltonian. Actually the issue is similar to the fact that the states of a hydrogen atom in a constant field are not discrete: strictly speaking, stationary states do not exist, yet there is no confusion in treating the problem in terms of transitions between stationary states. In fact, Krieger and Iafrate sidestepped the issue of stationary states. Instead, they showed that optical transitions involving a selection rule, which is consistent with the notion of the SL.

To eliminate some of the confusion, for example, in the relationship between localization and Bloch oscillation what is the degree of localization, the effects of tunneling into other bands, finite length and finite mean free path, relationship to quantum well structure, and so on? We apply the approach used in Sections 1.3 and 1.4 to include a time-varying electric field. As Lebwohl and Tsu (1970) pointed out, Chambers's path integral method for the Fermi–Dirac distribution leads to the same result as Eq. (1.15). Therefore, Chambers's path integral method will be used to treat a more general excitation, having an electric field consisting of a constant term as well as a time-dependent term.

Starting with the distribution function for a constant relaxation time τ given by Eq. (1.14) and with $\mathbf{v}_x = (1/\hbar)(\partial E/\partial k_x)$, then:

$$\langle v_x \rangle = \frac{1}{4\pi^3} \int f \mathbf{v}_x \, d\mathbf{k} = \frac{1}{4\pi^3} \int_k f_0 \, d\mathbf{k} \int_{-\infty}^{t'} \frac{1}{\hbar} \left(\frac{\partial E(t')}{\partial k_x} - \frac{\partial E(t)}{\partial k_x} \right) e^{-(t-t')/\tau} \frac{dt}{\tau}. \quad (1.17)$$

Taking a general field F,

$$F = F_0 + \sum_n F_n e^{-i\omega_n t}, \quad \omega_n = n\omega_1.$$

From $\hbar(dk_x/dt) = eF$,

$$k_x(t) = k_x(t') + \frac{eF_0}{\hbar}(t - t') + \sum_{n>0} \frac{2eF_n}{\hbar n\omega_1}(\sin n\omega_1 t - \sin n\omega_1 t'). \qquad (1.18)$$

For $n = 1$,

$$k_x(t) = k_x(t') + \frac{eF_0}{\hbar}(t - t') + \frac{2eF_1}{\hbar\omega_1}(\sin \omega_1 t - \sin \omega_1 t'). \qquad (1.19)$$

For the tight-binding $E_x - k$, $E_x = E_0 - E_1\cos(2\pi k_x/K)$ and $E_x(k_x) = E_x(-k_x)$, putting Eq. (1.18) in Eq. (1.17) with part integration, the expectation value of the velocity becomes

$$\langle v_x \rangle = nHv_0 \int_{-\infty}^{t} \sin(g(t, t')d)e^{(t-t')/\tau}dt', \qquad (1.20)$$

where $v_0 = E_1 d/\hbar$, $nH = (1/4\pi^3) \int \cos(k_x d)f_0 d\vec{k}$ and $g(t, t')$ is defined by the sum of the second and third terms on the right side of Eq. (1.19), or,

$$gd = \frac{eF_0 d}{\hbar}(t - t') + \frac{2eF_1 d}{\hbar\omega_1}(\sin \omega_1 t - \sin \omega_1 t'). \qquad (1.21)$$

The integration of Eq. (1.20) is readily done using the expansion $\exp(iz \sin \theta) = \sum_{n=-\infty}^{+\infty} e^{in\theta}J_n(z)$, where J_n is the Bessel function, then the expectation value of the velocity becomes

$$\langle v_x \rangle = v_0 H \sum_{m,n=-\infty}^{\infty} J_m\left(\frac{\omega_{B1}}{\omega_1}\right) J_n\left(\frac{\omega_{B1}}{\omega_1}\right) \frac{\sin(m-n)\omega_1 t + (\omega_B + n\omega_1)\tau \cos(m-n)\omega_1 t}{(\omega_B + n\omega_1)^2\tau^2 + 1},$$

$$\qquad (1.22)$$

where $\omega_B \equiv eF_0 d/\hbar$ and $\omega_{B1} \equiv eF_1 d/\hbar$. For $H = 1$, $\langle v_x \rangle$ in Eq. (1.22) is same as that was first obtained by Tsu (1990). For $\omega_{B1}/\omega_1 \ll 1$ (small F_1),

$$\langle v_x \rangle = \langle v_x \rangle_0 + \text{Re}\langle v_x \rangle_1 \cos \omega_1 t + \text{Im}\langle v_x \rangle_1 \sin \omega_1 t, \qquad (1.23)$$

where

$$\langle v_x \rangle_0 = v_0 H \frac{\omega_B \tau}{(\omega_B \tau)^2 + 1}, \qquad (1.24)$$

which is identical, as it should be, to the previous results. Let us discuss the case for $H = 1$, then $v_0 H = E_1 d/\hbar$ and the maximum extent $\langle x \rangle_m = \langle v_x \rangle_m \tau = E_1/2eF_0$.

Since length is measured by nd, with n being the number of periods, the maximum number of periods covered is given by

$$n = E_1/2eF_0d. \tag{1.25a}$$

For $\omega_B\tau \gg 1$, $\langle v_x \rangle \to 0$, $\langle x \rangle = E_1/eF_0$,   and $\langle x \rangle = 2\langle x \rangle_m$. $\tag{1.25b}$

The electrons will now oscillate with a period $T = 2\pi/\omega_B$, which was known to Bloch (1928) and discussed by Houston (1940).

Without collision, an electron will oscillate at a frequency of ω_B and cover a distance of E_1/eF_0. Note that the extent of an electron without collision is twice the maximum distance given by $\omega_1\tau = 1$. We shall make this point clearer with respect to the degree of localization. Rabinovitch and Zak (1971) cast some doubts about the existence of this oscillation. In order to find out whether a Bloch electron can oscillate, we shall examine Eqs. (1.22) and (1.23) in more detail. First of all, it is obvious that without both F_0 and F_1, it is pointless to discuss whether a Bloch electron oscillates or not. I would like to point out what constitutes basic understanding and what is needed in engineering optimization. It turns out that all models apply within the limits set by the models. Often, a good model is one that has an interrelationship with others familiar to a large number of researchers even though the models themselves maybe quite limited. On the other hand, a less limited model, by virtue of its complexity, maybe familiar to very few. Do we pick the former over the latter? Sidestepping these philosophical points, we shall take a look at the linear system. The in-phase component with time goes as $\cos \omega t$ which we abbreviate by writing $\mathrm{Re}\langle v_x \rangle$ and the out-of-phase component with time goes as $\sin \omega t$ denoted by $\mathrm{Im}\langle v_x \rangle$. Thus, we sum all the terms in Eq. (1.22) for $n - m = 1$. The equations describing the linear response, for convenience, are given as follows:

$$\mathrm{Re}\langle v \rangle \equiv \frac{\mathrm{Re}\langle v_x \rangle_1}{v_0 \cos \omega t}\left(\frac{2\omega}{\omega_{B1}}\right) \tag{1.26a}$$

and

$$\mathrm{Im}\langle v \rangle \equiv \frac{\mathrm{Im}\langle v_x \rangle_1}{v_0 \sin \omega t}\left(\frac{2\omega}{\omega_{B1}}\right). \tag{1.26b}$$

Equations (1.26a) and (1.26b) for various $\omega_B\tau$ are plotted in Figure 1.6. For simplicity, we take only the case of small electric field, i.e., $\omega_{B1} \ll \omega_B$.

In Figure 1.6, for $\omega_B\tau = 1$, $\mathrm{Re}\langle v \rangle$ is always positive indicating the lack of gain or self-oscillation. The $\mathrm{Im}\langle v \rangle$ has a maximum at $\omega = \omega_B$. For $\omega_B\tau = 2$, $\mathrm{Re}\langle v \rangle$ has a minimum at $\omega = \omega_B/2$ and is negative, but $\mathrm{Im}\langle v \rangle$ has a peak at $\omega = \omega_B$. With a further increase to $\omega_B\tau = 3$, $\mathrm{Re}\langle v \rangle$ has a maximum negative value at $\omega = 2\omega_B/3$ and the $\mathrm{Im}\langle v \rangle$ has a peak at $\omega = \omega_B$. Thus, the peak in Im part always appears at $\omega = \omega_B$, supposedly substantiating the intuitive understanding that the system is

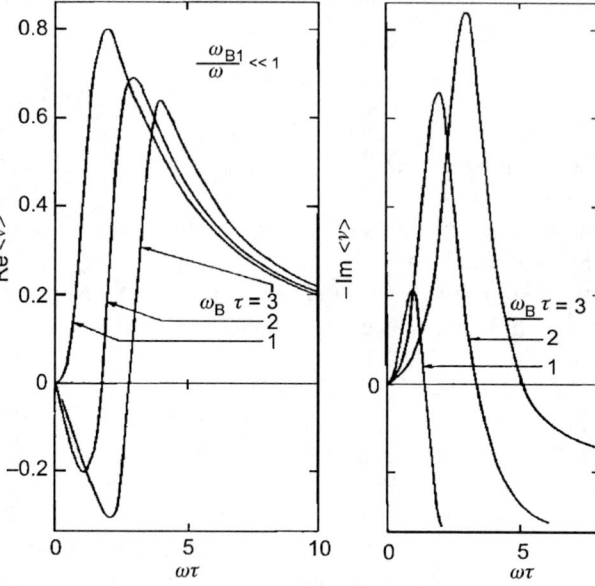

Figure 1.6 The in-phase, $\mathrm{Re}\langle v \rangle_1$ and out-of-phase $\mathrm{Im}\langle v \rangle_1$ components of the linear response function for a superlattice with an applied electric field of $F = F_0 + 2F_1\cos \omega t$, $\omega_B \equiv eF_0 d/\hbar$ and $\omega_{B1} \equiv eF_1 d/\hbar$.
Source: After Tsu (1993), with permission.

oscillating at the Bloch frequency. The question of gain or loss is another matter as we need to focus on $\mathrm{Re}\langle v \rangle$. Note that $\mathrm{Re}\langle v \rangle$ always has a maximum negative value below ω_B, indicating that self-oscillation that occurs at the maximum gain is never at the Bloch frequency. Only as $\omega_B\tau \to \infty$ does the maximum gain coincide with the Bloch frequency. For both $\omega_B\tau \gg 1$ and $\omega\tau \gg 1$, it is seen that $\mathrm{Re}\langle v \rangle_3$ can have a substantial region that is negative, indicating that in the region of nonlinear optics, an intense optical field is needed for gain.

What is happening is that higher energy photons cause transitions between minibands, providing additional nonlinear response. This is because k is conserved to within multiples of the reciprocal lattice vector, as in *umklapprozesse*. In the usual solids, optical nonlinearity arises from small nonparabolicity of the $E-k$ relation as treated by Jha and Bloembergen (1968), as well as from the optical phonons in a multilayer dielectric medium treated by Bloembergen and Sievers (1970). However, in man-made superlattices, nonparabolicity is huge, leading to substantial 2nd and 3rd harmonics shown in Figure 1.6, taken from Tsu and Esaki (1971), where the nonlinear effect is a factor of 20 greater than that calculated by Wolf and Pearson (1966).

Experimentally one needs to arrange the polarization with a component of the electric field in the superlattice direction. For $\tau = 0.5$ ps and $\omega_B\tau = 3$ gives $eF_0 d \sim 4$ meV corresponding to $F_0 \sim 4 \times 10^3$ V cm^{-1} for $d = 10$ nm. Therefore, the condition for self-oscillation at $\omega = 2\omega_B/3$ or $\omega \sim 4 \times 10^{12}$ Hz, should be quite accessible. The model calculations presented have answered the question that was raised about whether or not a Bloch electron can oscillate. In fact, our results show that not only can a Bloch electron go into oscillation, it can also serve as an amplifier. From device viewpoint, one wonders whether it is better to use a superlattice

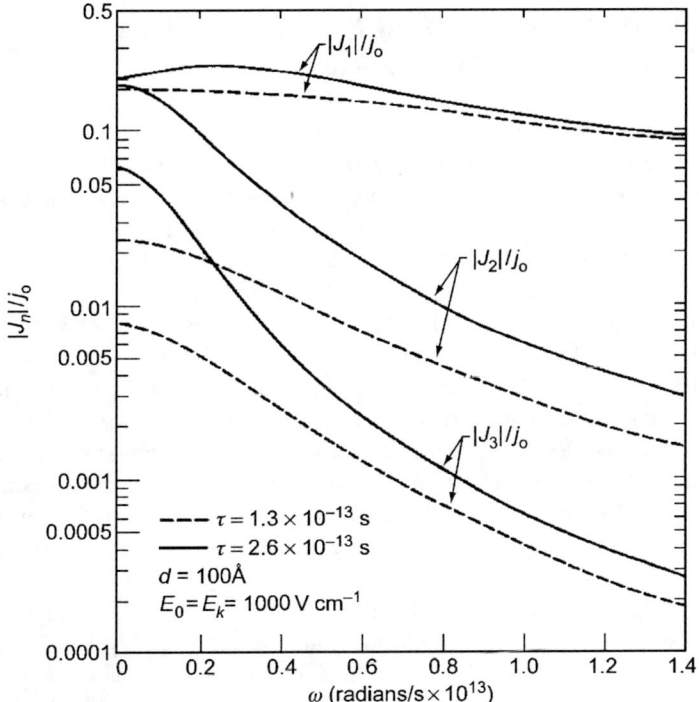

Figure 1.7 Calculated $|J_n|/j_0$ versus ω for two values of τ, with $J = en_0 \langle v \rangle$, where n_0 is the usual sum over the DOS.
Source: After Tsu and Esaki (1971), with permission.

as a "Bloch amplifier" or one with a single quantum well with external feedback. At this point, I remember years ago at BTL that a GaAs pn-junction transducer could operate without an external cavity, rather different from the usual barium titanate transducer. However, it was discovered that although the pn-junction acts as an electrical resonating system, owing to the small value of the electromechanical coupling, the Q value of the transducer is rather low. In short, the GaAs pn-junction transducer is not very efficient, without, somehow, introducing a resonating system. Also, similar to the traveling wave amplifier, the superlattice may have larger bandwidth and higher power-handling capacity. Additional considerations related to the effects of electric field—induced localization will be treated in a later section relating to the NDC involving phonon-assisted hopping, where the effects of a superlattice of finite length will be discussed. The physical picture maybe simply summarized that the periodicity of the solid serves as a distributed feedback. Therefore, one must recognize that the validity of these results presented as applying to a finite number of periods must be taken with additional considerations (Figure 1.7).

1.6   NDC from the Hopping Model and Electric Field—Induced Localization

At a sufficiently high applied electric field, the energy gain by an electron reaching the minizone boundary can exceed the energy bandwidth of the miniband. The band model ceases to be valid. The hopping model of Tsu and Döhler (1975) is based on the model that phonons provide the transfer of this energy to the lattice to reach equilibrium. The mathematical derivation is based on the Kane function (Kane 1959). Some highlights duplicated here are essential for the discussion, particularly in relation to the discussions on electric field—induced localization, which was only mentioned in the original work. Figure 1.8 shows possible transitions between electrons localized in a given cell to adjacent cells. Under the application of a field F_0, a constant potential energy difference exists between adjacent cells and a ladder structure for the energy state appears. If eF_0d is such that level 1 coincides with level 2', electrons may tunnel resonantly from 1 to 2' marked by (a), followed by an inelastic scattering marked by (b) to level 1', in order to repeat the process to the next cell. This process has been treated by Kazarinov and Suris (1972).

The hopping model of Tsu and Döhler, as in the treatments of Saitoh (1972) and Fukuyama et al. (1973), considers transitions between energy levels of an SL via phonon-assisted transitions. The main motivation is to bypass the objections to using a band model for the case where the energy gain by an electron in an electric field in a period exceeds the energy bandwidth, i.e., $\hbar\omega_B \geq 2E_1$. Following Callaway (1964), the wave function in the crystal momentum representations:

$$\psi(\mathbf{r}) = \sum \int \varphi_n(\mathbf{k})\psi_n(\mathbf{k})d\mathbf{k}, \tag{1.27}$$

where $\psi_n(\mathbf{k}) = U_n(k, r)e^{i\mathbf{r}\cdot\mathbf{r}}$ and the Kane function (Kane, 1959).

$$\varphi_n(\mathbf{k}) = \left(\frac{d}{2\pi}\right)^2 \exp\left\{-\frac{i}{eF}\int_0^{k_x}[\varepsilon - \varepsilon_n^{(1)}(\mathbf{k}')]dk_x'\delta(k_y - k_y')\delta(k_z - k_z')\right\}, \tag{1.28}$$

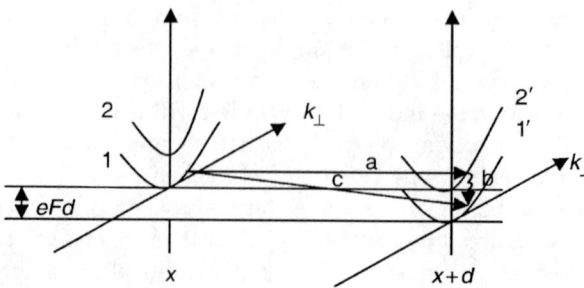

Figure 1.8 Energy states under an applied electric field F. Process (a) involves direct tunneling followed by an inelastic scattering (b), and process (c) describes inelastic scattering with emission and absorption of phonons. *Source*: After Tsu and Döhler (1975), with permission.

where

$$\varepsilon_n^{(1)}(\mathbf{k'}) = eFX_{nm} + \varepsilon_n(\mathbf{k}) \quad \text{and} \quad X_{nm} = \frac{(2\pi)^3}{V_0} i \int U_n^* \frac{dU_n}{dk_x} d\mathbf{r},$$

in which $\varepsilon_n(\mathbf{k})$ is the energy of the nth band. The requirement that $\varphi_n(\mathbf{k})$ is periodic in $\mathbf{k}$ results in the quantized levels,

$$\varepsilon_{v,n}(\mathbf{k}) = eFd + \frac{d}{2\pi} \int_{-\pi/d}^{\pi/d} [\varepsilon(\mathbf{k}) + eFX_{nm}] dk_x, \tag{1.29}$$

with v being any integer and $X_{nm} \equiv L$, a purely real constant because of the periodic function U_n. Using the tight-binding $E-k$,

$$\varepsilon(\mathbf{k}) = \hbar^2 k_t^2 / 2m + \varepsilon_0 - (1/2)\varepsilon_1 \cos k_x d, \tag{1.30}$$

then

$$\varepsilon_v(\mathbf{k}) = \hbar^2 k_t^2 / 2m + \varepsilon_0 + eFd(v + L/d). \tag{1.31}$$

This energy forms the basis of the SL, having equally spaced levels for different v. The function,

$$\varphi_v(\mathbf{k}) = \left(\frac{d}{2\pi}\right)^{1/2} \exp\left\{-i\left[k_x(vd + L) + \left(\frac{\varepsilon_1}{2eFd}\right) \sin k_x d\right]\right\}. \tag{1.32}$$

For one-dimensional periodic in x, with k restricted to the first minizone, $-\pi/d$ to π/d,

$$\psi(\mathbf{r}) = \left(\frac{d}{2\pi}\right)^{1/2} e^{ik_t\rho} \int_{-\pi/d}^{\pi/d} U(k_x, x) \exp\left\{i\left[k_x(x - vd - L) - \left(\frac{\varepsilon_1}{2eFd}\right) \sin k_x d\right]\right\} dk_x. \tag{1.33}$$

For large F such that $eFd \gg 2\varepsilon_1$, this wave function is highly localized with $\Delta x \sim \varepsilon_1/eF$, centered at $x = L + vd$. Therefore, the mean free path of the electrons must be greater than Δx. With the transition probability due to phonons using the golden rule,

$$\omega_{vv'}^{\pm}(k_t - k_{t'}) = \frac{2\pi}{\hbar} \sum_q |\langle v', k_t', n_q \pm 1 | H_{ep} | v, k_t, n_q \rangle|^2 \delta(\varepsilon_v(\mathbf{k}_t) - \varepsilon_v(\mathbf{k}_t') \mp \hbar\omega_q), \tag{1.34}$$

with (+) for phonon absorption and (−) for phonon emission. Owing to small inter-cellular transitions compared to intracellular relaxation, prior to and after each tran-sition, both the phonon and electron populations are governed by the equilibrium distribution functions, Bose−Einstein and Fermi−Dirac functions, respectively. This is a major assumption that allows us to proceed with the calculation. Skipping some detailed calculations involving quite complicated numerical integration found in Tsu and Döhler (1975), the hopping current j_γ between v and v' cells, or $\gamma = v - v'$, becomes

$$j_\gamma = \frac{em\gamma}{\hbar^2} \int_0^\infty \left[\omega_\gamma^+(\varepsilon_t) + \omega_\gamma^-(\varepsilon_t)\right] f(\varepsilon_t)\left[1 - \exp(-\gamma eFd/k_BT)\right] d\varepsilon_t, \qquad (1.35)$$

where ε_t is the energy transverse to the superlattice direction. Some of the steps used in computing j_γ, involve first computing $\omega_{vv'}^\pm (k_t - k_{t'})$ by integration on the angle ϑ and q_x, and summing over k_t for $v - v' = 1, 2, 3, \ldots$. To compare with the results derived for the time-varying electric field presented in Section 1.5, let us return to the defini-tions for the parameters used in the band model, i.e., energy bandwidth $2\varepsilon \to 2E_1$, the applied electric field $F \to F_0$, and the position of the miniband at E_0, then

$$E_v(\mathbf{k}) = \frac{\hbar^2 k_\perp^2}{2m} + E_0 + eF_0 d(v + L/d), \qquad (1.36)$$

and the wave function

$$\psi_v(\mathbf{r}) = \left(\frac{d}{2\pi}\right)^{1/2} e^{ik_\perp \rho} \int_{-\pi/d}^{\pi/d} u(k_x, x) \exp\left\{i\left[k_x(x - vd - L) - \frac{E_1}{eF_0 d} \sin k_x d\right]\right\} dk_x. \qquad (1.37)$$

For large F_0, the wave function is highly localized at a distance $\Delta x \sim E_1/eF_0$ and centered at $x = L + vd$. Using such a wave function to calculate the current due to an electron in the cell v (energy state v) making a hop to cell v' (energy state v'), we can examine the important role of τ, the scattering time. First of all, the locali-zation distance $\Delta x \sim E_1/eF_0$ agrees with that for the band model for $\omega_B \tau \gg 1$. Our previous results have pointed out that within the model of a constant scattering time τ, the effect of scattering is to reduce this localization length to half the value without scattering, as shown in Eq. (1.25a). Although in this treatment and most SL calculations in general, finite scattering time has not been included, it is neces-sary to see how the effect of the finite scattering time can be understood. To take scattering into account, the total current should be a sum of individual hopping between v and v', in other words, τ determines the range of γ in

$$J = \sum_{\gamma=1}^n j_\gamma, \quad \text{with} \quad \gamma = v' - v. \qquad (1.38)$$

This is an extremely important point, one that we shall go into great detail after the presentation of a physical model involving hopping using a coupled two-well model with a voltage applied across the two adjacent wells. If the barrier is such that the tunneling probability from one well to the next-nearest cell maybe neglected, we may consider only the net phonon-assisted transitions between the neighboring cells. Our purpose is to derive the hopping current for this simple model and compare the results with those in Section 1.6. A general treatment for large tunneling between cells and long mean free path will be discussed later.

1.6.1 Two-Well Model

Figure 1.9 shows a section of the superlattice potential profile. We assume that all the energy states other than the lowest levels denoted by λ_1 and λ_2 are far away, so that it is meaningful to consider the two lowest states only.

When the barrier width l is so large that electron wave functions do not overlap,

$$H_0|1\rangle = \lambda_1|1\rangle \quad \text{and} \quad H_0|2\rangle = \lambda_2|2\rangle \tag{1.39}$$

and $\lambda_1 - \lambda_2 = V$. Obviously for very high field, the next level Λ_2 maybe brought to the vicinity of λ_1, so that level λ_1 will be coupled to Λ_2 more so than with λ_2. This belongs to the resonant tunneling case. As l is reduced, electrons in the states λ_1 and λ_2 are coupled represented by $\alpha \equiv \langle 1|H_1|2\rangle$, in which H_1 is the coupling operator, so that

$$\varepsilon_{1,2} = \lambda_0 \pm \left[\left(\frac{V}{2}\right)^2 + \alpha^2\right]^{1/2} + \frac{\hbar^2 k_\perp^2}{2m}, \tag{1.40}$$

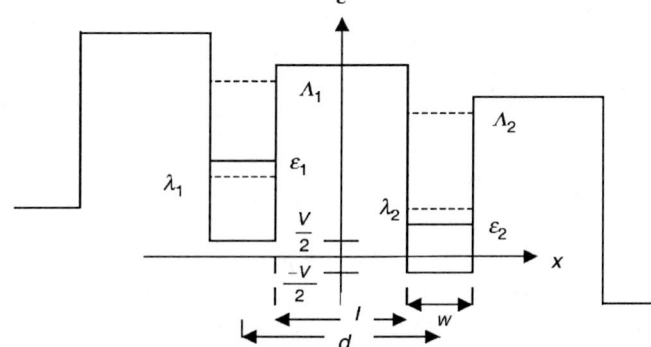

Figure 1.9 Section of the superlattice potential profile. λ_1, λ_2, Λ_1, and Λ_2 are the energies of the uncoupled wells with width W. ε_1 and ε_2 are the energies of the coupled wells when the coupling is increased by decreasing the barrier width l.
Source: Tsu and Döhler (1975), with permission.

where λ_0 is the longitudinal energy when $V = 0$ and $\alpha = 0$. Note that 2α is the splitting for $V = 0$. These two states have wave functions

$$\varepsilon_1 : \psi_1 = \left(\frac{1}{1+b^2}\right)^{1/2}(|1\rangle + b|2\rangle), \quad \varepsilon_2 : \psi_2 = \left(\frac{1}{1+b^2}\right)^{1/2}(b|1\rangle - |2\rangle), \quad (1.41)$$

where

$$b \equiv \left[1 + \left(\frac{V}{2\alpha}\right)^2\right]^{1/2} - \frac{V}{2\alpha}. \quad (1.42)$$

Most people that I had discussions with, because of their familiarity with the molecular orbitals, would quickly point out that the higher coupled state is anti-symmetric and the lower of the two is symmetric. Note that in Eq. (1.41) the symmetry is reversed because we have positive potential for the barrier, while in atomic systems, the positive nuclear charge gives rise to negative potential. If we take

$$|1\rangle \propto \sin(p\pi/W)(x + l/2)e^{ik_\perp \rho}$$

and

$$|2\rangle \propto \sin(p\pi/W)(x - l/2)e^{ik_\perp \rho}, \quad (1.43)$$

the matrix element

$$\langle \psi_1 | e^{i\mathbf{q}\cdot\mathbf{r}} | \psi_2 \rangle = \frac{i2b}{1+b^2}\frac{1}{q_x W}\sin\frac{q_x d}{2}\sin\frac{q_x W}{2} \times \frac{1}{1 - (q_x W/2\pi p)^2}\delta_{k_\perp', k_\perp + q_\perp'}. \quad (1.44)$$

For $p = 1$ (ground state) and $V \gg 2\alpha$,

$$\langle \psi_1 | e^{iq_x X} | \psi_2 \rangle = \frac{i2\alpha}{V}\sin\left(\frac{q_x d}{2}\right)\frac{1}{q_x W} \times \frac{\sin\left(\frac{1}{2}q_x w\right)}{1 - (q_x W/2\pi)^2}, \quad (1.45)$$

which is identical to the result of using Kane function for $eFd \gg \varepsilon_1$, if we identify 2α, the splitting with the width of the $E-k$ ε_1, $eFd = V$ and $J_1(\delta) \sim 1/2(\delta)$. The transition probability via electron–phonon interaction is then

$$w_{12}^{\mp} = \frac{2\pi}{\hbar}\sum_q |\langle \psi_1 || \psi_2 \rangle|^2 \delta(\varepsilon_1(\vec{k}_\perp) - \varepsilon_2(\vec{k}_\perp) \mp \hbar\omega_q). \quad (1.46)$$

The rest of the calculation is identical to the previous treatment. The validity of the two-well model depends on the extent of the localization of the wave functions.

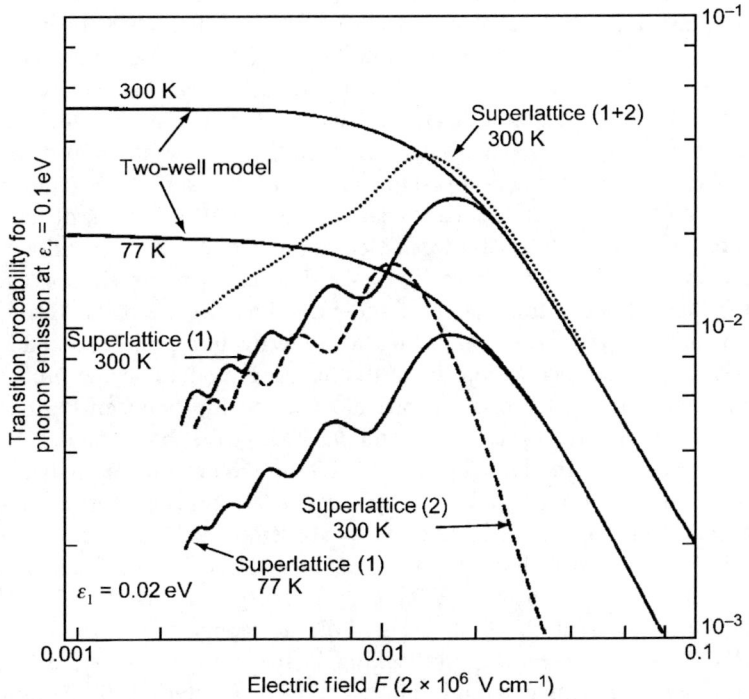

Figure 1.10 Comparisons between the hopping model using the Kane function with the two-well model for $\omega^{-}(4\pi C_s \rho \hbar^2/C_1^2)$ versus F at 300 and 77 K.
Source: Tsu and Döhler (1975), with permission.

Therefore, it is generally not applicable for low electric fields. Although the two-well model is only correct at high fields, the approach maybe used to treat the resonant tunneling case more easily, because in principle, many higher levels maybe incorporated into the formulation.

Figure 1.10 gives the comparisons between the superlattice case using the Kane function with the two-well model for $\omega^{-}(4\pi C_s \rho \hbar^2/C_1^2)$ versus F, at 300 and 77 K, with ρ, C_1, and C_s being the density, the deformation potential and the longitudinal sound velocity, respectively. Superlattice (n) denotes $\gamma = n$. For the case shown, there are two terms, $\gamma = 1$ and 2, added to give the resultant. Note that oscillations at low fields are almost gone, $j = j_1 + j_2$. With large τ, coupling to remote wells becomes meaningful, so that $j = j_1 + j_2 + j_3 + \cdots$ moving the peak toward low field. We thus see that there is a role for the relaxation time even we did not specifically include it. This is an important point to cause misunderstanding, not mentioned in the original paper by Tsu and Döhler (1975), but by Tsu and Esaki (1991). Therefore, physically there is not much difference between the hopping model and the band model, including the Bloch oscillation.

How many terms we need to sum depends on n, and in turn depends on the applied electric field F_0 in Eq. (1.25a). Therefore, in the extreme localization case where hopping can only take place between adjacent cells, $n = 1$, or in the preceding expression, $v = 1$, then the Esaki and Tsu (1969, 1970) expression $\omega_B \tau > 1$ is not necessary because implicitly we have already assumed that the mean free path is greater than the localization distance. This point maybe made quite clear by an example. Taking a regime where $\hbar\omega_B \sim E_1$, such that we cannot confidently use the band model, but the localization distance $\Delta x = nd$, the sum of j_γ requires that all these terms up to n be included. Note that in Figure 1.10, the peak of the total current progressively shifts toward lower field, while the number of individual hopping currents increases. How many terms we take depend on τ, which is not taken into account in any of these treatments of the SL. Thus, we need the Esaki and Tsu (1970) condition, $\omega_B \tau > 1$, to tell us how many terms in Eq. (1.38) should be taken. Since scattering is not accounted for in the hopping model and the limit of sum is set by the band model, there is an intricate relationship between the band model and the SL treatment. For $\omega_B \tau \gtrsim 1$ and $\hbar\omega_B \ll E_1$, the band model Esaki–Tsu applies. For $\hbar\omega_B \sim E_1$, the hopping model of Tsu–Döhler model applies, however, the Esaki–Tsu condition, $\omega_B \tau \gtrsim 1$, is still needed to tell us how many terms are required in the sum in Eq. (1.38). Only in the extreme localization case can the hopping model of Tsu and Döhler (1975) be taken without the condition $\omega_B \tau \gtrsim 1$ and then NDC is always given by $eFd > 2E_1$, the energy bandwidth of the band. As noted in Tsu and Esaki (1991), it is futile to apply the hopping model for $eFd \ll 2E_1$. A quantitative theory is lacking in the regime where the band model does not apply, and without taking into account the scattering, the hopping model also cannot apply. The situation is analogous to localization by a magnetic field.

Figure 1.11 summarizes the discussion taken from Tsu and Esaki (1991). As $\tau_3 > \tau_2 > \tau_1$, the shaded region gives the limits $\omega_B \tau \gtrsim 1$ and the peak current progressively moves toward larger F. On the other hand, the calculated current from the hopping model, j_n progressively moves toward lower F. Therefore, the two models practically describe the same basic phenomenon. This is a curious point

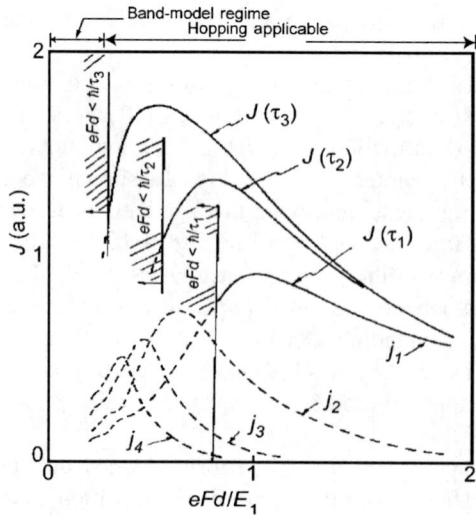

Figure 1.11 Current J versus eFd/E_1. As $\tau_3 > \tau_2 > \tau_1$, the shaded region gives the limits $\omega_B \tau \gtrsim 1$, and the peak current progressively moves toward larger F, where the same trend applies for the current j using the hopping model.
Source: After Tsu and Esaki (1991), with permission.

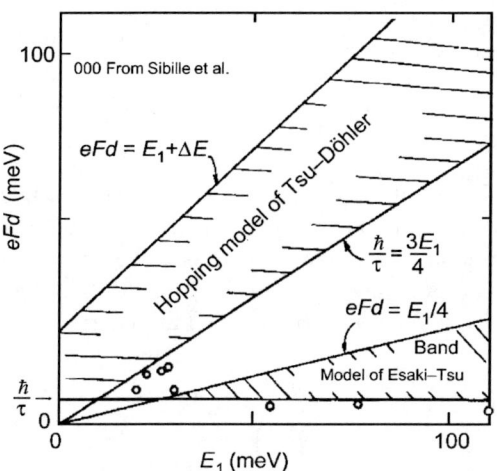

Figure 1.12 SL energy eFd versus miniband bandwidth E_1. The shaded region shows the two models: band and hopping.
Source: Tsu and Esaki (1991) (erratum), with permission.

shared by the lack of significant difference between the coherent resonant tunneling of Tsu and Esaki (1973) and the sequential model of Luri (1985), when scattering is taken into account. We shall come back to these points later after the treatment of tunneling through a quantum well structure.

Figure 1.12 shows the SL energy eFd versus miniband bandwidth E_1 taken from Tsu and Esaki (1991) (erratum). The shaded region shows the two models: band and hopping. ΔE is the separation between minibands. In between the shaded regions, there is a lack of a quantitative theory.

1.6.2 Effects of Finite Length

Some concerns about the existence of the SL in a periodic system of finite length led to the discussion by Shockley (1972) and Fukuyama et al. (1973). Fukuyama considered the Hamiltonian

$$H = \sum_n H_n + H'_n, \tag{1.47}$$

with $H_n = H_0 + V_n$, where $V_n \equiv eFdn \equiv nV_0$ and the nearest neighbor interaction such that $\langle n'|H'_n|n\rangle = 0$ for $n' = n$, and α for $n' = n + 1$. With a wave function

$$\psi = \sum_{n=1}^{N} C_n a_n^+ |0\rangle. \tag{1.48}$$

The expansion coefficient C_n satisfied the recursion relation

$$\left(\frac{E - E_0}{V_0} - n\right) C_n = \frac{\alpha}{V_0}(C_{n+1} + C_{n-1}). \tag{1.49}$$

Using a rigid wall boundary condition, i.e., $C_0 = C_{N+1} = 0$, an equation for the eigenstate E is obtained

$$J_{-\varepsilon}(2\alpha/V_0)Y_{N+1-\varepsilon}(2\alpha/V_0) = J_{N+1-\varepsilon}(2\alpha/V_0)Y_{-\varepsilon}(2\alpha/V_0), \qquad (1.50)$$

in which $\varepsilon \equiv (E - E_0)/V_0$. Fukuyama showed that $\varepsilon = n$ for $\alpha = 0$, which is a statement of Stark quantization. For $\alpha \ll V_0$, he found that only states near the band edges are affected $\sim N^{-1}$. We shall take only three coupled wells and examine the question of localization, where the separations are given by V_0 (Tsu, 1992, 1993). With $V_0 \equiv eFd$ and the nearest neighbor given by $\langle n|H'_n|n' \rangle = \alpha$, the energy states are given by

$$\frac{E}{E_0} = 1 + \left(\frac{\alpha}{E_0}\right)b, \quad \psi_+ = \frac{1}{b}\left(\frac{|1\rangle}{b-\beta} + |2\rangle + \frac{|3\rangle}{b+\beta}\right),$$

$$\frac{E}{E_0} = 1, \qquad\qquad \psi_c = \frac{1}{b}(|1\rangle - \beta|2\rangle - |3\rangle)), \qquad (1.51)$$

$$\frac{E}{E_0} = 1 - \left(\frac{\alpha}{E_0}\right)b, \quad \psi_- = \frac{1}{b}\left(\frac{-|1\rangle}{b+\beta} + |2\rangle - \frac{|3\rangle}{b-\beta}\right),$$

where $b = (2 + \beta^2)^{1/2}$ with $\beta \equiv V_0/\alpha$.

Figure 1.13 shows the plot of $|\psi|^2$ for several values of β with $\alpha/E_0 = 0.1$, and Table 1.1 gives the values of the energy E/E_0.

It is clear from these energies that, whenever $\beta > 1$, the energy separation approach V_0, the applied voltage, and the corresponding wave functions start to be localized. For example, the case for $\beta = 4$, i.e., $V_0 = 4\alpha$, $|\psi|^2$ shows almost total

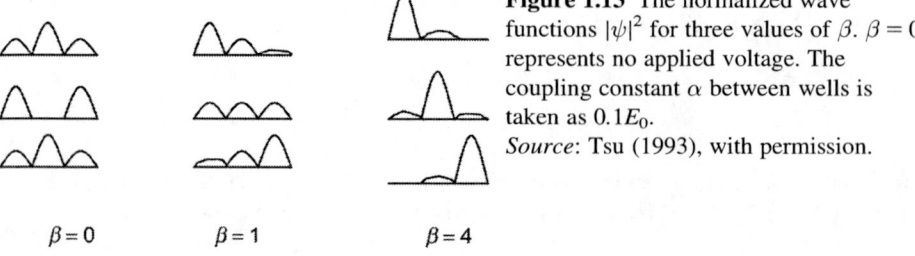

Figure 1.13 The normalized wave functions $|\psi|^2$ for three values of β. $\beta = 0$ represents no applied voltage. The coupling constant α between wells is taken as $0.1E_0$.
Source: Tsu (1993), with permission.

$\beta = 0$ \qquad $\beta = 1$ \qquad $\beta = 4$

Table 1.1 Values of E/E_0 for Various β and $\alpha = 0.1E_0$

	$\beta = 0$	$\beta = 1$	$\beta = 4$
E_+/E_0	1.141	1.173	1.424
E_c/E_0	1	1	1
E_-/E_0	0.859	0.827	0.576

localization in the sense that the peak moves from the left to the right and the separation is dominated by the applied voltage. The formation of the SL is based on the translational symmetry in a constant applied field. Although Esaki and I used the model of an infinite structure to calculate the minibands quite simply, we knew in reality one deals with finite chains, perhaps no more than three as pointed out in the very first paper (Esaki and Tsu, 1970). In an infinite system under an applied voltage, with $+V$ on the right, the band profile displays the familiar slope down, shown in Figure 1.14, and the minibands slope down accordingly. However, we must be very careful not to misuse such a potential profile. As soon as the energy gain eV_d from the applied voltage per period exceeds the bandwidth of the miniband, we are not in the band regime, but rather, in the hopping regime where transport as we normally understand it does not apply, because the energy states break up into localized states as in a quantum well. In such a finite structure consisting of two or three coupled wells, energy states that are flat and horizontal across the coupled wells do not slope down unlike the drawing applied to normal transport involving bulk semiconductors. Take a system of two coupled quantum wells forming a repeated chain. The energy of each period is flat across the period. Now, let us extend to three or four and we end up with the picture represented by the coupled wave function shown in Figure 1.13. When the voltage drop per period exceeds the bandwidth of the miniband, localization dominates, the wave function is localized within each well and a ladder structure appears. In a normal bulk semiconductor, the high-field region is actually limited by the velocity saturation with a field of $\sim 10^5$ V cm^{-1}. In a normal solid having a lattice constant of few angstroms, eV_d is less than few millielectron volts, which is minuscule in comparison to the energy bandwidth. The failure to distinguish cases with finite length contributed to the debate on the existence of the SL (Rabinovitch and Zak, 1971).

The interesting domain formation in a superlattice (Esaki and Chang, 1974) is in the regime of localized hopping conduction. In their structure of a 50 period—GaAs/AlAs superlattice, with an energy band $\Delta E_1 \sim 10$ meV and at an applied voltage of 10 meV per period, the energy states are certainly localized. Similarly, one of the three samples in the photoconductivity study reported by Tsu et al. (1975) is also within the localized regime. In essence, most of these structures

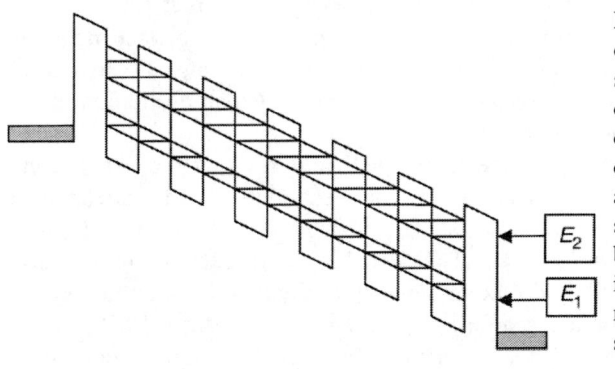

Figure 1.14 Conventional way of representing a periodic structure with energy bands denoted by E_1 and E_2, sloping down from the left to the right contact with a positive V. With a voltage drop per period of V_d such that eV_d exceeds the bandwidth ΔE_1, such a profile is incorrect, because the minibands stay flat instead of sloping down.

were in the regime of multiple quantum wells, rather than true superlattices where conduction is dominated by the conventional band model. An account dealing with these points will be given later.

1.6.3   Origin of NDC from the Band Model and Hopping Model

Thus far, using the band model (Esaki and Tsu, 1970), the condition for NDC is $\omega_B \tau \geq 1$, and in physical terms this is the presence of a Bloch oscillation as electrons are driven to the minizone boundary by the applied electric field F. On the other hand, Tsu (1975) pointed out in an unpublished paper that NDC in the hopping model is due to the decrease in overlap of the wave functions between the SL energy levels as the applied field is increased.

Let us point out some of the salient features described by Tsu (1975). The condition for Bloch oscillation is $\omega_B T = 2n\pi$, with $T = d/\langle v \rangle$. For the tightly binding $E-k$ relationship, $\langle v \rangle = E_1 d/\pi\hbar$, so that

$$eFd = E_1/n. \tag{1.52}$$

And the maximum excursion $x_m = E_1/eF$, which is same as Eq. (1.25b) with F replaced by F_0, because in Section 1.5, F is reserved for a time-varying electric field. At x_m, the corresponding maximum k is $k_m = \pi/d$, so that electrons can, at most, gain an energy of E_1 from the field F. When E_1 is greater than energy of the optical phonon, interaction with phonons reduces the scattering time, requiring an increase of the field to reach Bloch oscillation.

On the other hand, in the picture of hopping using the Kane function (1959), the transition of a hop to the next state from ε_1 to $\varepsilon_1 - eFd$, or to the next nearest energy state from ε_1 to $\varepsilon_1 - 2eFd$, and so on, can be expressed by

$$\langle 1|e^{i\mathbf{q}\cdot\mathbf{r}}|1+n\rangle^2 \to \left(\frac{E_1}{V}\right)^{2n}, \quad \frac{E_1}{V} \to 0. \tag{1.53}$$

Thus, the hopping model also gives the condition $neFd = E_1$, which is same as Eq. (1.52). Therefore, the onset of NDC for the hopping model is described by the same physical principles. Furthermore, the decrease of overlap between neighboring states leads to the same effects even with impurity scatterings (Döhler et al., 1975). In summary, NDC is nothing but the condition for Bloch oscillation as first described by the simple band model.

Let us calculate the mobility of the GaAs well region using the experimental data for the $I-V$ curves (Chang et al., 1973) for a superlattice with barrier width $B = 1$ nm, well width $W = 6$ nm, and a barrier height $V_1 \sim 0.4$ eV. From the Penny–Krönig model, $E_1 = 70$ meV. The measured current maximum gives $eFd = 18$, or $n = 70/18 = 3.9$. The electron mean free path is $\sim 4d$, four periods. This allows us to obtain for the scattering time $\tau = nd/\langle v \rangle \sim 1.2 \times 10^{-13}$ s corresponding to $\mu = 4650$ (cm^2 V^{-1} s^{-1}) using the tight-binding model. Other

Table 1.2 Comparison of Calculated Scattering Time and Mobility from Measured Data

W	B	E_1 (meV)	n	nd (nm)	τ (10^{-13} s)	μ (cm^2 V^{-1} s^{-1})	References
6	1	70	4	28	1.2	4650	Chang et al. (1973)
5	5	38	2	20	5.9	4650	Tsu et al. (1975)

measurements from Tsu et al. (1975), with different W and B values, led to the same value for the mobility. These are given in Table 1.2.

Note that the mobility is very close to the typical value of 5000 for molecular beam epitaxy (MBE) GaAs epitaxial layers. The exact equality for the two cases is due to rounding off the decimals. Nevertheless, it is remarkable that they agree. Therefore, these results indicate that it is simple to characterize the mobility using only the position of the onset of the NDC. Let us ask what happens to real solids. If we assume that the anomaly observed by Maekawa (1970) maybe due to the same physics as sketched preceding, then indeed the SL in real solids has already been observed. More recently, the existence of the SL has been experimentally verified (Leo et al., 1991; Feldmann et al., 1992; Shah, 1992).

The interesting domain formation in a superlattice (Esaki and Chang, 1974) is in the regime of localized hopping conduction. In their structure of a 50 period–GaAs/AlAs superlattice, with an energy band $\Delta E_1 \sim 10$ meV and at an applied voltage of 10 meV per period, the energy states are certainly localized. Similarly, one of the three samples in the photoconductivity study reported by Tsu et al. (1975) is also within the localized regime. In essence, most of these structures were in the regime of multiple quantum wells, rather than true superlattices where conduction is dominated by the conventional band model. An account dealing with these points will be given later.

1.7 Experiments

1.7.1 Domain Oscillation in a Superlattice

Esaki and Chang (1974) reported the oscillation observed in a superlattice due to a domain of high-field regions induced by the NDC at an electric field beyond the inflection point. Because of unavoidable inhomogeneity, the high field–induced domain appears and is followed by oscillatory behavior. The superlattice structure used in their experiments comprises 50 periods of 4.5-nm GaAs wells and 4 nm of AlAs barriers. The calculated energy: $E(\Delta E)$, both in millielectron volts, is 80(8). NDC appears at $\sim$0.8 V, corresponding to 16 mV per period. About half the voltage per period $\sim$16 mV is across the well region. Thus, NDC appears at $\sim$8 meV, which is close to the bandwidth $\Delta E \sim 8$ meV of the calculated miniband for a 4.5-nm well width. Therefore, the instability starts at the threshold of localization, rather than the NDC in the superlattice. In short, this superlattice is closer to a

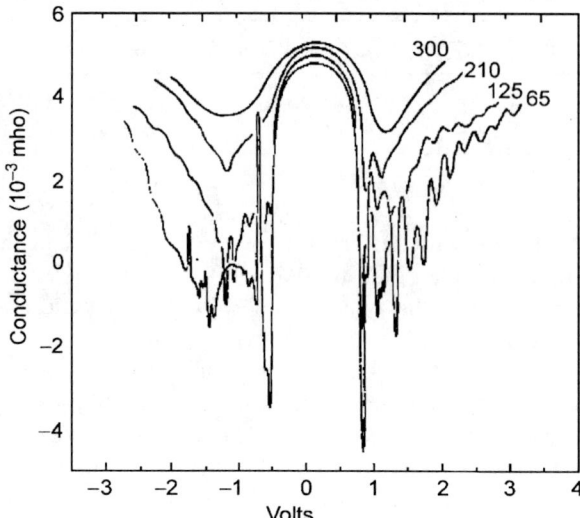

Figure 1.15 Difference conductance versus applied voltage at 65–300 K. Note the NDC at 125 and 65 K. *Source*: After Esaki and Chang (1974), with permission.

multiple quantum well rather than a superlattice. Figure 1.15 shows the conductance from 65 to 300 K.

Figure 1.16 shows energy profiles with the domain formation on the left and the corresponding conductance at points: (a) initial low field band conduction, followed by the appearance of NDC shown on the right in (b), where domain formation is initiated. On further increase of the applied voltage, a domain is established in (c). And finally conductance is restored when the current in the lower band E_1 can tunnel through the higher E_2 of the adjacent double barrier quantum well section, resulting in the establishment of sections, and the domain throughout the length of the structure. This experiment has established without doubt the presence of man-made quantum states. Depending on the applied voltage, conduction is transformed from initial band-like conduction to hopping conduction. Does the appearance of oscillation signify instability? Not at all! Any system with an NDC will result in oscillation. Even if there is none internally, the system will oscillate at the maximum gain dictated by the overall circuit in the measurement. Shall we apply bias with a large resistor to prevent oscillation? The answer is yes, purely for observing the slope of the NDC. However, for the amplifier, we need all the NDC possible for positive feedback.

1.7.2   Experiments on the SL

Koss and Lambert (1972) reported the effect of a constant electric field on optical absorption in GaAs. Figure 1.17 is a replot of their data with the lines removed. Without the guide from the calculated lines, the data do not clearly indicate jumps in $\hbar\omega = eFa$. The situation is quite different in a superlattice. Mendez et al. (1988) observed field-dependent photoluminescence and photocurrent with results consistent with field-induced localization. They observed an increase in localization with

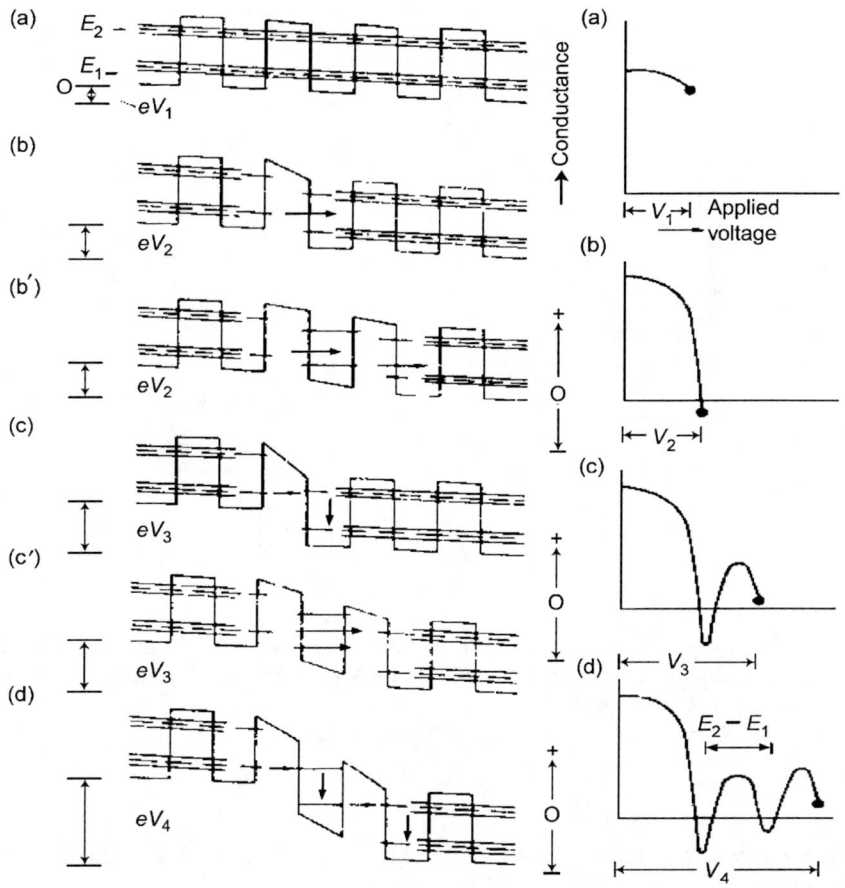

Figure 1.16 Energy profile with domain formation (left). Conductance at points: (a) band conduction, (b) initial domain formation, (c) expansion of domains, and (d) fully developed domains.
Source: After Esaki and Chang (1974), with permission.

an increase in the applied electric field. Voisin et al. (1988) observed the effect of Stark quantization in a GaAs/AlGaAs superlattice on electroreflectance and Lyssenko et al. (1997) observed the spatial displacement of the Bloch oscillating electrons. Leavitt and Little (1990) put a quantum well in between superlattices serving as a notch for reference in their study of optical transition involving SL. It is interesting to note that Dignam and Sipe (1991) considered the treatment by Emin and Hart (1987) as the ultimate proof of the existence of the SL. In this work, they take the constant electric field as a superposition of a sawtooth ramp and a step. Actually this approach is no more valid a proof than previous work. This is the proper place to comment on the many theoretical investigations.

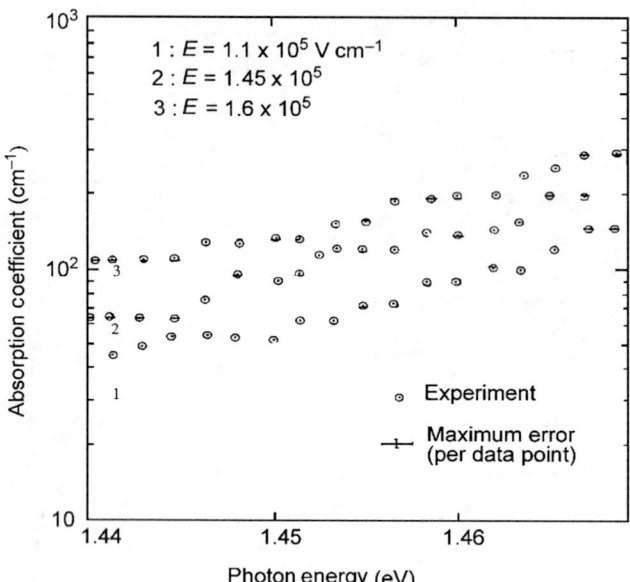

Figure 1.17 Absorption coefficient versus photon energy taken from Koss and Lambert (1972). The figure is replotted with comparison to theory removed. The stepwise increase at $\Delta\hbar\omega = neFa$ is not so obvious.
Source: After Koss and Lambert (1972), with permission.

Personally, we feel that the many body approach utilizing a linear chain model is basically pedagogical (Chen and Zhao, 1991).

Recently, the NDC first predicted by Esaki and Tsu (1969) was observed by several teams (Beltram et al., 1990; Sibille et al., 1990). These are very important experimental confirmations of Stark quantization. As we discussed earlier, for $\omega_B\tau \gg 1$, an electron oscillates by virtue of repeated Bragg reflections, leading to localization. Again, an analogy to the magnetic quantization is in order. The observation of cyclotron resonance or the de Haas—van Alphen effect is a direct manifestation of magnetic quantization. The observation of NDC is a direct manifestation of electric field quantization. Beltram et al. (1990) were quite correct in pointing out that localization starts at $\omega_B\tau > 1$, the condition for NDC. SL theories predict that SL exists at any field because scattering is not considered. Obviously, it is meaningless to consider quantized energy level for $\omega_B\tau < 1$. This author tried to consider the effect of incoherent scattering on the energy states of a quantized system (Tsu, 1989). It is common first to find the eigenstates using perturbation calculations. Scatterings are considered and energy broadening is obtained from the scattering cross section. When scattering is large, we know that not only levels are broadened, but there should be an accompanied level shift. These considerations together with the discussion of the effects of scattering on localization should induce some theorists to formulate a realistic theory instead of redoing what has been done many times. The reason is obvious: most semiconductors do not have sufficiently long τ, or mean free path, so that the effects of scattering are dominant. In other words, experiments with real solids instead of superlattices have a marginal chance of success. For this reason, scattering must be taken into account in any realistic theory. This is also why this author took the Boltzmann transport approach.

Apart from what has been discussed, there is another important difference between a superlattice and real solids. In superlattices, if the applied electric field is misoriented with respect to the periodic axis there are few consequences. However, in real solids, there is a built-in smearing effect. If the applied electric field is slightly off the major symmetry axes, theory predicts that there will be no Stark quantization, and yet, intuitively, no major problem should arise because there is always some spread in the field axis. Perhaps, a wave packet consisting of a solid angle of **k** vector centered about the field axis should be taken.

The coherent oscillations of a wave packet in a double-well structure have been recently observed by Leo et al. (1991). Furthermore, Bloch oscillation in a semiconductor superlattice has also been observed by Feldmann et al. (1992). Actually, superlattice and coupled quantum wells are not all that different theoretically and experimentally. As discussed earlier, an electric field induces localization in a periodic system, as well as coupled wells, as shown in Figure 1.18, by the same mechanism, separating the energies between adjacent cells resulting in a decoupling of wave functions. Therefore, these experimental observations confirm localization induced by the application of a constant electric field.

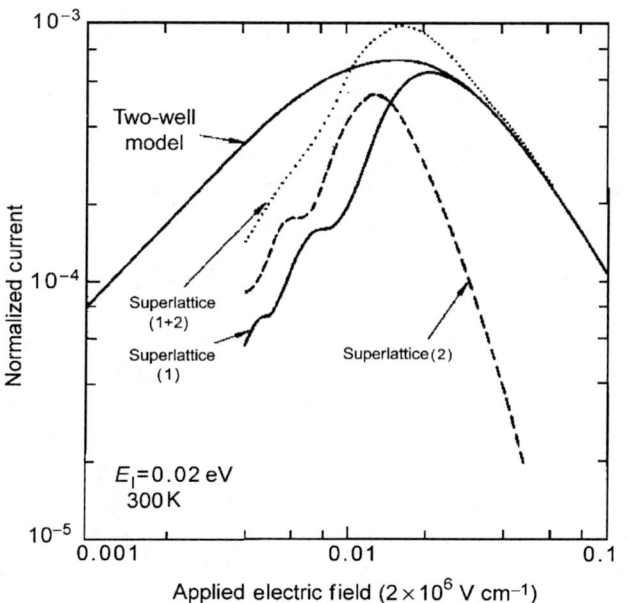

Figure 1.18 Current versus applied voltage in the hopping regime. *Source*: After Tsu and Döhler (1975), with permission.

1.7.3 Comparison with Cyclotron Resonance

I spent some time during the summer of 1991 at IBM Research working with Leo
Esaki. He asked me to look into the confusion that had surfaced between the band
model and the hopping model and which resulted in our paper (Tsu and Esaki,
1991), in particular, making a comparison between cyclotron resonance and the SL.
Since the analogy is much more than casual, the situation will be discussed here in
more detail.

Before the establishment of energy quantization in units of $\hbar\omega_c$ with $\omega_c = eH/mc$,
the dynamics of electrons in a magnetic field was treated in the traditional equation

$$\mathbf{k} = (e/\hbar c)\mathbf{v} \times \mathbf{H}. \tag{1.54}$$

The situation is entirely analogous to our situation, with the use of $\mathbf{k} = (e/\hbar)\mathbf{F}$
for NDC, by Esaki and Tsu (1970), establishing the condition for the Bragg reflection. Without scattering, owing to the oscillatory nature, the electrons are entirely
localized within a distance $\langle x \rangle$ given by Eq. (1.25b). Similarly, in cyclotron resonance, without scattering, electrons are entirely localized within a cyclotron orbit.
The quantization of cyclotron orbit is treated without scattering (Ziman, 1988), in
much the same way as the treatment of the SL. Experimentally, taking scattering
into account, the condition $\omega_c\tau > 1$ must be observed. In addition, as $\hbar\omega_c$
approaches the bandwidth of the energy band in question, simple transport theory
using the preceding equation of $\mathbf{k}$ cannot apply. And yet, a formal quantization
treatment by solving the Schrödinger equation with multiple bands lacks the effect
of scattering. To treat it properly, strictly speaking Green's function should be
used, as discussed in the section on damping in resonant tunneling. Therefore, we
must examine closely the validity of the band model and the hopping model.
However, from an experimental point of view, as in the case of cyclotron resonance, solution of the Boltzmann equation is certainly the key to understanding the
experimental results. As in most experiments on superlattices, the range of validity
of the band model is exceeded, and therefore, in most situations, the concept of
transitions between SL levels needs to be considered. There is one subtle point that
needs to be clarified—the role of scattering in localization. Again, we shall first
discuss magnetic quantization. When an electron undergoes a collision before completing one circular orbit, cyclotron resonance cannot be important. Similarly,
whenever an electron suffers a collision before the Bragg reflection, the SL cannot be
dominant. Intuitively, if a wave function is localized in one cell, scattering can couple a
wave function to wave functions in the adjacent cells, resulting in "delocalization," similar to the case of an electron in one cyclotron orbit scattered into another orbit. Why
then is the maximum excursion $\langle x \rangle_m$ reduced because of scattering? The answer to this
paradox lies in the fact that localization in quantum mechanics is only defined in terms
of stationary eigenstates, i.e., the extent of the eigenfunction, and scattering brings about
incoherent transitions among stationary states. Therefore, scattering reduces the coherent length—in other words, "delocalizes" the wave function—resulting in spreading.
Since this spreading is incoherent, we should not consider that localization is

reduced. On the contrary, localization, as defined by the extent of the coherent part of the wave function, is in fact increased. We can thus clearly state that localization is increased by scattering. This is certainly true in amorphous quantum wells. We shall close this discussion with a definitive statement regarding the effects of scattering. Scattering increases localization. And this increase in localization in turn accompanies broadening of the SL. Therefore, we have reached an important understanding that should be remembered—*When keeping the scattering fixed, increases in the applied electric field increase localization and sharpen the SL levels. When keeping the field fixed, increases in scattering will also increase localization, but broaden the SL levels!*

To summarize, the relationship between cyclotron resonance and magnetic quantization is identical qualitatively to the Bloch oscillation and electric field—induced quantization. At present, before scattering is included in the quantization schemes, the Boltzmann transport equation provides a better guide for experiments. At extremely high field where $\hbar\omega_\mathrm{B} \sim E_1$, scattering considerations are less important in hopping models because localization is sufficiently complete that the wave functions are almost decoupled between adjacent cells. This extreme quantization is unlikely to be observable in a real solid except in molecular crystals. Nonetheless, this extreme localization is in fact easier to observe in superlattices and quantum well structures. Since the DOS obtained from SL formulation is not obtainable from a semiclassical band model, effects such as optical transitions, involving these Stark levels, and resonant tunneling when these levels are lined up between adjacent cells, require a quantum mechanical approach. This aspect is, of course, apparent in the case of magnetic quantization—transitions involving Landau levels must be treated with the full Hamiltonian. Unlike in magnetic quantization, $\omega_\mathrm{c}\tau \gg 1$ is generally achievable, but in the SL, $\omega_\mathrm{B}\tau \gg 1$ is generally difficult to obtain. Theorists willing to tackle the scattering in the SL can indeed provide an important contribution to the field of electric field quantization.

From a device point of view, which was what provided the rationale in this field, with the advent of hypermobility materials (Pfeiffer et al., 1989), many new effects such as NDC, material for nonlinear optics (Tsu and Esaki, 1971), parametric light amplifier (Monsivais et al., 1990; Tsu, 1990), and ultimately a terahertz Bloch oscillator, are experimentally realizable or already proven.

1.8   Type-III Superlattice (Historically Type-II Superlattice)

Superlattices treated thus far involve modulations of the conduction band or valence band. In some special heterostructures of A/B the conduction band of A lies close to the valence band of B, forming the so-called type-III superlattice. In InAs/GaSb, the top of the valence band of GaSb is above the bottom of the conduction band of InAs. It turns out that this type of superlattice has a unique feature— the band gap of the miniband, the energy separation between the miniconduction band and minivalence band depends on the length of the periodic variation of the two materials, A and B. For small separations, these two bands interact, resulting in a new band gap. However, for large separations, interaction of these bands vanishes so that the minigap tends to zero resulting in a zero-gap superlattice. Since part of the purpose of this book is to stimulate ideas, I would like to describe how the idea came to Esaki and me. Esaki asked me if I had some insight into how to move the point of inflection, when the mass changes sign at some value of $E-k$ of the miniband, closer to the minizone center, in order to lower the electric field necessary for NDC. While we were examining various combinations of compound semiconductors, Esaki took a copy of *Handbook of Electronic Materials* by Neuberger (1971), and asked what would happen if the top of the valence band lies below the bottom of the conduction band. Although we did not come up with a clear-cut physical insight regarding whether this system would result in moving the point of inflection closer to the Γ point, we agreed that the idea was definitely worth exploring. I decided that Kane's two-band k·p model (Kane, 1959) would be most suited for this problem, partly because I had used this method before.

1.8.1   Material Parameters for the $In_{1-x}Ga_xAs/GaSb_{1-y}As_y$ System

If we select a pair of alloy semiconductors, $In_{1-x}Ga_xAs/GaSb_{1-y}As_y$, and choose $y = 0.918x + 0.082$, then a perfect lattice match is obtained for the A/B system. The material parameters used for a suitable A/B system (Sai-Halasz et al., 1977) are tabulated in Table 1.3 taken from Neuberger (1971).

Three pairs—the (a), (b), and (c) with band-edge energies shown in Figure 1.19 by bracketed arrows—are plotted against the alloy composition x or y. The energy ordinate scale is referenced to the vacuum level. A possible bowing is shown as a dashed line. However, Vegard's law is used for the calculation.

1.8.2   Kane k·p Two-Band Model

Because the adjacent layers involve the coupling of the conduction and valence bands, unlike type-I superlattices, Bloch waves instead of plane waves must be

Table 1.3 Material Parameters Selected for the Type-II Superlattice

	Energy gap E_g (eV)	Electron affinity χ (eV)	Electron mass m_e/m_0	Light hole mass m_{lh}/m_0	Lattice constant a (nm)
(a) InAs	0.36	4.9	0.024	0.024	0.6058
GaSb	0.70	4.06	0.048	0.056	0.6094
(b) InGaAs	0.44	4.838	0.027	0.028	0.6028
GaSbAs	0.81	4.062	0.051	0.060	0.6028
(c) InGaAs	1.078	4.344	0.053	0.063	0.5786
GaSbAs	1.211	4.067	0.062	0.074	0.5786

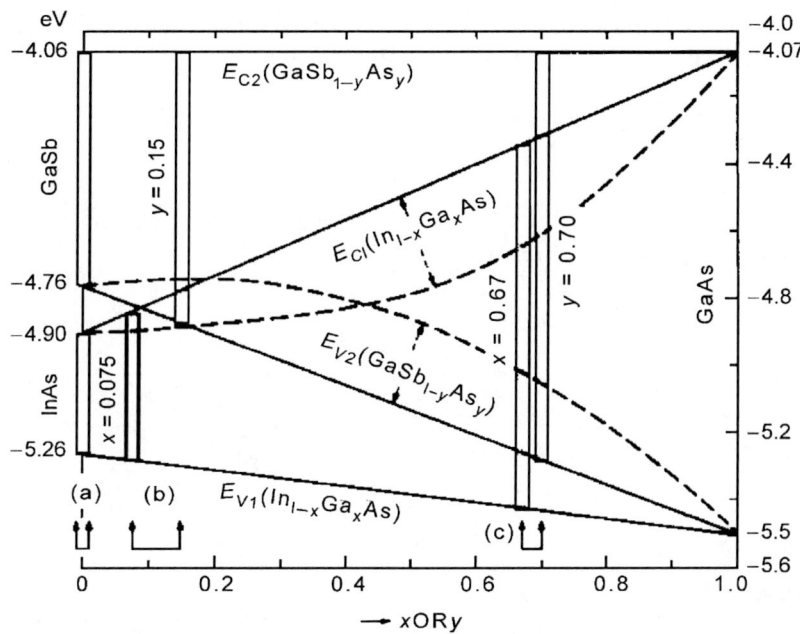

Figure 1.19 Changes in band-edge energies with composition x or y in the $In_{1-x}Ga_xAs$ and $GaSb_{1-y}As_y$ quaternary system. Pairs (a), (b), and (c) bracketed by arrows are used for the calculations.
Source: Sai-Halasz et al. (1977), with permission.

used. The plane wave solution is adequate in general whenever the conduction band minimum of one layer is far from and above the valence band maximum of an adjacent layer. In the quaternary system, the top of the valence band of GaSb is 0.14 eV above the bottom of the conduction band of InAs. The k·p two-band model of Kane (1959) is ideal for treating this case, although in general the method can be extended to include other bands, for example, the calculation of the energy band for GeTe by Tsu et al. (1968). The Hamiltonian matrix for the two bands is given by

$$
H = \begin{bmatrix} E_g + \dfrac{\hbar^2 k^2}{2m_0} & \dfrac{\hbar}{m_0} \mathbf{k \cdot p} \\[2ex] \dfrac{\hbar}{m_0} \mathbf{k \cdot p} & \dfrac{\hbar^2 k^2}{2m_0} \end{bmatrix},
\tag{1.55}
$$

where p is the momentum matrix element taken, for convenience, to be real as originally assumed by Kane, and the zero energy is at the valence band maximum.

Then the $E-k$ relationships of the conduction and valence band, ε_c and ε_v, respectively, obtained from the secular determinant are given by

$$\varepsilon_{c,v} = \frac{\hbar^2 k^2}{2m_0} + \frac{E_g}{2}\left[1 \pm \sqrt{C}\right]. \tag{1.56}$$

where $C \equiv 1 + (\hbar^2 k^2/\mu E_g) \equiv \kappa^2$, in which the reduced mass $\mu^{-1} = m_c^{-1} + m_v^{-1}$. The eigenfunctions diagonalize equation (1.55), are

$$U_c = (1 + |b|^2)^{-1/2}[|1\rangle + b|2\rangle] \tag{1.57a}$$

and

$$U_v = (1 + |b|^2)^{-1/2}[b|1\rangle - |2\rangle], \tag{1.57b}$$

in which

$$b \equiv [(\sqrt{C} - 1)/(\sqrt{C} + 1)]^{1/2},$$

where $|1\rangle$ and $|2\rangle$ represent the wave functions for $k=0$. For $k=0$, $C=1$, and $b=0$, there is no mixing of the two bands, whereas at large k, $C \gg 1$ so that $b \to 1$, the U functions are the usual $(|1\rangle \pm |2\rangle)/\sqrt{2}$. At the branch point, $\hbar^2 k^2 = -\mu E_g$, b is purely imaginary, so that $U = (|1\rangle + i|2\rangle)/\sqrt{2}$. In general for $\mathbf{k} \neq 0$, the states $|1\rangle$ and $|2\rangle$ represent the wave functions at $\mathbf{k} = \mathbf{k}_0$, then

$$\varepsilon_{c,v} = \frac{\hbar^2 (\mathbf{k} - \mathbf{k}_0)^2}{2m_0} + \frac{E_g}{2}\left\{1 \pm \left[\frac{\hbar^2}{E_g}(\mathbf{k} - \mathbf{k}_0)m^{-1}(\mathbf{k} - \mathbf{k}_0)\right]^{1/2}\right\}, \tag{1.58}$$

where the mass tensor m^{-1} has $m_{11} = m_{22} = \mu_t^{-1}$ and $m_{33} = \mu_t^{-1}$. The coupling parameter in terms of $\kappa^2 \equiv \hbar^2 k^2/\mu E_g$ becomes $b = (\sqrt{C} - 1)/\kappa$, when it is evident that the coupling parameter, b depends on the sign of κ. A wave traveling in the $+x$ direction with $(+)$ and in the $-x$ direction with $(-)$ should be used in U_c of Eqs. (1.57a) and (1.57b) for $b(\pm\kappa)$ or $b(\pm k)$. Furthermore, for the valence band, the signs are reversed such that electrons traveling in the $+x$ direction with $(-)$ and $-x$ direction with $(+)$ should be used in U_v of Eq. (1.58) for $b(\pm\kappa)$ or $b(\pm k)$. In particular, the next section describes the reflection and transmission coefficients for waves incident on an interface, when the Bloch functions instead of the plane waves for the electrons must be used. The use of the two-band k·p model is ideal for this situation.

1.8.3 When Are the Full Bloch Waves Needed?

Figure 1.20A shows a type-I superlattice, i.e., an electron in a conduction band incident to the left of another conduction band separated by an interface and a type-III superlattice in (B) where the right side is a valence band at the same energy.

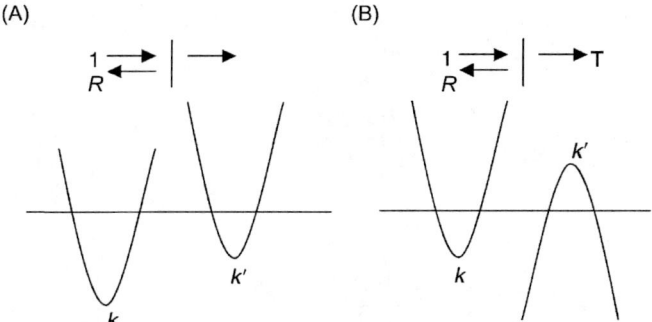

Figure 1.20 The $E-k$ for (A) type-I and (B) type-III superlattices. The horizontal line indicates the energy of the electron.

It is generally unnecessary to include the full Bloch functions for conduction band to conduction band, whereas for conduction band to valence band shown in (B), serious error would generally prevail without the Bloch functions. Explicitly, the superscripts $(+)$ and $(-)$ denote the waves moving to the right and left, respectively and the subscripts c and v denote the conduction and valence bands, or the upper and lower bands:

$$\psi_c^+ = U_c(k, x)e^{ikx}, \quad \psi_c^- = U_c(-k, x)e^{-ikx}, \tag{1.59a}$$

$$\psi_v^+ = U_v(-k', x)e^{ik'x}, \quad \psi_v^- = U_v(k', x)e^{-ik'x}, \tag{1.59b}$$

It is important to note the signs of the exponentials for the waves $(+)$ and $(-)$ here, as well as in U_c and U_v, where the signs originate from $b(k)$. Incorrect signs in Eqs. (1.59a) and (1.59b) are often the source of errors in calculations involving incident and reflected waves.

Let us proceed with the reflection problem, with an electron from the left conduction band and emerging from the right of the interface into the conduction band with $(+)$ for k_2, and valence band with $(-)$ for k_2. For convenience, for the conduction band electron incident from the left onto an interface located at $x = 0$, we use $U_1 \equiv U_c(k_1, x)$, $V_1 \equiv U_c(-k_1, x)$; and for the transmitted electron to the right, $U_2 \equiv U_v(\mp k_2', x)$, $(-)$ for movement to the right and $(+)$ for movement to the left, then

$$\psi_1 = U_1 \exp(ik_1 x) + R V_1 \exp(-ik_1 x) \tag{1.60a}$$

and

$$\psi_2 = T U_2 \exp(ik_2 x). \tag{1.60b}$$

Matching these wave functions and their derivatives (equal effective masses are taken for this example), there results

$$R = \frac{(k_1 - k_2) - i[(U_1'/U_1) - (U_2'/U_2)]}{(k_1 + k_2) - i[(U_2'/U_2) - (V_1'/V_1)]} \tag{1.61a}$$

and

$$T = \frac{2k_1 - i[(U_1'/U_1) - (V_1'/V_1)]}{(k_1 + k_2) - i[(U_2'/U_2) - (V_1'/V_1)]}. \tag{1.61b}$$

Obviously for plane waves, these U functions do not appear in R and T. Note that these U functions, for example, U_1, are given by Eqs. (1.57a) and (1.57b). Therefore, some algebra is involved resulting in R and T once the basis states $|1\rangle$ and $|2\rangle$ are chosen. As an example leading to useful and simple results, two ortho-normal functions satisfying the periodicity of the lattice $g \equiv 2\pi/a$, where a is the lattice constant, maybe taken as

$$|1\rangle = (g/\pi)^{1/2}\cos(gx) \quad \text{and} \quad |2\rangle = i(g/\pi)^{1/2}\sin(gx), \tag{1.62}$$

for type I, conduction band to conduction band, so we have

$$R = \frac{(k_1 - k_2) + [b_1(k_1)g_1 - b_2(k_2)g_2]}{(k_1 + k_2) + [b_1(k_1)g_1 + b_2(k_2)g_2]} \tag{1.63a}$$

and

$$T = \frac{2[k_1 + b_1(k_1)g_1]}{(k_1 + k_2) + [b_1(k_1)g_1 + b_2(k_2)g_2]}. \tag{1.63b}$$

For type II with an incident electron in the conduction band and a transmitted electron in the valence band, R and T are

$$R = \frac{(k_1 - k_2) + [b_1(k_1)g_1 + b_2^{-1}(k_2)g_2]}{(k_1 + k_2) + [b_1(k_1)g_1 - b_2^{-1}(k_2)g_2]} \tag{1.64a}$$

and

$$T = \frac{2[k_1 + b_1(k_1)g_1]}{(k_1 + k_2) + [b_1(k_1)g_1 - b_2^{-1}(k_2)g_2]}. \tag{1.64b}$$

We shall take a special case using Eq. (1.64a) for the case shown in Eq. (1.56), a typical case where the incident electron from the conduction band is transmitted into the valence band on the right with the top of the valence band lying above the bottom of the conduction band, for $k_1 \approx k_2$, $g_1 \approx g_2 \gg k_1$ and $b_1 \approx b_2 \sim 1/\sqrt{2}$, $R \sim 1/3$, whereas without these U functions, $R \sim 0$. Physically, the wave functions for the conduction band and valence band are orthogonal, but are coupled at the same energy, in this case, the $\mathbf{k} \cdot \mathbf{p}$ term that couples them results in a linear combination of these orthogonal functions. We recognize that when the reflectivity increases from 0 to 1/3 it indicates that an effective barrier is present between the electron in the conduction band and valence band. To get some idea about this effective barrier, one may simply substitute the U functions from Eqs. (1.57a) and (1.57b) into the Schrödinger equation after canceling the plane wave parts for the effective V. The seemingly complicated problem for the overlapping type-III superlattice can be treated quite simply using the principles established in this section. A more direct explanation in terms of an effective potential experienced by an electron in the case of InAs/GaSb, when an electron from the conduction band aligns in energy to an electron in the valence band, will be further elaborated after the next section.

1.8.4 Type-III Superlattice with the Kane Two-Band Model

Writing the Bloch function $\psi(k, x) = U(k, x)e^{ikx}$ for the wave going to the right, and $\psi(-k, x) = U(-k, x)e^{-ikx}$, for the wave going to the left, then the waves in each sections are

$$\psi_1(k_1,x) = [AU_1(k_1,x)\exp(ik_1 x) + BU_1(-k_1,x)\exp(-ik_1 x)], \quad 0 < x < d_1,$$
(1.65a)

$$\psi_2(k_2,x) = [CU_2(k_2,x)\exp(ik_2 x) + DU_2(-k_2,x)\exp(-ik_2 x)], \quad d_1 < x < d_1 + d_2 = d,$$
(1.65b)

$$\psi_3(k_3,x) = \psi_1(k_1, x - d)\exp(ikd), \quad d < x < d_1 + d.$$
(1.65c)

Assuming each layer thickness is an integer multiple of the lattice constant a in Table 1.3, the $E-k$ relationship for k as a function of E results,

$$\cos(kd) = \cos(k_1 d_1)\cos(k_2 d_2) - F\sin(k_1 d_1)\sin(k_2 d_2),$$
(1.66)

in which

$$F = \frac{1}{2}\left[\frac{ik_1 + U_1'/U_1}{ik_2 + U_2'/U_2} + \frac{ik_2 + U_2'/U_2}{ik_1 + U_1'/U_1}\right],$$
(1.67)

where $U_i = U_i(k_i, 0)$ and $U_i' = (dU_i(k_i,x)/dx)|_{x=0}$, and the allowed bands correspond to energy with real k. Equations (1.66) and (1.67) reduce to the well-known Krönig–Penney solution if the logarithmic derivatives in Eq. (1.67) are zero. Some of the details in calculating the $E-k$ from the preceding equations have been left out in the published version (Sai-Halasz et al., 1977). A one-dimension derivation without including self-consistent potential and spin will be treated in what follows. The LCAO by Sai-Halasz et al. (1978) did take three dimension into account, but left out the self-consistency and spin. In fact, LCAO led to the same results as the use of k · p.

From the viewpoint of physical insight, the k · p model is more transparent to numerical computations. Using the U functions in Eq. (1.57b), and the $|1\rangle$ and $|2\rangle$ in Eq. (1.62), the energy versus k relations for several cases are shown in Figure 1.21, taken from Sai-Halasz et al. (1977). A word of caution on this simple orthonormal function is appropriate.

The momentum matrix element $\langle 1|P|2\rangle = \hbar g = 2\pi\hbar/a$, with a being the lattice constant. However, the Kane model also gives $\langle 1|P|2\rangle = (m_0/2)(E_g/\mu)^{1/2}$. Therefore, the use of these simple functions puts a restriction on the lattice constant

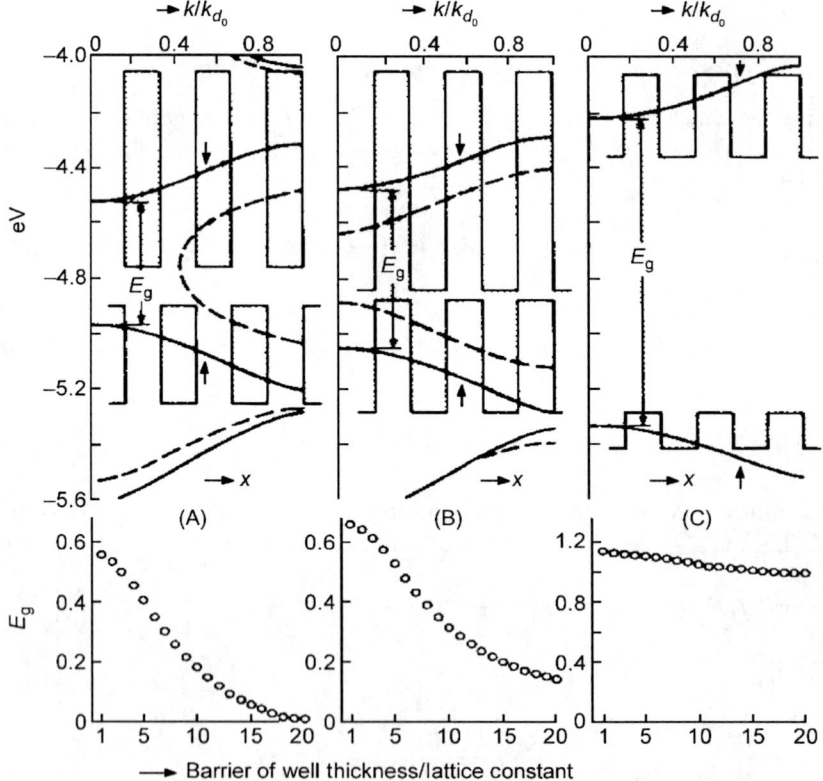

Figure 1.21 The $E-k$ relationships calculated by the two-band k · p model, top of figures and energy gap, bottom of figures, versus the reduced $k/(\pi/d)$, for three pairs (A), (B), and (C), together with their respective band-edge profiles. Dashed curves are derived from plane waves.
Source: Sai-Halasz et al. (1977), with permission.

$a = (4\pi\hbar/m_0)\mu/E_g$. Nevertheless, in a computational exercise, we have tried various orthonormal functions for $|1\rangle$ and $|2\rangle$ and discovered that results are not very sensitive on the particular forms used, partly because U functions come in as logarithm derivatives, U'/U, somewhat akin to variational solutions where a ratio is involved. Several cases were calculated by Tsu (1976), an unpublished work containing much of the presented material in this Section 1.8 and which formed the basis of the published version by Sai-Halasz et al. (1977).

The bottom part of the figure shows the reduction of the new E_g versus the period. The smaller the period, the larger is the interaction resulting in coupling. At a large period, simply there are not too many interactions per unit volume, leading to zero band gap. In this coherent picture, the system tends toward a zero band gap, but is not semimetallic. As the period $d > l$, the coherence length of the electrons, the system returns to the semimetallic state with overlapping of bands. This is really a point of fundamental importance, in that as long as coherent interactions dominate, even with long periods, only zero-gap behavior should be expected.

Figure 1.22 shows various U functions used. Before explaining the details of Figure 1.22, let us see what was done to avoid the restriction on the lattice constant a. By adding another term to Eq. (1.62),

$$|1\rangle = \left(\frac{2}{a(1 + B^2)}\right)^{1/2} [\cos gx + B \cos 2gx], \tag{1.68a}$$

$$|2\rangle = i\left(\frac{2}{a(1 + B^2)}\right)^{1/2} [\sin gx - B \sin 2gx]. \tag{1.68b}$$

Now the momentum matrix

$$\langle 1|P|2\rangle = \hbar g\left(\frac{1 - 2B^2}{1 + B^2}\right) = (m_0/2)(E_g/\mu)^{1/2} \tag{1.69}$$

for the determination of B. With B, these basis functions are then used in Eqs. (1.68a) and (1.68b) for the computation of $E-k$ for the InAs/GaSbAs systems shown in Figure 1.22, where $E - k/k_d$, as well as the variation of the effective band gap, are shown. The notation (3,3) is for three unit cells of InAs and three unit cells of GaSb$_{0.915}$As$_{0.085}$ and (5,1) is for five InAs and one GaSbAs unit cells, and so on. Solid lines apply to (3,3) using the basis states given by Eqs. (1.68a) and (1.68b). Solid line segments with three dots apply to (3,3) using the simple basis states given by Eq. (1.62). Note that the results are different though not too significant. Short of expanding this approach with minimization of the total energy, there is really no way for us to tell which is more correct. Therefore, we must view this approach as giving a semiquantitative guide to the design of type-II SL. Returning to Figure 1.22, we note that using plane waves, i.e., with $U = 1$, the $E-k$ relationship is multivalued and obviously incorrect. We have also computed $E-k$ by interchanging $\sin gx$ and $\cos gx$ in Eq. (1.62) and no change was found. This should not be

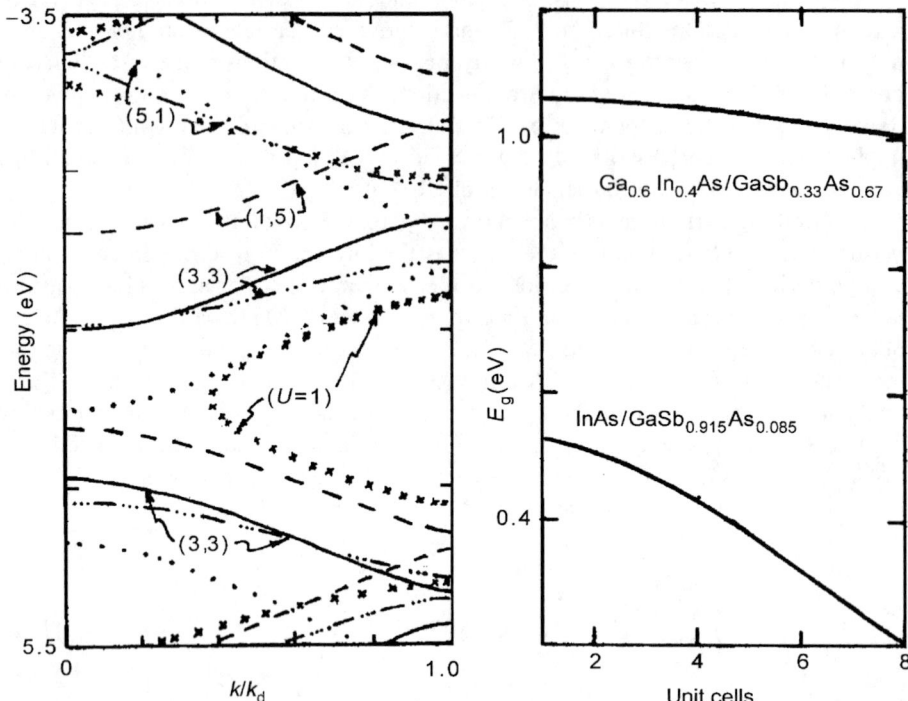

Figure 1.22 Energy versus momentum k/k_d for the InAs/GaSb$_{0.915}$As$_{0.085}$ system taken from Tsu (1976, unpublished), left, and energy versus unit cell thickness, right. The notation (3,3) is for three unit cells of InAs and for three unit cells of GaSbAs. Solid lines and lines with three dots are for U functions using basis sets given by Eqs. (1.68a), (1.68b), and (1.62), respectively. For $U = 1$ or plane waves shown by xxx, $E-k$ is multivalued thus the use of plane waves is obviously incorrect.

surprising because interchanging $\sin gx$ and $\cos gx$ is nothing more than shifting the interface between the two solids by $gx = \pi/2$. In summary, the k·p model may serve as a guide to understanding the type-II superlattice. For a specific design, the computed $E-k$ for the SL serves as semiquantitative guide for the design.

Let us make a comparison between the k·p calculation and the LCAO calculation by W.A. Harrison. Figure 1.23 is a replot of figure 3 of Sai-Halasz et al. (1978), where band structures of InAs and GaSb are zone-folded and superimposed on the central figure for the superlattice. For clarity, the portion outside the two basic overlapping bands is not included. The circle is drawn centered at the crossing of the conduction band of InAs with the valence band of GaSb, where the interaction is the strongest. The interaction of these two bands opens up an energy gap.

The new band gap $E_g = 0.44$, 0.54, and 0.9 eV, at $k = 0$, $0.3k_d$, and $1.0k_d$, respectively calculated with the k·p as shown in Figure 1.21A. The corresponding $E_g = 0.45$, 0.54, and 0.86 eV from the LCAO are shown in Figure 1.23. It is remarkable that the two calculations, one with k·p and the other with LCAO, give essentially the same results. I would like to point out that both calculations may not be unquestionably correct, however, the computed results are essentially the same.

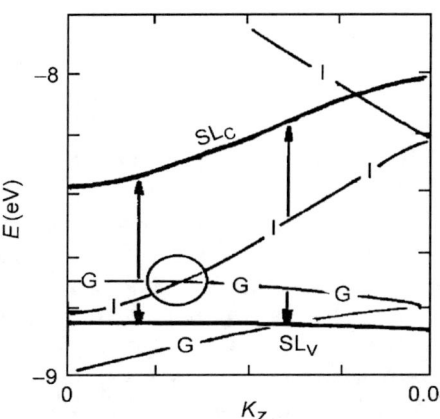

Figure 1.23 Band structures for superlattices consisting of alternate layers of 12 atomic planes. K_z is in units of $2\pi/a$, with the lattice constant a assumed to be the same for both cases.

This prompted me to present an "effective potentials" for the type-III SL, different from the type-I case. Let us substitute the U functions from the two-band $\mathbf{k} \cdot \mathbf{p}$ model back to the Schrödinger equation:

$$-\frac{\hbar^2}{2m}\nabla^2 U_\pm -i\frac{\hbar^2}{m}\mathbf{k}\cdot\nabla U_\pm + \left[V - \left(E - \frac{\hbar^2 k^2}{2m}\right)\right]U_\pm = 0. \qquad (1.70)$$

In one dimension, using zone folding into the reduced BZ, the terms inside the bracket in the preceding equation become

$$\left(E - \frac{\hbar^2 k^2}{2m}\right) = \left(E - \frac{\hbar^2(k+g)^2}{2m}\right) \equiv \varepsilon', \qquad (1.71)$$

then

$$V^\pm = \varepsilon' + \frac{\hbar^2 k}{m}(g + iU'_\pm/U_\pm) = \varepsilon' + \frac{\hbar^2 k}{m}g\frac{1+b}{1-b^{-1}}. \qquad (1.72)$$

We define an effective barrier by $\Delta V \equiv V_2 - V_1$. For conduction band to conduction band at the same energy ε' we use $V_2^+ - V_1^+$, and for conduction band to valence band at the same energy, we use $V_2^- - V_1^+$, then

$$\Delta V(\text{type I}) = \frac{\hbar^2 k}{m}[g_2(1 + b_2) - g_1(1 + b_1)], \qquad (1.73a)$$

$$\Delta V(\text{type III}) = \frac{\hbar^2 k}{m}\left[g_2(1 + b_2^{-1}) - g_1(1 + b_1)\right]. \qquad (1.73b)$$

Let us take a drastic case where $k_1 \sim k_2$, $g_1 \sim g_2$, $b_1(k) \sim b_2(k)$, then ΔV (type I) = 0, but ΔV (type II) = $(\hbar^2/kg)/(m) (b^{-1} - b)$, which is obviously not equal to zero. I think this is the best way to explain why there is reflection even though the **k** vectors are equal on both sides. I want to emphasize that this derivation applies to electrons transported between two bulk materials, which is rather different from the superlattice shown in Figures 1.21 and 1.22.

Before we close the discussion on this subject, I would like to point out that the k·p calculation is somewhat overshadowed by *ab-initio* techniques nowadays. Nevertheless, as pointed out by Cardona and Pollak (1966), the parameters that diagonalize a 15×15 k·p Hamiltonian matrix are obtained quite directly from experiments, partially because the momentum matrix element in the k·p theory is the same momentum matrix element for optical transitions. One may further ask if in fact an even simpler way maybe developed by taking two states at the point of crossing in k, as in elementary techniques forming the symmetric and anti-symmetric states of a molecule, using the k·p model as coupling. This is precisely the route we took.

We shall summarize the difference between type-I and type-III SL. In the usual GaAs/GaAlAs superlattices, the modulation of the band-edge energy of the conduction or valence bands results in the formation of a confinement. In the case of an InAs/GaSb-based system with overlapping conduction−valence bands, the orthogonality of the states gives rise to an effective confinement. This shows very directly that the reflectivity from Eq. (1.61a) in the overlapping region of energy is high and in some instances, close to unity, as if a fairly high barrier were separating the well regions. This feature presents a wide range of applications of technological interest.

What happens to two bulk solids, InAs and GaSb, forming an interface? At the interface, two things happen. First, alignment of the Fermi level results in the transference of electrons in the valence band of GaSb to the conduction band of InAs. When both are undoped, near the interface, electrons form an accumulation in InAs and holes form an accumulation in GaSb, a situation that cannot really be described as semimetal where the overlapping of the conduction and valence bands is in the *k*-space, as in graphite and bismuth. Since both InAs and GaSb are undoped, far away from the interface, they behave as an intrinsic semiconductor, but having a large conduction along the interface. Second, with thin alternate layers of InAs and GaSb forming a type-III superlattice, something very different results, i.e., a semiconductor with a variable band gap. Only because Esaki and I tried to move the point of inflection toward the center of the mini-BZ, did we stumble on this new and rich physical phenomenon.

1.9 Physical Realization and Characterization of a Superlattice

1.9.1 First Attempt: GaAs/GaAsP Vapor Phase Epitaxy Superlattice

The first superlattice constructed with a periodic variation in the phosphorus content the $GaAs_{1-x}P_x$ alloy system using vapor phase epitaxy (VPE) was attempted by A.E. Blakeslee (Esaki et al. 1970). Figure 1.24 shows the transmission electron micrograph (TEM) of a 30-period structure with a 20-nm period.

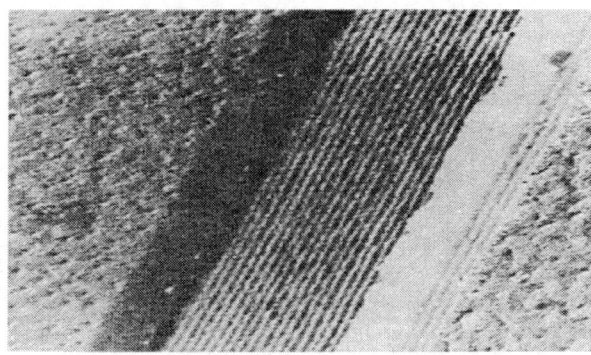

Figure 1.24 TEM of a GaAs/ GaAs$_{0.5}$P$_{0.5}$ 30-period structure with 20 nm for each period. *Source*: Esaki et al. (1970), with permission.

To characterize the structure further, X-ray diffraction was used to find the modulation parameters Δ_1 and Δ_2 of an atom located at x_p given by

$$x_p \sum_{n=0}^{p}(a = \Delta_1 \cos(2\pi na/d) + \Delta_2 \cos(4\pi na/d)), \qquad (1.74)$$

in which $a = 0.56$ nm, is the lattice constant and d is the modulation of the alloy. Figure 1.25A shows the X-ray intensity versus degree angle of the (400) diffraction peak. The $|A_0|$ peak, a doublet of the $K_{\alpha 1}$ and $K_{\alpha 2}$ lines of Cu at 67.5° and 67.7° is from GaAs, with a strong left peak from the GaAs substrate. The remaining structure, including the center peak, with satellites $|A_L|$ and $|A_R|$, belongs to the GaAs and GaAs$_{0.5}$P$_{0.5}$ superlattice. The parameters Δ_1 and Δ_2, the variation of the lattice constant from the mean value, were chosen to fit the diffraction data. Without the second harmonic term in Eq. (1.74), the asymmetry of the two satellites could not have been explained. The X-ray characterization is in good agreement with rough values from the TEM. We could not explain the X-ray data if instead it was assumed that the satellites originated from the difference in the amplitude of the scattering between the P and the As atoms. Further results of characterization using cathode luminescence (CL) are shown in Figure 1.25B, with two samples of 20- and 140-nm periods compared with pure GaAs at 4.2 K excited by 15 keV electrons. The 1.82 eV peak of the 20-nm sample may indicate the formation of a superlattice; however, the 1.65 eV peak of the 140-nm sample is most likely to be due simply to the alloy material because it is unlikely that the mean free path is longer than 140 nm.

Transport measurements on the $n-n$ heterojunction of a GaAsP superlattice and GaAs with a donor concentration of $n \sim 10^6 - 10^7$ cm^{-3} indicates that the energy discontinuity is ~0.05 eV, agreeing with what is already known (Davis et al., 1969) that most of the difference in the effective energy gap lies in the valence band.

The failure to observe NDC led to the conclusion that the fabricated GaAs/ GaAsP superlattices are basically metallurgical, i.e., it is only a superlattice structurally. However, electronically, owing to the poor overall mean free path from a poor lattice match and for other reasons, there is no clear-cut evidence of the formation of minibands. At this point, in spite of the results of Alferov et al. (1971),

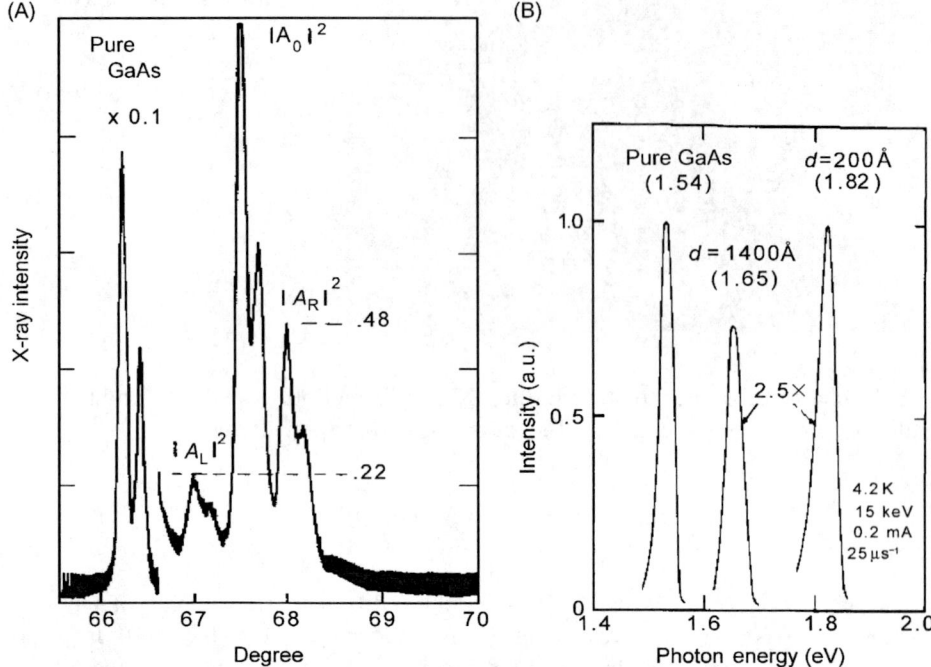

Figure 1.25 (A) X-ray diffraction of a GaAs/GaAs$_{0.5}$P$_{0.5}$ superlattice showing a periodic structure with satellite peaks of 100 periods of 20 nm per period and (B) CL of two samples: 20- and 140-nm periods.
Source: Esaki et al. (1970), with permission.

which may indicate some hope of continuing with the GaAs/GaAsP system, Esaki led our effort toward examining other systems, mainly the GaAs/GaAlAs system, where a major effort has been devoted for some time to the DH lasers (Hayashi et al., 1970; Hayashi, 1984).

The GaAs/GaAlAs superlattice will be covered in the next section; however, the X-ray results are shown in Figure 1.26 for high angle with a 8.8-nm period and in Figure 1.27 for low-angle scattering with a 9.05-nm period taken from Chang et al. (1974a,b). For the center peak, the zero order is much stronger than the weak superlattice reflections, ±1, ±2, and ±3 satellites. The relative intensities of these reflections give a measure of strain, in good agreement with the lattice constant mismatch (Segmuller et al., 1977).

1.9.2 The GaAs/GaAlAs Superlattice: Determination of the Alloy Concentration

The most important parameter for the formation of superlattices is the band-edge alignment of the two adjacent materials. Because the band-edge offset is proportional to the energy band gap, this information is of primary importance

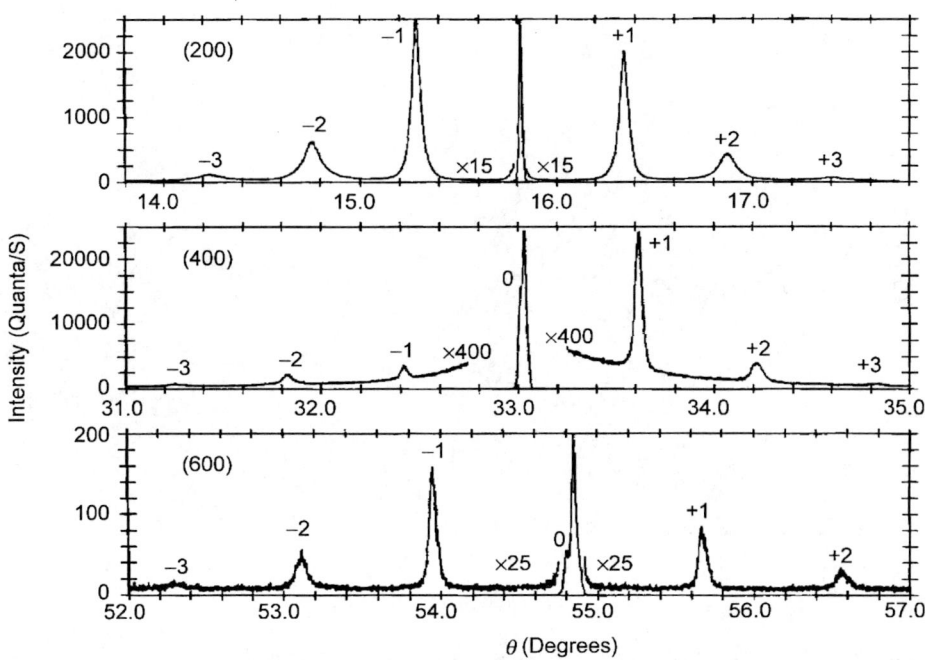

Figure 1.26 High angle X-ray diffraction of GaAs/AlAs superlattice with an 8.8-nm period near the (200), (400), and (600) reflections.
Source: Segmuller et al. (1977), with permission.

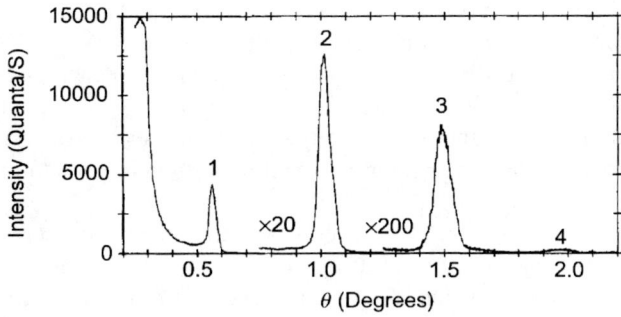

Figure 1.27 Low-angle X-ray scattering for a GaAs/AlAs SL with a 9.05-nm period.
Source: Chang et al. (1974a,b), with permission.

for the design of a superlattice. Moreover, lattice matching is of primary importance for epitaxial growth at low strain, and therefore, low defect densities at the heterojunction interface of two adjacent materials. Figure 1.28 shows the dependence of the energy gap versus the lattice constant for a variety of

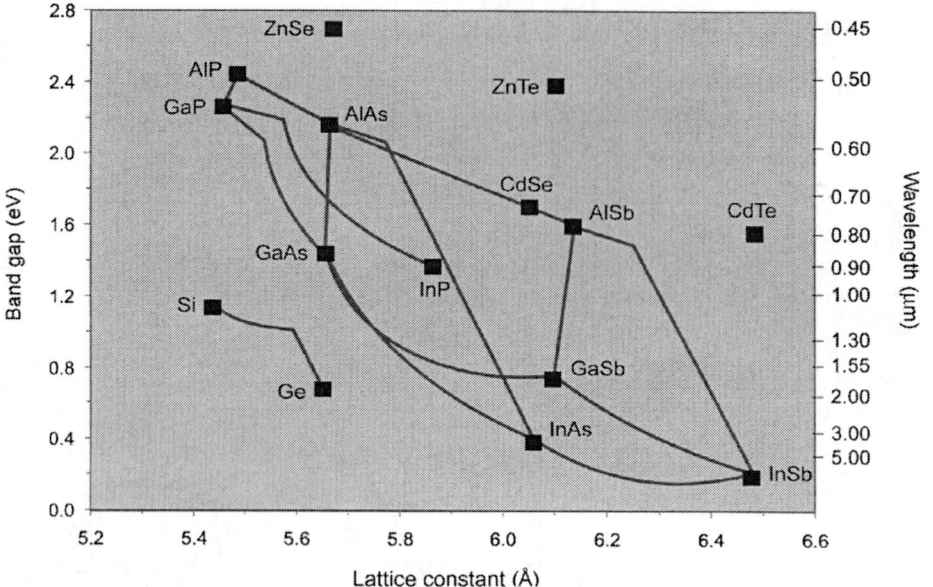

Figure 1.28 Energy band gap versus the lattice constant for a number of systems. *Source*: Courtesy from EPI-MBE.

systems, Si/Ge, as well as a variety of Group II—VI and III—V compound semi-conductors. Note that the lattice constant matching for GaAs and AlAs is much better than GaAs and GaP. To achieve matching, as well as other considerations like the transition from direct gap to indirect gap, we need to select the proper molar fraction for the alloy system. For this reason, we need to control the deposition for a precise molar fraction which requires characterization of this molar fraction and the stoichiometry of the particular pseudo-binary or even pseudo-quaternary alloys.

As mentioned, apart from lattice constant matching, another parameter, the band-edge offset, is needed for the design of the miniband of the superlattice. And here we need first to have some idea of the variation of the band gap with composition. Moreover, the transition from direct to indirect band gap is sometimes crucial for proper design. Figure 1.29 shows this variation for GaAs and AlAs, with measurements taken from Casey and Panish (1969), and lines from Eqs. (1.75a) and (1.75b) taken from Onton et al. (1974). Notwithstanding that, as the molar fraction of Al increases beyond $\sim 40\%$, the lowest fundamental band gap is switched from the BZ center, the Γ-point, to the X-point. This is the reason why the molar fraction of Al must be less than 40%.

$$E_{g\Gamma} = 1.44 + 1.04x + 0.47x^2 \qquad\qquad (1.75a)$$

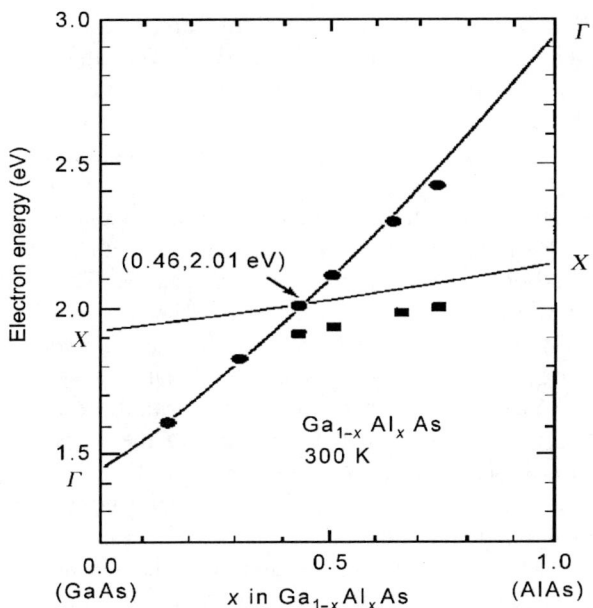

Figure 1.29 The band structure of $Ga_{1-x}Al_xAs$. The lines are given by Eqs. (1.75a) and (1.75b).
Source: Onton et al. (1974), with permission.

and

$$E_{gX} = 1.92 + 0.17x + 0.07x^2. \tag{1.75b}$$

DH lasers using GaAlAs as a barrier to confine the space charge have been quite successful for several years. (We must not confuse charge confinement with quantum confinement involving the wave nature of electrons, which came much later as consequences of the introduction of the superlattice and quantum wells.) In addition to the determination by chemical means, such as ESCA which emphasizes composition without indicating whether the alloy system is stoichiometric, most researchers utilized luminescence data for indications of quality. I have been fascinated by Raman scattering for some time, thinking that the phonon structure alone should give indications of both composition and stoichiometry, with which Esaki agreed. IBM Poughkeepsie terminated research on photonic systems leaving the use of many lasers available for experiment. At the time, Esaki was concentrating on the design and installation of an MBE deposition system, leaving little capacity for anything else. Kawamura, a prominent professor from Japan who was spending a couple of years with us, also wanted to work on Raman scattering. There was just sufficient capital equipment budget to add a SPEX double spectrometer for high-resolution Raman scattering. I succeeded in getting the first photon counting system from SPEX as a tryout. After Professor Kawamura arrived, I learned that he, like me, also had no working knowledge of Raman scattering. G. Burn, who was able to offer us help, commented to me, "The two of you are like blind leading the blind." Well, new tools

always lead to new results. We discovered that the spectrum of GaAs taken by
Kawamura and myself consisted of the usual transverse-optical (TO) and longitudinal-optical (LO) phonons of GaAs, but, in addition, other structures, later
identified as the phonons at the BZ boundary caused by the breakdown of selection rule induced by disorder, a random distribution of Al in GaAs forming a
GaAlAs random alloy.

With the help of M. Lorenz and A. Onton, R. Chicotka was asked to help me
and we acquired a Bridgman-grown single crystal of $Ga_{1-x}Al_xAs$, with $x = 0$ at
one end and $x \sim 1$ at the other end of a 3-cm long by 1-cm wide sample. This sample was cut into two pieces lengthwise, with a further cut into 10 pieces for ESCA
and, later, for wet chemical analysis. The other half was used for Raman measurements. A typical spectrum is shown in Figure 1.30 taken from Tsu et al. (1972),
showing the weak longitudinal acoustic (LA) modes that were normally forbidden
at the BZ zone boundary. I like to point out that the discovery of these normally
forbidden modes was due to a sensitive photon counting system with an
adjustable discriminator that was not possible with the Keithley meter normally in
use at the time. See the additional remarks in the preface.

The variations of the optical local (OL) mode, and the acoustical local (AL)
mode for various compositions are shown in Figure 1.31. The results of everything
are summarized in Figure 1.32, with the molar fraction obtained from both ESCA
and wet chemical analysis.

This Raman spectrum was given to L. Chang for refinement of the Knudsen cell
calibration used in MBE growth of the $Ga_{1-x}Al_xAs$ barriers for the GaAlAs superlattices and quantum wells.

Unlike ESCA, the phonon modes are similar to those used in luminescence characterization, and molar fraction is determined from the mode spectra rather than
from composition. Therefore, one is assured of stoichiometry. It is generally agreed
that luminescence peak intensity and/or lifetime gives a measurement of the quality
in terms of defects, mobility, and so on. Nevertheless, the phonon linewidth also
gives information about the quality, although not directly. This point will be further
discussed in the later section dealing with Raman scattering. The quantum well

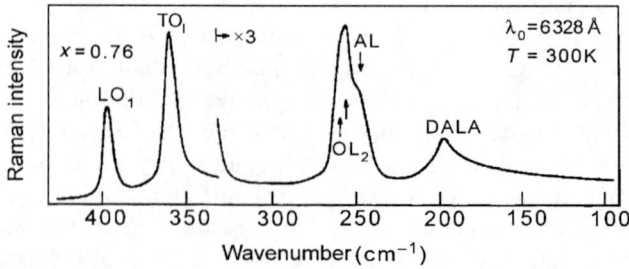

Figure 1.30 Raman spectrum for the $Ga_{0.24}Al_{0.76}As$ alloy with polarization ($\parallel, \parallel$). DALA is
disorder-activated LA phonon mode.
Source: Kawamura et al. (1972), with permission.

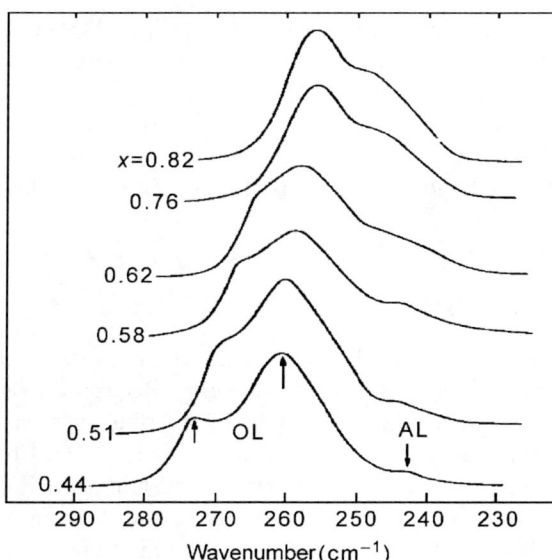

Figure 1.31 Mode frequency versus molar fraction x for the $Ga_{1-x}Al_xAs$ alloy system. *Source*: Kawamura et al. (1972), with permission.

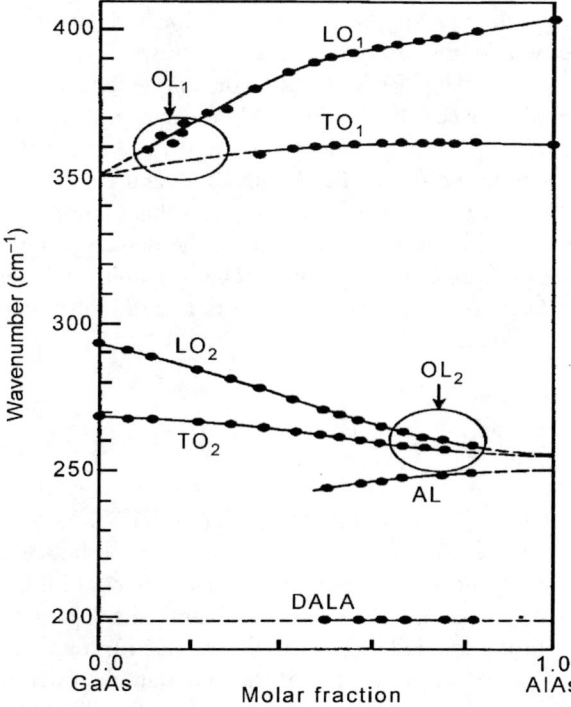

Figure 1.32 Raman phonon modes versus molar fraction for the $Ga_{1-x}Al_xAs$ alloy system. *Source*: Tsu et al. (1972).

structures (Dingle et al., 1974) and superlattices (Dingle et al., 1975) fabricated at BTL seemed to be better quality in terms of optical data, although these reports came nearly 2 years after the work presented by IBM, owing, perhaps, to several years of work on heterojunction lasers. However, I think the main reason was that the alloy composition used at IBM had nearly 50% [AL], unlike [Al] ~30% at BTL with better lattice matching. Obviously the mistake involved a lack of experience.

1.9.3   Other Structural Characterizations

In the initial learning stage at IBM GaAs depositions on GaAs substrates were monitored for epitaxy during growth using reflection high-energy electron diffraction (RHEED) mounted *in situ* (Dove et al., 1973). I asked John Arthur, who supposedly coined the term "molecular beam epitaxy," what distinguishes MBE from older deposition schemes with a heated molecular oven. He replied that apart from having a "good ring," it is really the use of RHEED *in situ* that ensures an instant check of epitaxy. Scanning electron microscopy (SEM) was not relied upon early on because of the lack of reliable information below 10 nm. In fact, our samples of superlattices, after characterization by X-ray, were given to Alex Broar as calibration for his high-intensity SEM. On the other hand, Rutherford back scattering (RBS) had been sufficiently developed and used for an attempt to determine the actual structure. The sample used for the RBS study by Mayer et al. (1973) has a periodicity of ~100 nm, which can be easily analyzed by SEM and Raman to determine the spatial periodicity and the alloy composition, respectively. The agreement of composition determined using RBS and Raman was better than 5%; however, the agreement of period between RBS and SEM was generally greater than 10%. Although the period of 50−100 nm is much too large for any quantum effects, it was designed to corroborate the use of RBS with SEM and Raman as a learning process. In particular, the barrier thickness is usually less than 2 nm; therefore, these techniques are not sufficiently good to characterize the device potential of these superlattices and quantum wells. The situation today is quite different from what Figure 1.33 portrays because monolayer thick barriers can be routinely realized in man-made quantum structures as discussed in Chapter 6.

1.10   Summary

Esaki and I decided that the Bloch oscillation and its manifestation in NDC constitutes the most direct evidence of the formation of a man-made superlattice as means of broadening the class of semiconductors for electronic devices. Initially we obtained the condition that whenever $\omega_B \tau > 1$, with ω_B being the Bloch frequency, the current decreases with the applied electric field, or NDC. In our original IBM report (Esaki and Tsu, 1969), we stated that if the superlattice structures are formed in such a manner that most scattering centers such as foreign atoms,

(A)   (B)   (C)

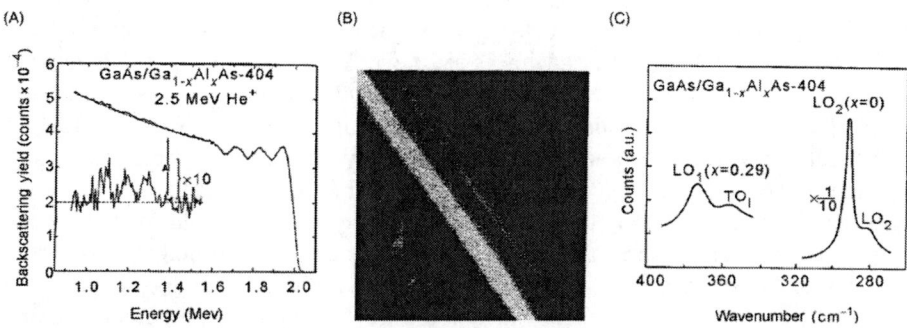

Figure 1.33 Evaluation of a GaAs/GaAlAs periodic structure. (A) RBS at 2.5 MeV He⁺ with onset of the Al portion. Oscillations show three periods of GaAs/GaAlAs with a period of ~120 nm. (B) From SEM. (C) Alloy composition determined from Raman Scattering to be 29% [Al].
Source: Mayer et al. (1973), with permission.

imperfections, and so on, are concentrated in the potential hills, that the matrix element for scattering is favorably reduced. *This prediction (Tsu, 2007) led to modulation doping, a widespread techniques for high mobility.* The first observation of NDC was 2 years later for a GaAs/GaAlAs superlattice. In an effort to distinguish domain oscillation from Bloch oscillation, a quantum phenomenon, we started on a path that involved only one period, a structure with two GaAlAs barriers on each side of a GaAs quantum well, the now-famous resonant tunneling diode (RTD), a subject that will be taken up in the next chapter. Meanwhile, an optical response to the time-dependent electric field of a superlattice was obtained. Two types of NDC, a band-like and a hopping-like NDC for the highly localized case when the energy gain exceeds the bandwidth of the superlattice, including the argument in favor of the existence of the SL, are considered. Initial experimental results in support of the observation of quantum effects are presented. As pointed out, originally Esaki and I tried to find a way to move the point of inflection toward the BZ center, by creating a superlattice of two materials with the bottom of the conduction band of one below the top of the valence band of the other. In this section we gave details of the use of the k·p model which is more than adequate. The effective band gap depends on how close the two materials forming the heterostructure are put together. The last section is devoted to characterization, including by X-ray, RHEED, RBS, SEM, and Raman. Compositional characterization by Raman, unlike SIMS, is capable of determining the stoichiometry via phonon modes. The characterization techniques used in those days have improved by a large margin and nowadays, STM and even X-TEM can reveal atomic resolution. When I left China, my great-uncle told me an ancient Chinese proverb on his death bed: *To excel in work needs to take advantage of the best tools.*

The concept of the superlattice has opened the door to devices such as quantum wire and quantum dots, as well as serving as a test bed for modern nanotechnology involving the wave nature of electrons. New ideas and technical breakthroughs are being added to the huge list of advances at an ever-increasing rate. Superlattices were originally based on minibands in energy bandwidth, however, most superlattices developed into operating units actually belong to tight-binding systems with narrow bandwidth so that conduction is due to phonon-assisted hopping. The ramification of our ideas plays an important role in research relating to the fundamental concepts of solids and slowly but surely developing into new devices. Although modulation doping, Bloch oscillation, and so on, came from man-made heterostructures, no superlattice devices appear in any electronic systems even after more than 30 years of intense research and development, except in optoelectronics such as the quantum cascade laser (Faist 1994) and THz oscillators (Williams 2007). It is interesting to note that officially QCL appeared in 1994, however, as early as 1990, Capasso (Capasso 1990) discussed the possibility of using superlattice for infra-red lasers, pointed out by Kazarinov and Suris (1971). Economically, devices utilizing superlattices, even with technological complexity in place, are simply not ready to be incorporated in any major way in the world of devices. One may argue that many technical breakthroughs such as the input/output (I/O) problems are needed before quantum dots can play a role in the electronic systems, but man-made one-dimensional superlattice structures have no I/O problems. Thus, to my mind, it is complexity and economic reasons that have prevented the wide spread usage. Nevertheless, the "genie" is out of the bottle (Tsu, 2007) it is only a matter of time.

As noted earlier, without the tenacity of Leo Esaki, this project would have been "dead-on-arrival." I, for one, quite likely would have given up.

References

Alferov, Zh.I., et al., 1971. Fizika Tekhnika Poluprovo Dnikov 5, 196.

Beltram, F., Capasso, F., Sivco, D., Hatchinson, A.L., Chu, S.N., Cho, A.Y., 1990. Phys. Rev. Lett. 64, 3167.

Bloch, F., 1928. Z. Phys. 52, 555.

Bloembergen, N., Sievers, A.J., 1970. Appl. Phys. Lett. 17, 483.

Budd, H., 1963. J. Phys. Soc. Japan 18, 142.

Callaway, J., 1964. Energy Band Theory, Academic Press, New York.

Cardona, M., Pollak, F.H., 1966. Phys. Rev. 142, 532.

Casey Jr., H.C., Panish, M.B., 1969. J. Appl. Phys. 40, 4910.

Chambers, R.G., 1952. Proc. Phys. Soc. (Lond.) A65, 458.

Chang, L.L., Esaki, L., Howard, W.E., Ludeke, R., Schul, G., 1973. J. Vac. Sci. Technol. 10, 655.

Chang, L.L., Esaki, L., Segmuller, A., Tsu, R., 1974a. Proceedings of the 12th International Conference of Physics on Semiconductors, B.G. Teubner, Stuttgart, p. 688.

Chang, L.L., Esaki, L., Tsu, R., 1974b. Appl. Phys. Lett. 24, 593.

Chen, S.G., Zhao, X.G., 1991. Phys. Lett. A 155, 303.

Choyke, W., Hamilton, D.R., Patrick, L., 1964. Phys. Rev. A 133, 1163.

Capasso, F., Mohammad, K., Cho, A.Y., 1986. QE 22, 1853.

Davis, M.E., Zeidenbergs, G., Anderson, R.L., 1969. Phys. Status Solidi. 34, 385.

Dignam, M.M., Sipe, J.E., 1991. Phys. Rev. Lett. 64, 1797.

Dingle, R., Gossard, A.C., Wiegmann, W., 1975. Phys. Rev. Lett. 34, 1327−1330.

Dingle, R., Wiegmann, W., Henry, C.H., 1974. Phys. Rev. Lett. 33, 827−830.

Dove, D.B., Ludeke, R., Chang, L.L., 1973. J. Appl. Phys. 44, 1897.

Döhler, G.H., Tsu, R., Esaki, L., 1975. Solid State Commun. 17, 317.

Emin, D., Hart, C.F., 1987. Phys. Rev. B 36, 7353.

Esaki, L., 1958. Phys. Rev. 109, 603.

Esaki, L., Chang, L.L., 1974. Phys. Rev. Lett. 33, 495.

Esaki, L., Tsu, R., 1969. In IBM Research note RC-2418.

Esaki, L., Tsu, R., 1970. IBM Res. Develop. 14, 61.

Esaki, L., Chang, L.L., Tsu, R., 1970. Proceedings of the 12th International Conference on Low Temperature in Physics, Kyoto, Japan, September, 1970, Keigaku Publishing Company, Tokyo, pp. 551−553.

Faist, J., Capasso, F., Sivco, D.L., Sirtori, C., Hutchinson, A.L., Cho, A.V., 1994. Science 264, 533.

Feldmann, J., Leo, K., Shah, J., Miller, D., Cunningham, J.E., Meier, T., 1992. Phys. Rev. B 46, 7252.

Fukuyama, H., Bari, R.A., Fogedby, H.C., 1973. Phys. Rev. B 8, 5579.

Hayashi, I., 1984. IEEE Trans. Electron Dev. ED-31, 1630.

Hayashi, I., Penish, M.B., Foy, P.W., Sumski, S., 1970. Appl. Phys. Lett. 17, 109.

Houston, W.V., 1940. Phys. Rev. 57, 184.

James, H.M., 1949. Phys. Rev. 76, 1611.

Jha, S.S., Bloembergen, N., 1968. Phys. Rev. 171, 891.

Kane, E.O., 1959. J. Phys. Chem. Solids 12, 181.

Kawamura, H., Tsu, R., Esaki, L., 1972. Phys. Rev. Lett. 29, 1397−1400.

Kazarinov, R.F., Suris, R.A., 1971. Sov. Phys. Semicond. 6, 120.

Kazarinov, R.F., Suris, R.A., 1972. Sov. Phys. Semicond. 5, 707.

Koss, R.W., Lambert, L.M., 1972. Phys. Rev. B 5, 1479.

Krieger, J.B., Iafrate, G.J., 1986. Phys. Rev. B 33, 5494.

Krömer, H., 1958. Phys. Rev. 109, 1856.

Leavitt, R.P., Little, J.W., 1990. Phys. Rev. B 41, 5174.

Lebwohl, P.A., Tsu, R., 1970. J. Appl. Phys. 41, 2664.

Leo, K., Shah, J., Gobel, E.O., Damen, T.C., Schmitt-Rink, S., Schafer, W., 1991. Phys. Rev. Lett. 66, 201.

Luri, S., 1985. Appl. Phys. Lett. 47, 490.

Lyssenko, V.G., Valuis, G., Löser, F., Hasche, T., Leo, K., Dignam, M.M., et al., 1997. Phys. Rev. Lett. 79, 301.

Maekawa, S., 1970. Phys. Rev. Lett. 24, 1175.

Martini, R., Klose, G., Roskos, H.G., Kurz, H., Grahn, H.T., Hey, R., 1996. Phys. Rev. B 54, 14325.

Mayer, J.W., Ziegler, J.F., Chang, L.L., Tsu, R., Esaki, L.I., 1973. Appl. Phys. 44, 2322−2325.

McLachlan, N.W., 1947. Theory and Application of Mathieu Functions, Oxford University Press, London.

Mendez, E.E., Agullo-Rueda, F., Hong, J.M., 1988. Phys. Rev. Lett. 60, 2426.

Monsivais, G., Castillo-Mussot, M., Claro, F., 1990. Phys. Rev. Lett. 64, 1433.

Neuberger, M., 1971. Handbook of Electronic Materials, III−V Semiconducting Compounds, vol. 2. IFI/Plenum, New York.

Onton, A., Lorenz, M.R., Woodall, J.M., Chicotka, R.J., 1974. J. Cryst. Growth 27, 166−176.

Pfeiffer, L., West, K.W., Starmer, H.L., Baldwin, K.W., 1989. Appl. Phys. Lett. 55, 1888.

Physics of Quantum Electron Devices, ed. by Capasso, F., 1990, Springer-Verlag.

Pippard, A.B., 1965. Dynamics of Conduction Electrons, Gordon & Breach Science Publication Inc., New York.

Rabinovitch, A., Zak, J., 1971. Phys. Rev. B 4, 2358.

Sai-Halasz, G.A., Esaki, L., Harrison, W.A., 1978. Phys. Rev. B 18, 2812.

Sai-Halasz, G.A., Tsu, R., Esaki, L., 1977. Appl. Phys. Lett. 30, 651.

Saitoh, M., 1972. J. Phys. C. Solid State 5, 914.

Segmuller, A., Krishna, P., Esaki, L., 1977. J. Appl. Cryst. 10, 1.

Shah, J. (Ed.), 1992. Hot Carriers in Semiconductor Nanostructures: Physics and Applications. Academic Press, Boston.

Shockley, W., 1972. Phys. Rev. Lett. 28, 349.

Sibille, A., Palmier, J.F., Wang, H., Mokot, F., 1990. Phys. Rev. Lett. 64, 52.

Smith, R.A., 1961. Wave Mechanics of Crystalline Solids, Wiley, New York.

Tsu, R., 1975. Unpublished.

Tsu, R., 1976. Unpublished containing much of the presented material in Section 1.8.

Tsu, R., 1989. J. Non-Cryst. Solids 114, 708.

Tsu, R., 1990. SPIE 1361, 231.

Tsu, R., 1992. Proceedings of 21st ICPS Beijian, Vol. 2, Eds. Jiang, P. & Zheng, H.-Z., p. 1198.

Tsu, R., 1993. In: Feng, Z.C. (Ed.), Semiconductor Interface, Microstructures and Devices. IOP Publishers, Bristol, Philadelphia, chapter 1.1.

Tsu, R., 2007. Microelectronics J. 38, 959−1012.

Tsu, R., Döhler, G., 1975. Phys. Rev. B 12, 680.

Tsu, R., Esaki, L., 1971. Appl. Phys. Lett. 19, 246.

Tsu, R., Esaki, L., 1973. Appl. Phys. Lett. 22, 562.

Tsu, R., Esaki, L., 1991. Phys. Rev. B 43, 5204, Erratum Phys. Rev. B 44, 3495.

Tsu, R., Chang, L.L., Sai-Halacz, G.A., Esaki, L., 1975. Phys. Rev. Lett. 24, 1175.

Tsu, R., Howard, W.E., Esaki, L., 1968. Phys. Rev. 172, 779.

Tsu, R., Kawamura, H., Esaki, L., 1972. Proceedings of the 11th International Conference in Physics Semiconductors, Warsaw, Poland 1972, PWN—Polish Scientific Publishers, Warsaw, Poland, pp. 1136−1141.

Voisin, P., Bleuse, J., Bouche, C., Gaillard, S., Alibert, C., Regreny, A., 1988. Phys. Rev. Lett. 61, 1639.

Wannier, G.H., 1960. Rev. Mod. Phys. 34, 645.

Wannier, G.H., 1961. Phys. Rev. 117, 432.

Waschke, C., et al., 1993. Phys. Rev. Lett. 70, 3319.

Williams, B.S., 2007. Nat. Photonics 1, 577.

Wolf, R.A., Pearson, G.A., 1966. Phys. Rev. Lett. 17, 1015.

Ziman, J.M., 1988. Principles of the Theory of Solids, Cambridge University Press, Cambridge, p. 313.

2 Resonant Tunneling via Man-Made Quantum Well States

2.1 The Birth of Resonant Tunneling

When Gunn pointed out to Esaki and me that the observed negative differential conductance (NDC) in a GaAlAs superlattice may originate from domain oscillations in GaAlAs bulk alloys (Gunn, 1963, 1976), we extended our study to tunneling via man-made quantum wells (QWs). Particularly, we were focused on a single QW of GaAs with two GaAlAs barriers on each side, to avoid any possibility of domain formation, as well as several wells coupled as shown in Figure 2.1. Before the theoretical calculation is presented, let us touch on some general issues.

The energy E is the sum of longitudinal and transverse energies,

$$E = E_l + E_t \quad \text{with} \quad E_t = \frac{\hbar^2 k_t^2}{2m^*}, \tag{2.1}$$

and the wave function is expressed as the product, $\psi = \psi_l + \psi_t$. For an n-period structure, the electron wave functions in the left- and right-hand contacts are, respectively

$$
\begin{aligned}
\psi_1 &= \psi_t[\exp(ik_1 x) + R \exp(-ik_1 x)], \\
\psi_2 &= \psi_t[A_1 \exp(ik_1 x) + B_1 \exp(-ik_1 x)], \\
&\vdots \\
\psi_n &= \psi_t[T \exp(ik_n x)],
\end{aligned}
\tag{2.2}
$$

where $\psi_n = \psi_{n+1}$ and $(1/m_n^*)\psi_n' = (1/m_{n+1}^*)\psi_{n+1}'$ at the boundaries separating the regions n and $n+1$. In the original derivation, however, the effective masses are taken to be equal, then

$$\begin{pmatrix} T \\ 0 \end{pmatrix} = M_1 \cdots M_p \cdots M_n \begin{pmatrix} 1 \\ R \end{pmatrix}, \tag{2.3}$$

Superlattice to Nanoelectronics. DOI: 10.1016/B978-0-08-096813-1.00002-3

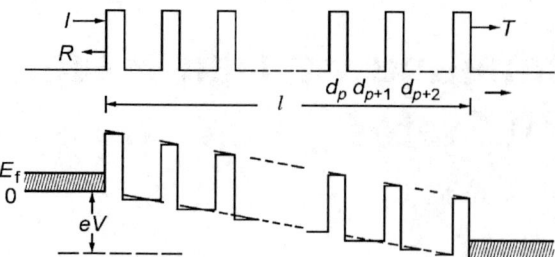

Figure 2.1 Top: a superlattice of length l with barrier eV_0. Bottom: the solid line shows a good approximation of the potential profile used in the calculation. *Source*: After Tsu and Esaki (1973), reprinted with permission.

where

$$M_p = \frac{1}{4} \begin{pmatrix} \exp(ik_{p+2}d_{p+2}) & \exp(ik_{p+2}d_{p+2}) \\ \exp(-ik_{p+2}d_{p+2}) & -\exp(-ik_{p+2}d_{p+2}) \end{pmatrix}$$

$$\times \begin{pmatrix} \exp(ik_{p+1}d_{p+1}) & \exp(-ik_{p+1}d_{p+1}) \\ -i(k_{p+1}/k_{p+2})\exp(-ik_{p+1}d_{p+1}) & -i(k_{p+2}/k_{p+2})\exp(-ik_{p+1}d_{p+1}) \end{pmatrix}$$

$$\times \begin{pmatrix} 1 + i(k_p/k_{p+1}) & 1 - i(k_p/k_{p+1}) \\ 1 - i(k_p/k_{p+1}) & 1 + i(k_p/k_{p+1}) \end{pmatrix},$$

(2.4)

in which $k_p = [2m^*(V_p - E_1)]^{1/2}/\hbar$. The reflection amplitude R and the transmission amplitude T are given by $R = -M_{21}/M_{22}$ and $T = M_{11} - M_{12}M_{21}/M_{22}$. In the original version, following Duke (1969), with $|T|^2 = (k_1/k_1')D(E_1)$, the net current density from the left at E, to the right at E' is

$$J = \frac{e}{4\pi^3\hbar} \int_0^\infty dk_1 \int_0^\infty dk_t [f(E) - f(E')]T * T(k_1'/k_1)\frac{\partial E}{\partial k_1}.$$

(2.5)

Note that there is a term (k_1'/k_1) in Eq. (2.5) that was left out in the original by Tsu and Esaki (1973). However, we shall follow the original treatment for the time being without this (k_1'/k_1) term, partly because the main features are not affected by this term. A full discussion of this point is given later. With the separation of variables due to the planar geometry, the transverse momentum k_t is conserved so that T^*T depends only on the longitudinal direction. After integration over the transverse dk_t,

$$J = \frac{em^*kT}{2\pi^2\hbar^3} \int_0^\infty dE_1 T^*T \ln\left(\frac{1 + \exp[(E_f - E_1)/k_bT]}{1 + \exp[(E_f - E_1 - eV)/k_bT]}\right).$$

(2.6a)

For $k_BT \rightarrow 0$,

$$J = \left(\frac{em^*}{2\pi\hbar^2\hbar^3}\right) \int_0^{E_f} (E_f - E_1)T * T \, dE_1 \quad eV \geq E_f$$

(2.6b)

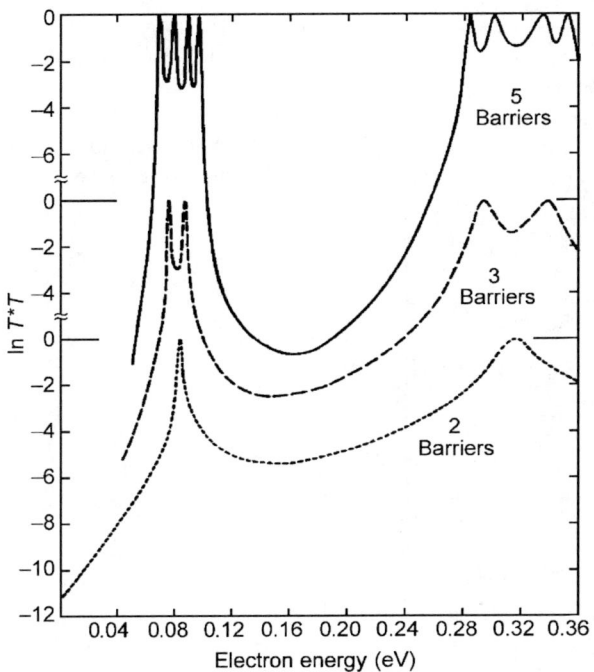

Figure 2.2 T^*T versus incident electron energy in eV, the barrier and well width are 2 and 5 nm, and the barrier height is 0.5 eV.
Source: After Tsu and Esaki (1973), with permission.

$$J = \left(\frac{em^*}{2\pi\hbar^2\hbar^3}\right)\left\{\int_0^{E_f - eV} (eV)T^*T \, dE_1 + \int_{E_f - eV}^{E_f} (E_f - E_1)T^*T \, dE_1\right\}, \quad eV \le E_f.$$

$$(2.6c)$$

Figure 2.2 shows the calculated transmission coefficient versus the incident electron energy and Figure 2.3 shows the total tunneling current versus the applied voltage V.

Let us discuss something important. Although T^*T reaches unity at the resonance shown in Figure 2.2 at $V = 0$, T^*T may be very much reduced at $V \ne 0$, owing to the loss of symmetry, a point that was not appreciated at first. To achieve a large NDC, one needs to design the structure such that it is nearly symmetrical at the operating point V. We shall go into detail on this point as well as some other important points later. But first we offer a physical picture of why NDC appears. Figure 2.4 shows an electron with energy E, incident from the left and transmitted to the right: (a) through a single barrier with barrier height V_b; (b) for a QW with a state denoted by E_1, an "eigenstate" (resonant state to be precise) created by a QW, as in typical elementary quantum mechanics; and (c) with transmission $|T|^2$. For a symmetrical structure, constructive interference gives rise to resonance with $T^*T = 1$, centered at E_1, similar to the optical Fabry–Perot interferometer.

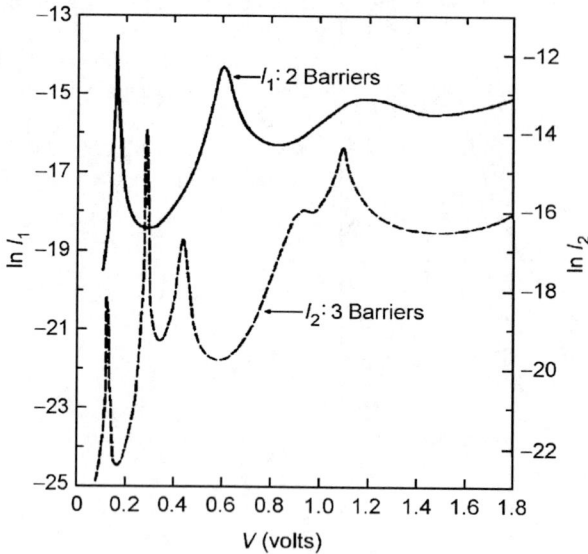

Figure 2.3 $J[(em*/2\pi\hbar^2 \ \hbar^3)^{-1}$ $\times (1.6 \times 10^{-12})^2]$ versus the applied voltage V showing NDC.
Source: After Tsu and Esaki (1973), with permission.

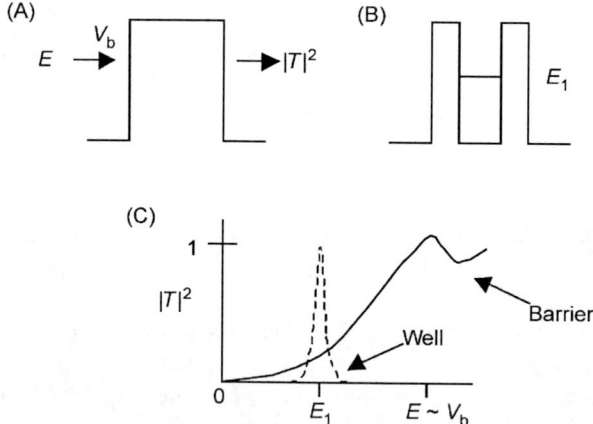

Figure 2.4 (A) Electron transmission at an energy E incident from the left and transmitted to the right through a single barrier. (B) The same barrier with a "cut," representing a QW between two barriers, the DBRT. (C) A resonant transmission peak at energy E_1 appears and is superposed onto the original barrier without the "cut."

In Figure 2.5, the same structure is subjected to an applied voltage between the left and right contacts. A current peak appears at $V \sim 2E_1$. The approximation sign is due to the loss of symmetry with an applied voltage. If symmetry can be preserved, the resonant tunneling peak appears exactly at $V = 2E_1$. The width of the current peak is approximately the width of E_F. As the voltage is such that the state E_1 moves into the forbidden gap, the supply of electrons from the left contact vanishes, resulting in a rapid decrease in the transmitted current. Thus, even without the use of a formula for the resonant tunneling current, the mechanism for NDC is easily understood.

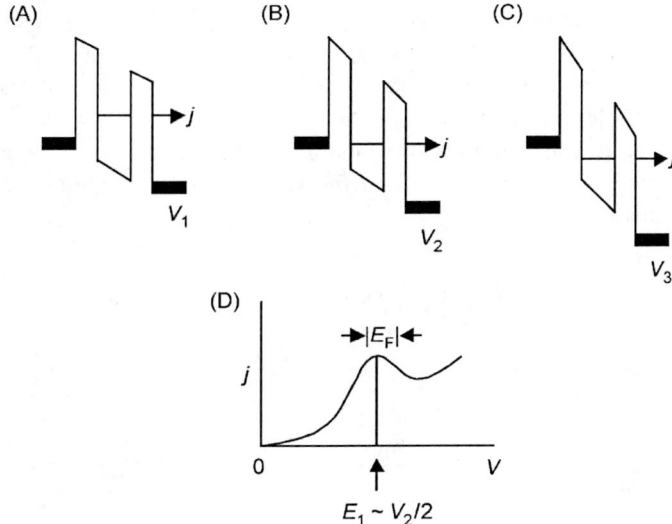

Figure 2.5 The same DB structure is subjected to an applied voltage between the left and right contacts. The low current is shown in (A), reaching a maximum value when the applied voltage is such that $E_F \geq E_1 - V_2/2 \geq 0$ shown in (B), and finally the current is much decreased at V_3 when the E_1 state moves into the forbidden gap of the left contact. j versus V is shown in (D). The background is direct Fowler–Nordheim tunneling through both barriers.

2.2 Some Fundamentals

Before we deal with the (k_1'/k_1) term that was left out in the original Tsu–Esaki formula, we shall first deal with some fairly elementary, yet quite important principles. For example, most quantum mechanics texts do not emphasize that $1 - |R|^2$ is not equal to $|T|^2$. Let an electron incident at a barrier with a potential $V = V\theta(x)$, and the Heaviside function $\theta(x)$, so that the wave function for $x < 0$ is

$$\psi_L = e^{ikx} + Re^{-ikx},$$

and for $x > 0$ on the right is

$$\psi_R = T_{LR}e^{ik'x}.$$

After imposing the continuity of ψ, and continuity of the current $\hbar/2im$ $(\psi^* \nabla \psi - \nabla \psi^* \psi)$ at the boundary at $x = 0$, there results $R = (k - k')/(k + k')$ and $T_{LR} = 2k/(k + k')$, so that

$$k(1 - |R|^2) = |T_{LR}|^2 k'. \tag{2.7}$$

Eq. (2.7) is simply the statement of current continuity. Similarly, we take an electron incident from the left, with R' and T_{RL}, and we obtain $R' = -R$, and $T_{RL} = 2k'/(k + k')$, so that $|R'|^2 = |R|^2$, but $|T_{RL}|^{2'} = |T_{LR}|^2(k'/k)^2$. Therefore, the reflection coefficient is the same regardless of whether it is going to the right or to the left, but the transmission coefficients are not the same. Now, we can calculate the net current going from the left to the right for our double-barrier (DB) QW tunneling structure.

$$j_{LR} = \frac{e}{4\pi^3 \hbar} \int dk_1 |T_{LR}|^2 \frac{\partial \varepsilon_1}{\partial k_1} \left(\frac{k_1'}{k_1} \right) f(\varepsilon)[1 - f(\varepsilon')] d\mathbf{k}_t, \quad \text{with } \varepsilon' = \varepsilon + eV. \quad (2.8)$$

Using $dk_1 = (m^*/\hbar^2)(d\varepsilon_1/k_1)$,

$$j_{LR} = \frac{em^*}{2\pi^2 \hbar^3} \int_0^\infty |T_{LR}|^2 \left(\frac{\varepsilon_1 + eV}{\varepsilon_1} \right)^{1/2} d\varepsilon_1 \int_0^\infty d\varepsilon_t [f(\varepsilon) - f(\varepsilon)f(\varepsilon + eV)], \quad (2.9a)$$

$$j_{RL} = \frac{em^*}{2\pi^2 \hbar^3} \int_0^\infty |T_{LR}|^2 \left(\frac{\varepsilon_1 + eV}{\varepsilon_1} \right)^{1/2} d\varepsilon_1 \int_0^\infty d\varepsilon_t [f(\varepsilon + eV) - f(\varepsilon)f(\varepsilon + eV)], \quad (2.9b)$$

and the net current from the left contact to the right contact becomes

$$j = j_{LR} - j_{RL} = \frac{em^*}{2\pi^2 \hbar^3} \int_0^\infty |T_{LR}|^2 \left(\frac{\varepsilon_1 + eV}{\varepsilon_1} \right)^{1/2} d\varepsilon_1 \int_0^\infty d\varepsilon_t [f(\varepsilon) - f(\varepsilon + eV)]. \quad (2.10)$$

The original Tsu−Esaki expression has left out a term, $(k_1'/k_1) = ((\varepsilon_1 + eV)/\varepsilon_1)^{1/2}$, referred to as the "kinematic term" by Coon and Liu (1985). Theories dealing with resonant tunneling appeared before and even after Coon and Liu used the Tsu−Esaki expression instead of the kinematic factor. The magnitude of this factor is $\sim\sqrt{3}$, because near resonance $eV \sim 2E_1$. Although this term generally does not affect the overall physical picture, let us look at what happens to the transmitted electron at an energy eV above the Fermi energy at the right contact. An electron, $|k_1'\rangle$, resonantly appears as a hot electron at the right contact at an energy $\varepsilon_1 + eV$, and is readily scattered to the equilibrium state $|k_1\rangle$ at energy ε_1 inside the right contact within perhaps few angstroms. Early on I struggled with this difficulty, even trying to create a transitional layer inside the contact, without any positive results. My idea was to create a transitional region in the contact somewhat similar to what has been discussed by Datta (1995). But any attempt to create two planes, 1 and 2, inside the contacts presents a formidable task with dubious consequences. From the intuitive pictures, we know that a large current in solids almost certainly originates from a large carrier density, such as in metals, or via multiplication mechanisms. Thus, we may view the process as the following sequence of events: (1) Electrons gain energy under an applied voltage and exit into a contact defined by the definition of a contact being an equal potential. (2) The extra energy is readily lost via scattering toward equilibrium with the final product being the creation of heat. (3) Such a process may be described by using Fermi's golden rule in a rate equation rather than a transport equation, allowing the establishment of equilibrium. Sometimes it is even necessary to define a hot electron temperature if elastic scattering dominates over inelastic events. Within the crude assumption that the Fermi levels between the two contacts are similar, electrons enter the collector at $E + eV$, and after relaxation return to the same Fermi distribution of the contact. Inclusion of a self-consistent potential such as by Cahay et al. (1987) and Bandara and Coon (1989) is a movement in the right direction, but not sufficient without taking scattering inside the contact into account. Nevertheless, the net result may very well be creating a term just canceling the kinematic term.

There was another nontechnical issue that I had to deal with. I have stated previously that the main reason for going to the double-barrier resonant tunneling (DBRT) was to model NDC without the possible domain oscillation pointed out to us by Gunn. Note that the very title of our paper, "Tunneling in a Finite Superlattice" (Tsu and Esaki, 1973), indicated my frame of mind. I wanted to produce the basis for extending the tunneling results as a model for transport in a superlattice showing NDC, where clearly equilibrium distribution should dominate every few tunneling events, or at least within a length limited by the mean free path. I also want to share with the reader the ideas that guided my thinking. When I embarked on this problem, a book by Duke (1969) treated everything except the feature I was looking for, allowing resonance to occur as in Fabry–Perot optical filter. I know that the DB structure is nothing more than an electron filter allowing electrons at the state approximated by $k_1 = n\pi/w$. Without an elaborate scattering formulation, it is clear that the transmitted electrons simply lose the extra energy while cascading down. I noticed that taking the velocity term as $\hbar k_1/m^*$ instead of $\hbar k_1'/m^*$ accomplishes just that. By doing so, I circumvented a complicated scattering formulation inside the right contact. I was convinced that whatever one does to take into account the scattering in the contact, the end result will involve giving back the extra energy. From my perspective, the Tsu–Esaki formulism represents a crude approximation. I erroneously chose to multiply $|T|^2$ by k_1, rather than by k_1', the correct way. However, as events have shown, the original work of Tsu and Esaki has led to an enormous number of improvements, including many body and hot electron effects (Tsu, 2007).

Now I think we need to recognize that the left out term may be important in some cases, preserving fundamental consistency within simple assumptions. Therefore, I want to take this opportunity to urge the inclusion of the "kinematics term," however, with recognition of the complexity I have pointed out. I also want to point out that I was disappointed that the extra correction, including the electron accumulation to the left and depletion to the right, made by Bandara and Coon (1989) further significantly increased the tunneling current, contrary to my intuition, where it decreased, as shown by Cahay et al. (1987), in a full self-consistent calculation. I think most of those involved in experimental work may prefer to see a correction factor decreasing the peak-to-valley ratio. The theory of resonant tunneling is still being developed after some 30 years. But before I leave this subject, I would like to comment briefly on several other treatments related to the discussion here. Duke's (1969) treatment is correct, but lacking the specific means to deal with coherent resonances. Vassell et al. (1983) were the first to point out the omission of this "kinematic" term from the Tsu−Esaki version. However, their presentation was quite confusing because they used k_0 for the incident electron from the left and k for the transmitted electron to the right, and yet, in their formula (Eq. (2.21)), the velocity operator is $v(k)$ instead of $v(k_0)$. In any case, the treatment of Coon and Liu (1985) is at least clear and their results are consistent as far as they go. Nonetheless, as noted by Noteborn (1993), the orthogonality process they introduced was somewhat incorrect. Where do we stand now? The self-consistent calculation by Cahay et al. (1987) is most appealing, because their results fit simple logic, where the tunneling current goes down rather than going up as reported by Bandara and Coon (1989). In a later section, I shall show how the self-consistent results of Cahay et al. are obtained with simple inclusion of the space charge in the QW.

Before I close this section, I want to emphasize that the sequential tunneling (ST) model first conceptualized by Luryi (1985) represents a subset of the coherent tunneling (CT) model, because, with unavoidable elastic as well as inelastic scattering in the QW, relaxation must be present in the well. I find it hard to imagine why it is surprising that the ST model developed by Payne (1986) and Weil and Vinter (1987) also shows NDC. As long as tunneling is into or out of a discrete state in the QW, outside the energy of this discrete state, tunneling cannot occur. As long as the mean free path is greater than the well width, scattering can only broaden the linewidth of this state, usually denoted by Γ, because there is no state available outside of this discrete state for the electron to be scattered into. The use of a non-Hermitian Hamiltonian, to be treated later, can naturally express this line broadening. Nonetheless, such an approach is embraced by very few because quantum mechanics, for most people, is synonymous with Hermitian operators. One often finds casual remarks that the use of the ST model is to facilitate the calculation of space charge in the well. I shall present a simple calculation of the space charge in the well using simply the wave function in the well. I think the most important value of looking at the problem concerned with the ST model is to allow a definite Fermi level inside the well region and an average between the Fermi levels on each side.

Since we have based on our physical picture of resonant tunneling on the important fact that the resonant energy $E_n \sim eV/2$, where the voltage V aligns the quantum states with the source of electrons, and generally $E_n \gg E_F$. Figure 2.6 shows the calculated $|T|^2$ using the Airy function versus the longitudinal electron energy for structures (b_1, w, b_2) at (3, 4, 3 nm), respectively, with barrier height $V_b = 0.4$ eV for GaAs well with GaAlAs barriers. For example, (0.355) refers to a

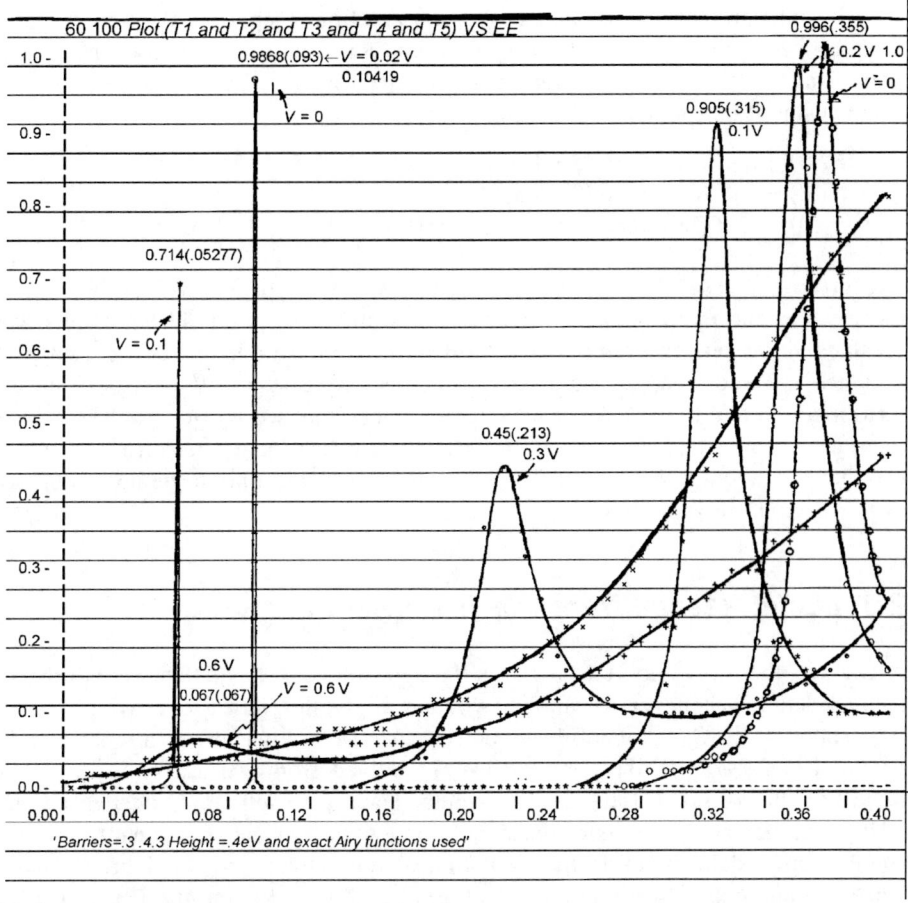

Figure 2.6 Calculated $|T|^2$ versus E for (3, 4, 3 nm) at $V_b = 0.4$ eV for the GaAs/AlGaAs DBRT structure, using the Airy function shortly after the Tsu and Esaki (1973) publication. Note that the peaks move very close to the $E_n = V/2$ rule.

peak located at 0.355 eV. $E_2(0.1)$ indicates the second energy state with an applied voltage of 0.1 eV. Specifically, we have the following:

$$E_2(0) - E_2(0.1) = 0.053 \sim V/2 \text{ at } V = 0.1,$$
$$E_2(0) - E_2(0.6) = 0.3015 \sim 0.6/2,$$
$$E_1(0) - E_1(0.1) = 0.10419 - 0.0528 = 0.051 \sim 0.1/2.$$

Note that the rule $E_n = V/2$ is followed very closely. Also, we want to point out that the height of the resonant peak is drastically reduced as $eV \geq V_b$, because the second barrier is depressed to the point where it cannot effectively confine the electron.

The same structure using the exponential function to approximate the triangular ramp by potential steps is also quite good, but not as close to the rule for the position of the resonant energy compared to the Airy function. Let us list the calculated results using the exponential function.

$$E_2(0) - E_2(0.1) = 0.3685 - 0.3185 = 0.005 \sim V/2 \text{ at } V = 0.1,$$
$$E_2(0) - E_2(0.6) = 0.3685 - 0.043 = 0.3255 > 0.6/2,$$
$$E_1(0) - E_1(0.1) = 0.10419 - 0.054 = 0.050 \sim V/2 \text{ at } V = 0.1.$$

Except for the case $eV = 0.6$, where the second barrier begins to lose its confinement effect, the exponential approximation is still very good. These results were obtained early on, enforcing our confidence in the simple picture presented in Figure 2.5. To summarize, we point out that as long as the QW is made from uniform material and the voltage is specified between the two barrier–well interfaces, regardless of the thickness of the barriers, symmetrical or not, electrons inside the QW have an average potential of $V/2$, commonly thought only to be applicable to a symmetrical DBRT structure.

2.3   Conductance from the Tsu–Esaki Formula

As we see, the most important result is the appearance of NDC that we will discuss later in detail. Whenever the applied voltage aligns the Fermi level with the resonant state of the QW, tunneling gives rise to NDC. However, at a small applied potential and $k_B T \sim 0$, the conductance G, derived from the use of Eq. (2.10) or Eq. (2.5), the original Tsu–Esaki formula, leads precisely to Landauer's (1957, 1970) conductance for one-dimensional tunneling. Basically, resonant tunneling comes from a delta function like T at finite temperatures and Landauer's conductance formula comes from a delta function like f where f is the the Fermi distribution function at very low temperature. Both are obtained from Eq. (2.10) or (2.5) taking the proper limits where these apply. To show this, it does not matter whether we include the kinematic term. We start from equation (2.182) of Mitin et al. (1999), which is identical to Eq. (2.5), for $k_B T \sim 0$. And far from resonant, $eV \ll E_n$, the conductance per electron (per spin) is

$$G = \frac{e^2}{h} \sum_{n,m} T(E_F, n, m), \tag{2.11}$$

where the sum is over the transverse degree of freedom (n, m), or integration in dk_t of the transverse channels and $T \equiv T * T(k_1'/k_1)$. The conductance per transverse channel becomes $G_{nm} = (e^2/h)T_{nm}$. If in each transverse channel $T_{nm} = 1$, then $G_{nm} = G_0 = e^2/h$, the so-called quantum conductance. This last assumption is frequently made; however, it is noted that the condition $T_{nm} = 1$ gives zero reflection, which happens near resonance and is contrary to the assignment of a contact conductance. In transmission line theory the only reflectionless contact is one with the input impedance exactly equal to the characteristic impedance of the line. Let us discuss this more in detail. First of all, an impedance function is merely a special case of a response function or a transfer function for the input/output. Therefore, there is no such thing unless two contacts are involved serving as input and output. The impedance or conductance has been referred to as contact conductance, for example by Datta (1995). In reality, it is not a contact conductance. If $T = 1$ is taken, then Eq. (2.11) applies to reflectionless contact. The real issue is why experimentally equal steps of G_0 appear? I think the answer lies in the fact that the transverse modes are not all coupled to a planar boundary. More precisely, G_0 is the conductance of the quantum wire with matched impedance at the input end, terminating in the characteristic impedance, and therefore also matched at the output end. We shall show later that G_0 represents the wave impedance of a closed system (a system with two contacts), the quantum conductance of a quantum electron waveguide, or a quantum wire, analogous to the treatment of photons, even in free space. For time dependence, we shall use the Laplace transform. Letting a wave bounce between the two reflectors, as adopted by Landauer for conductance, is a very special case of the general time-dependent solution.

2.4 Tunneling Time from the Time-Dependent Schrödinger Equation

I took a sabbatical at the Institute of Physics and Chemistry in the University of Sao Paulo, Brazil, between 1983 and 1985. During this period, I went back to Stark ladder (SL) and QW areas, working on doping and the dielectric function in an SL, as well as the old problem of damping to treat internal scattering in a QW, which is also closely tied to the residence time for the electrons inside a QW. Tunneling time is not only important in its own right, for example for device speed, it is also important in modeling involving charging time, scattering time, and so on, where one finds it in use in conjunction with the Matthiessen (1857) rule of adding inverse time constants. Part of the content of this section, a thesis by Subrata Sen, has been briefly summarized by Tsu (2001); however, some details, using the Laplace transform, are not only important from a technological point of view but also provide important concepts in fundamental physics and engineering, particularly for graduate students. Sen came to me to do an M.S. in Electrical Engineering in order to get a job in Silicon Valley, after having received a Ph.D. in theoretical nuclear physics from SUNY. I was quite satisfied with Sen's thesis, a little too mathematical perhaps, but worthy of publication. I take this opportunity to present it almost in its entirety, because, unfortunately, it has never even been submitted for publication when he graduated in 1989.

2.4.1 Stevens's Problem

The time-dependent Schrödinger equation has been treated like a stepchild by physicists and mathematicians. Expanding and following on the work of Stevens

(1983), time-dependent formulation of a transmission resonance in a DBRT structure is treated with Laplace transforms. Starting with Stevens's approach to the Schrödinger equation,

$$i\hbar\frac{\partial\psi}{\partial t} = -\frac{\hbar^2}{2m}\frac{\partial^2\psi}{\partial t^2} + V(x)\psi, \tag{2.12}$$

where $\psi(x, t) = 0$, for $|x| \to \infty$, and $\psi(x, 0) = r(x)$, and using the Laplace transform (p is used instead of the traditional s used in most engineering fields) gives

$$\phi'' - \beta^2\phi = \frac{i2m}{\hbar}r(x), \quad \text{with} \quad \beta^2 = \frac{2m}{\hbar^2}(V - i\hbar p), \quad \text{and}$$

$$\phi(x, p) = \int_0^\infty \psi(x, t)\exp(-pt)\mathrm{d}t = \mathcal{L}(\psi(x, t)) \tag{2.13}$$

Using Green's function $G(x, x', p)$, satisfying $G'' - \beta^2 G = \partial(x - x')$, then

$$\phi(x, p) = \frac{i2m}{\hbar}\int_{-\infty}^\infty G(x, x', p)r(x)\mathrm{d}x'. \tag{2.14}$$

In free space, $G(x, x', p) = (-1/2\beta)\exp(-\beta|x - x'|)$.
For a square wave, $r(x) = A_0\theta(a - |x|)\exp(ikx)$, leading to

$$\phi(x, p) = \frac{2maA_0}{\hbar\beta}\exp(\mp\beta x)\mathrm{sync}((\beta \pm ik)a), \quad \text{with}$$

$(-, +)$ for $x \geq a$, and $(+, -)$ for $x \leq -a$,

in which $\mathrm{sync}(y) = [\sin(y)]/y$. Taking the inverse Laplace transform,

$$2\alpha\psi(x, t) = \exp(ikx - \omega t)\{\mathrm{erfc}(\gamma_-) - \mathrm{erfc}(\gamma_+)\},$$

where $\alpha = \sqrt{m/\hbar}(1 - i)$ and $\gamma_\pm = ik(\sqrt{t}/\alpha) + \alpha(x \pm a)/\sqrt{t}$.
For Stevens's half-space problem, the boundary conditions become: $V(x) = 0$ for $x \geq 0$, $\psi(x, t) = 0$, for $x \to \infty$ and $\psi(0, t) = f(t)$. By specifying $f(t)$ we are in effect putting a current source j at the origin as $j(x, t) = i\hbar\mathbf{Im}(\psi\partial\psi^*/\partial x)/2m$. Note that direct specification of $j(0, t)$, which is nonlinear in ψ, would not have permitted the use of the Laplace transform in solving the initial value problems. Then,

$$\phi'' - \beta^2\phi = 0, \quad \text{and} \quad \phi(0, p) = F(p) = \mathcal{L}(f(x, t)); \quad \phi(x, p) = F(p)\exp(-\beta x).$$

With $f(t) = \exp(-i\omega t)$, then $F(p) = 1/(p + i\omega)$, and

$$2\psi(x,t) = \exp(-i\omega t)\{\exp(ikx)\text{erfc}(\eta_-) + \exp(-ikx)\text{erfc}(\eta_+)\},$$

where $\eta_\pm = x\sqrt{\dfrac{\alpha}{t}} \pm \sqrt{-i\omega t}$.

2.4.2   The DB Problem

We first consider the case of a semi-open system where the DB has one side at which we put a current source. Initially, $t < 0$, there is no charge without a source. At $t = 0$, the source is switched on for a time period of T. The charge $Q(t)$ is obtained by integrating $|\psi^*\psi|^2$ over the barriers and the well. For time $t > T$, $\psi(0, t) = 0$, the charge leaks out through the right side only. We shall also treat a system that is open at both ends of the barrier. The applied voltage allows the charge to flow through the system. However, we use an exponential function instead of an Airy function for convenience. To calculate the charge inside the QW, $Q(t)$ must be multiplied by $[n(\omega) - n'(\omega)]$ and integrated over all ω, using the distribution functions n for the input side and n' for the output side, the right side. Our goal is to study the decay time of the trapped charge as a function of excitation energy, represented by $\sim \exp(-i\omega t)$ whose transform $\sim 1/(p + i\omega)$.

The response in terms of an impedance function Z is given by

$$\phi(x, p) = F(p)Z(x, p). \tag{2.15}$$

For the half-space problem in the previous section, $Z(x, p) = \exp(-\beta x)$. Let us take the potential of a symmetrical DB for barrier width, b, and well width, w, of a potential

$$V(x) = \begin{cases} V, & a \le x \le a + b \quad \text{and} \quad a + b + w \le x \le a + 2b + w \\ 0, & \text{elsewhere.} \end{cases}$$

The boundary conditions and initial conditions (see Morse and Feshbach (1956) on under and over specifications of boundary and initial conditions) of the half-space problem are

$$\psi(x, t) = 0, \quad x \to \infty, \quad \text{and} \quad \psi(x, t) = \exp(-i\omega t)\{\theta(t) - \theta(t - T)\}, \quad \text{then}$$
$$F(p) = \{1 - \exp[-(p + i\omega)T]\}/(p + i\omega).$$

The system under consideration is a DB, on one side of which we put a current source. At $t < 0$, there is no charge. The source is turned on at $t = 0$ for a time T. The charge $Q(t)$ is obtained by integrating $|\psi(x, t)|^2$ over the well and the barriers. For $t > T$, $\psi(0, t) = 0$, so that the charge can only leak out through the right side only. Therefore, this is a typical semi-open system. The flow of charge through the structure requires a potential drop requiring the use of the Airy function; however, we approximate the potential profile by steps for approximating to exponential functions.

Let us define

$$\beta^2 = -\frac{2mip}{\hbar} \quad \text{and} \quad \beta'^2 = \frac{2m(V - i\hbar p)}{\hbar^2}$$

$$d_1 = (\beta + \beta')^2 - (\beta - \beta')^2 \exp(-2\beta'b)$$

$$d_2 = \frac{2mV}{\hbar^2}(1 - \exp(-2\beta'b))$$

$$d_3 = (\beta - \beta')^2 - (\beta + \beta')^2 \exp(-2\beta'b)$$

$$s = x - a, \quad y = x - a - b, \quad \text{and} \quad z = x - a - b - W$$

then, $Z(x, p)$, the Laplace transform of Green's function

$$\begin{aligned}
Z(x, p) = \ & -\Delta_0[\beta \sinh \beta'(b - s)(d_1 + d_2 \exp(-\beta W)) \\
& + \beta' \cosh \beta'(b - s)(d_1 - d_2 \exp(-\beta W))] \quad (a \leq x \leq a + b) \\
= \ & 2\beta' \Delta_0[\exp(-\beta y)d_1 - \exp(-\beta(2W - y))d_2] \quad (a + b \leq x \leq a + b + W) \\
= \ & 4\beta\beta' \Delta_0 \exp(-\beta W)[\exp(-\beta'z)(\beta + \beta') \\
& - \exp(\beta'(z - 2b))(\beta - \beta')] \quad (a + b + W \leq x \leq a + 2b + W),
\end{aligned}$$

$$(2.16)$$

where

$$\Delta_0 = \frac{\exp(-\beta a - \beta'b)}{\exp(-2\beta W)d_2(d_2 + d_3) - d_1(d_1 + d_2)}.$$

There is a branch cut along the Re p from $-\infty$ to 0 and a double pole at (Re p, Im p) = $(0, -\omega)$. The path of integration is along the Im p in the right half of the plane from $-\infty$ to $+\infty$. The inverse transform is performed numerically. For thick barriers, ≥ 1.5 nm, there are first-order poles on the negative imaginary axis where Z is inside the well. If these poles coincide with the excitation $F(p)$, the doubly singular response in $\phi(x, p)$ exhibits resonance behavior. We shall see that these resonances are characterized by a monotonic growth of the space charge inside the well and current through the well. The charge and the current diminish while oscillating rapidly as a function of energy separate from resonance, as shown in Figure 2.7 for $Q(t)$ and in Figure 2.8 for $j(t)$. For a numerical example, we have chosen $W = 6$ nm, $b = 2$ nm, with the barrier height $V = 0.3$ eV and $m = 0.067m_e$. The limits of integration for the inverse transform were -3 eV $<$ Im($p\hbar$) < 3 eV, more than sufficient for the energies involved. The transmission resonance occurs at $\omega = 0.0682$ eV. Note that at resonance, there is a slight overshoot, but basically growth is monotonic until the excitation is cut off at $t > T$, after which rapid decay takes place. *However, away from resonance, oscillatory behavior is evident.* These results are familiar to those working on transients, but those who use the steady-

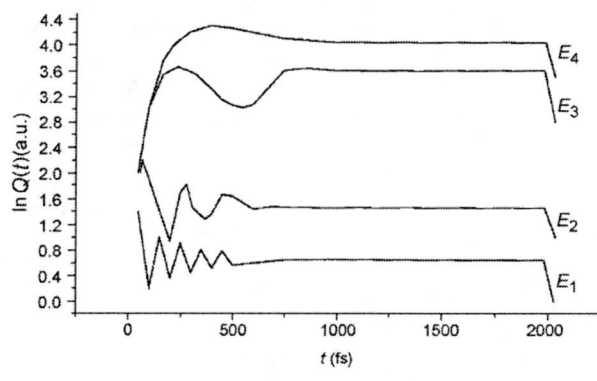

Figure 2.7 The trapped charge $Q(t)$ versus time (fs) for $w = 6$ nm, $b = 2$ nm with GaAs well and GaAlAs barriers. The energy at resonance is located at $E_4 = 0.0682$ eV, $E_3 = 0.079$ eV, $E_2 = 0.05$ eV, and $E_1 = 0.03$ eV.
Source: First appeared in published form in Tsu (2001) taken from Sen's unpublished M.S. thesis (1989). From Tsu (2001), with permission.

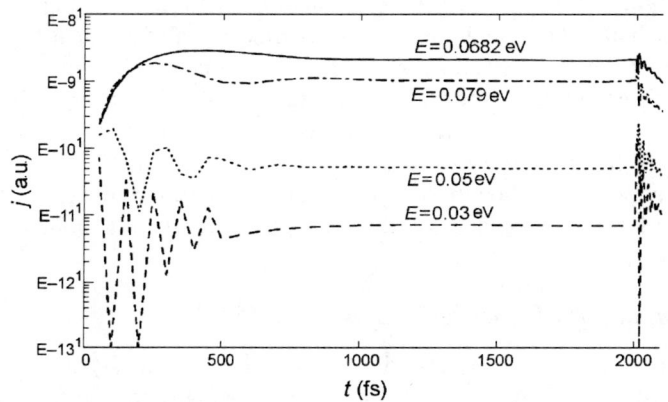

Figure 2.8 Transmitted current j versus t (fs) for the same case as Figure 2.7.

state time-independent model could possibly take these oscillations as showing instability (Goldman et al., 1987). For $t > T$, the excitation is not singular, the response disappears, and the confined charge starts decaying. We define the decay time τ by $Q(T + \tau) = Q(T)/e$. Similarly, we can define a buildup time τ_b by $Q(\tau_b) = (1 - e^{-1})Q_0$, where Q_0 is the steady-state value of the charge inside the well. Note that this definition is not unique when separated from resonance, because the trap charge is oscillatory. Therefore, we take the time to reach the first peak as the buildup time. In a way, we must recognize that decay time and buildup time can only be uniquely defined for monotonic variations. With this cautionary remark, the buildup and decay times are shown in Figure 2.9.

Next, we consider the case of a pulse of width $T = 100$ fs. The growth and decay is more or less similar to the case shown, however, the decay starts before reaching the steady state, resulting in a much wider peak than anticipated. In the calculation, we take $0 < a < 0.1$ nm. However, the position of the resonance transmission peak remains stable for any a.

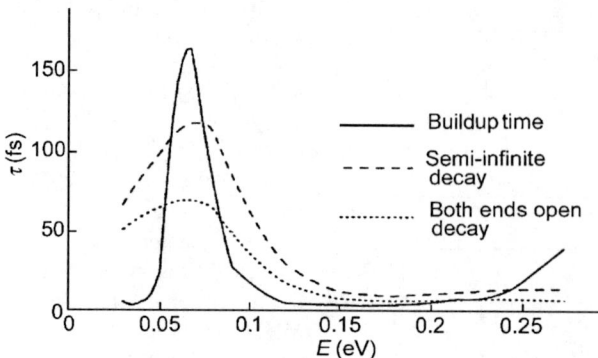

Figure 2.9 Buildup time and two cases of decay times. Dashed line, open at one end; dotted line, open at both ends, in τ (fs) versus energy. The rise at $E > 0.25$ eV shows the beginning of the next quantum state.

It is interesting to note that even the decay after the excitation is turned off is not monotonic; rather, it consists of very rapid oscillations as shown in Figure 2.8.

Note that both charge and current versus time are almost identical, a fact to be expected from our understanding of Fabry−Perot optical cavity. This fact negates the notion of instability caused by space charge buildup discussed in the last section. The buildup time has a maximum at resonance as in any resonating system. However, it is important to recognize that even the decay time peaks near resonance, a fact usually not much thought out. The maximum buildup time is $\sim$200 fs, which is the reason for the fast device speed of the resonant tunneling diode (RTD).

2.4.3   Spreading of a Wave Trapped in a Well

Instead of using Stevens's specification of the wave packet $r(x)$ at $t = 0$, we shall be concerned with cases where $r(x)$ is a steady state solution of the Schrödinger equation. We start with an excitation $\exp(-i\omega t)$ outside the well as in the previous section and compute $r(x)$. The response $\phi(x, p)$ can now be calculated using Eq. (2.14) with the appropriate Green's function. When the barrier is open at both ends, the charge naturally leaks out faster and the resonance is less pronounced. The resonance peak for buildup time is due to the doubly singular character of the response function for $t < T$. However, during decay, the excitation function ceases to be singular and the spectrum displays a wider peak. Next, we consider $\exp(-i\omega t)$ at $x = x'$ as an excitation within the well and calculate the steady-state wave packet. To exploit the symmetry of the barrier, we need to modify our coordinate system and formulate the problem as follows:

$$V(x) = \begin{cases} V, & -b - \dfrac{W}{2} \leq x \leq -\dfrac{W}{2} \\[2ex] V, & \dfrac{W}{2} \leq x \leq b + \dfrac{W}{2} \\[2ex] 0, & \text{elsewhere} \\ \psi(x, t) = 0, & x \rightarrow \infty \\ \psi(x', t) = \exp(-i\omega t). \end{cases}$$

Then,

$$\phi(x,p) = F(p)Z_1(x,p), \quad x' \le x \le \frac{W}{2}$$

$$\phi(x,p) = F(p)Z_2(x,p), \quad -\frac{W}{2} \le x \le x'$$

$$Z_1(x,p) = \frac{\exp(-\beta x) - H \exp(\beta x)}{\exp(-\beta x') - H \exp(\beta x')}$$

$$Z_2(x,p) = \frac{\exp(\beta x) - H \exp(-\beta x)}{\exp(\beta x) - H \exp(-\beta x)}$$

$$H = \exp(-\beta W)(d_1/d_2).$$

The impedances Z_1 and Z_2 on either side of the point of excitation closely resemble the impedances due to two single barriers. As the point of excitation moves toward the center of the well from the shadow of the barrier, the singular nature we had observed in the impedance of the system for an external excitation gradually disappears and with it the resonant character of the response. The time evolution of the charge confined in the well at steady state can now be calculated by using Green's function. The decay for different points of excitation located at x' inside the well are all different as shown in Figure 2.10. Note that the decay time for $x' = 0$ is nearly independent of energy. Unlike the case for external excitation, with internal excitation near the center of the well at $x' = 3$ nm, the decay time drops off slowly as the energy goes above the resonant energy. This feature was studied because I asked Subrata Sen what happened if charges were injected into the center of the QW as in a device with a "gate" or due to relaxation processes. He was quickly able to show that with a delta function as a source in the middle of the QW, the two barriers are decoupled and sharp resonance gives way to a soft peak owing to the two-dimensional nature of the QW states, resulting in sequential process, not a true resonant process. Subrata explained that the loss of the double pole leads to the loss of resonance. Mathematically, however, the space is divided into two half-spaces.

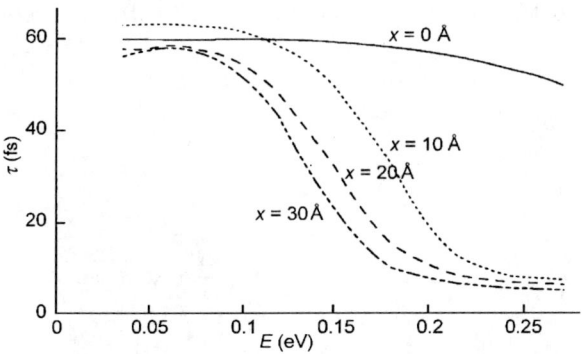

Figure 2.10 Decay time τ (fs) versus E (eV) for various points of excitation, x', in the well.

2.4.4 *The Series Expansion*

In the 1880s, Oliver Heaviside developed operational calculus, a powerful mixture of common sense, physical intuition, and mathematical insight/skullduggery, and applied it to solve linear differential equations. His paper containing series solutions of the wave equation and the heat equation in free space was rejected by the editors of the *Proceedings of the Royal Society* because of its lack of rigor (see Whittaker, 1928). Laplace transform was extensively used by Carslaw and Jaeger (1948, 1959), beginning the characterization of linear systems by their eigenvalues and poles in the complex plan, an approach adopted here. According to Carslaw and Jeager, the hyperbolic functions in the transform were expanded in a series of exponentials leading to solutions that have a convenient physical meaning in terms of reflected waves, which are described in text books on the Laplace transform (Carslaw and Jaeger, 1948). However, for the Schrödinger equation, the series expansion loses a definite meaning. Here, we shall demonstrate that in the asymptotic limit, a sinusoidal excitation produces a steady state in a QW with standing waves at resonance.

We shall approximate the DB by delta function potentials. For our half-space problem $(x \geq 0)$, $V(x) = V_0 \delta(x - a)$, and $\psi(0, t) = \exp(-i\omega t)$. The excitation, $F(p) = 1/(p + i\omega)$ and the impedance function

$$Z(x,p) = \frac{(V_1 + \beta)\exp(-\beta x) - V_1 \exp(-\beta(2a - x))}{V_1 + \beta - V_1 \exp(-2\beta a)},$$

where $V_1 = 2mV_0/\hbar^2$. Expanding $Z(x, p)$ in a series we obtain the response

$$\phi(x,p) = F(p) \sum_{n=0}^{\infty} \left[V_2^n \exp(-\beta(x + 2na)) - \frac{V_2^{n+1}}{V_1} \exp(-\beta(2(n + 1)a - x)) \right].$$

$$(2.17)$$

The first term in the summation asymptotically represents the direct wave while all other terms represent the reflected waves, the quantity $V_2 = V_1/(V_1 + \beta)$ being the reflection coefficient. The inverse transform of the first term $F(p)\exp(-\beta x)$ gives

$$\tfrac{1}{2}[\exp(i(kx - \omega t))\mathrm{erfc}(\gamma) - \exp(-i(kx + \omega t))\mathrm{erfc}(\delta)],$$

where

$$\alpha = \sqrt{m/\hbar}(1 - i), \quad \gamma = x\sqrt{\frac{\alpha}{\tau}} - \sqrt{-i\omega t}, \quad \delta = (2a - x)\sqrt{\frac{\alpha}{\tau}} + \sqrt{-i\omega t}.$$

As $t \to \infty$, $\mathrm{erfc}(\delta) \to 0$; we can treat the other terms similarly. The final asymptotic expansion is

$$\psi(x, t) = \sum_{n=0}^{\infty} \left[V_3^n \exp(ix_n)\mathrm{erfc}(\gamma_n) - \frac{V_3^{n+1}}{V_1} \exp(iy_n)\mathrm{erfc}(\delta_n) \right],$$

where

$$x_n = k(x + 2na) - \omega t, \quad y_n = k(2(n+1)a - x) - \omega t,$$
$$\gamma_n = x_n \sqrt{\frac{\alpha}{t}} - \sqrt{-i\omega t}, \quad \delta_n = y_n \sqrt{\frac{\alpha}{t}} - \sqrt{-i\omega t}.$$

As $V_3 = V_1/(V_1 + k)$ is less than 1, the series converges and a steady state is reached. As $t \to \infty$ the complementary error functions tend to 1. The terms in the series will then add up in phase whenever $ka = n\pi$, $n = 0, 1, 2, \ldots$. Hence, standing waves are possible in the asymptotic limit.

In conclusion, the Laplace transform is a powerful tool to study time-dependent problems with spatial boundary conditions. One can also use the derivative of the phase shift with frequency for the time delay using the time-independent approach. However, only a Gaussian packet leads to the same result as a time-dependent treatment (Tsu and Zypman, 1990). On the other hand, it is not really different from the use of Green's function, which we shall treat next. Although the Wigner function has been popularized recently (Datta, 1995), my preference is still for Green's function.

2.4.5 Delay Time in DBRT

An estimate of a signal delay time may be obtained with the uncertainty relation $\Delta E \Delta t \approx \hbar$, where ΔE is the linewidth of the transmission peak of the QW structure. Another estimate, generally more accurate, involves $\tau = Qt_0$, with t_0 being the transit time given by $t_0 = d/(\hbar k/m^*)$ and $Q = \Delta E/E$. A rigorous expression for the delay time of a signal propagating through a linear network (any book on signal processing) is

$$\tau = d\phi/d\omega = (d\phi/dk)(dk/d\omega), \quad \phi = kd + \theta,$$

where ϕ is the total phase shift and θ the phase of the transmission amplitude through the DBRT structure. The delay time τ for a structure with the barrier width and well width equal to 2.5 and 6 nm, respectively, is shown in Figure 2.11. Note that the approximate values shown as circles are close to the computed delay time τ. At resonance, the delay time is very long.

In Figure 2.12, the time delay using a Gaussian at the left and at a time τ later at the right is compared with the buildup time and decay time for a δ-function

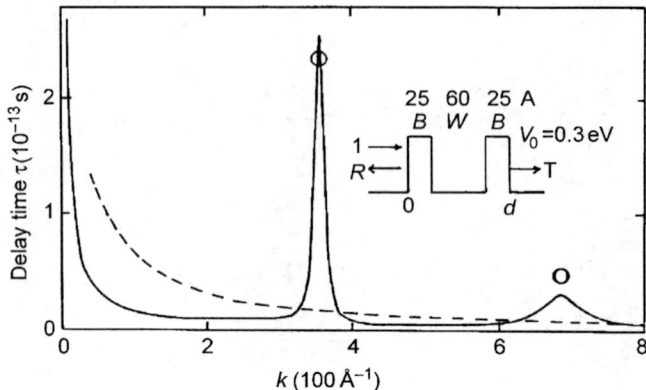

Figure 2.11 Delay time τ for the first two resonant peaks versus k for the structure shown, solid $t_0 = dm^*/\hbar k$. Dashed line and circles, $t_0 Q$.
Source: After Tsu and Zypman (1990), with permission.

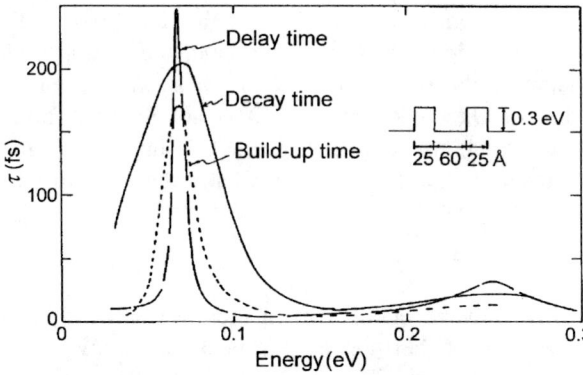

Figure 2.12 Comparison of delay time, decay time, and buildup time for the structure shown. Delay time was calculated by Zypman (1988, unpublished). The others were from Sen.

excitation within the well, taken from Subrata Sen's 1989 thesis. The delay time, peaking higher and narrower, is similar to Figure 2.11.

Basically, the time slows down by the factor Q, the quality factor of the resonating system similar to any resonating systems. However, as shown in Figure 2.12, the time scale is a couple of femtoseconds, still very fast. Thus, we predict that any space charge effects cannot be the reasons for the appearance of hysteresis or instability, which have been attributed by some, in a resonant tunneling device. There will be a special section devoted to this subject later.

2.5 Damping in Resonant Tunneling

2.5.1 The Quality Factor Q

The quality factor Q is useful for characterizing losses in a resonating system and analysis of linewidth in any spectrum. All experimentalists know about the linewidth at half maximum (LWHM), ΔE. And Q is $E/\Delta E$. Formally, Q is defined by:

(1) $Q \equiv 2\pi \times$ number of wavelengths in a given mean free path, ℓ or $Q = k\ell$.
(2) $Q \equiv 2\pi \times$ number of periods in a given mean free time, τ or $Q = \omega\tau$.
(3) $Q \equiv 2\pi \times$ (energy stored $\div$ energy loss per cycle, T or per wavelength, λ), or $2\pi \times$ (particle stored $\div$ particle loss per cycle, T or per wavelength, λ),

for

$$A = A_0 \exp\{i[(k_r + ik_i)x - \omega t]\},$$

then

$$Q = 2\pi/\exp(1 - \exp(-2q\lambda)) \quad \text{or} \quad Q = 2\pi/\exp(1 - \exp(-T/\tau)), \quad q \equiv k_i.$$

As $q\lambda \ll 1$, $Q \rightarrow k_r\ell$ and as $T/\tau \ll 1$, $Q \rightarrow \omega\tau$, the same as (1) and (2) above. But if $q\lambda \gg 1$, or $T/\tau \gg 1$, $Q \rightarrow 2\pi$. This last result is obviously meaningless. What we need to realize is that Q cannot be defined by definition (3) if the system damps before it goes through less than one cycle of oscillation or one wavelength. Therefore, we see that the first two definitions are always correct while the last is only meaningful for low loss. Unfortunately, definition (3) is used far more extensively, particularly in the engineering field. We shall use the first two definitions. Generally speaking, definition (1) is more useful in transport problems and (2) is used in the time—oscillatory situation such as in cyclotron resonance. In fact (1) and (2) are identical if we take the decay time $\tau = (2\nu)^{-1}$, with ν, the collision frequency. And (3) is popular with engineers, partly because we are not interested in a highly damped system. Since mobility is defined as $e\tau/m^*$, regardless of whether we use (1) or (2), the linewidth from resonant tunneling gives a measure of the mobility. Q is a single parameter figure of merit.

2.5.2 The Simplest Way to Account for Damping

Before I discuss where and how we can include damping, let us go back to the usual Green's function G, in Fourier transformed space, for free space, which is adequate for us to make the point. Take the free space Green's function

$$G_\pm(E) = \frac{1}{E - E_k \pm i\Sigma}, \quad \Sigma \rightarrow 0. \tag{2.18}$$

The reason we insist on $\Sigma \rightarrow 0$ is because we want to preserve the Hermitian nature of

$$(H - E_k)\psi_k = 0 \quad \text{and} \quad (H - E_k)G_k = \partial(x - x'), \quad \text{where } H = H^+. \tag{2.19}$$

If we do not let $\Sigma \to 0$, damping is involved, then Green's function does not satisfy Eq. (2.19), rather, it satisfies an equation

$$(\mathcal{H} - \mathcal{E}_k)\mathcal{G}_k(x - x') = \delta(x - x'), \quad \text{with } \mathcal{H} \neq \mathcal{H}^+. \tag{2.20}$$

It is interesting to note that Green's function need not be limited to Hermitian operators. For an open system, the operator is generally non-Hermitian, whether one deals with a lossy system such as in the optical Fabry—Perot interferometer, or with a microwave cavity that has dissipation. In fact, even an antenna radiating energy to infinity cannot be described by a closed system such as an atom or molecule. The simplest way to make any sense of the use of a Hermitian Hamiltonian for a closed system applying to an open system is to cut the system into parts, having the non-Hermitian parts sum to zero, where each individual part or parts are non-Hermitian. Such individual parts obviously do not have real eigenvalues, which is fine because we need the complex eigenvalues to represent dissipation. The real problem is that we can no longer assign occupation, not even the usual distribution functions, and certainly Pauli's exclusion principle cannot be rigorously enforced. The common denominator with all these problems lies in the fact that these solutions are not orthogonal, preventing us from representing the excitations as quasi-particles. To avoid these problems, quantum mechanics tells us to treat the dynamics by a scattering formulation, abandoning the eigenvalue approach. Very early on in 1973 when I first calculated resonant tunneling using complex $k = k_r + ik_i$, not long after my solution with real k was published (Tsu and Esaki, 1973), I showed my results to Rolf Landauer, expressing some satisfaction that the calculated $I-V$ shows the usual line-broadening feature and a reduction in the peak-to-valley ratio for the tunneling current. I was cautioned about putting a non-Hermitian term in the Schrödinger equation itself. Landauer suggested that I consider using the density matrix. I want to share with the reader here why I wanted to put the damping term in the wave equation itself. The density matrix operator, like the distribution function of the Boltzmann transport equation, has no phase. It is certainly possible to formulate the problem step-by-step, such as using the unitary operator for the solution from $\psi(t)$ to $\psi(t + \Delta t)$, calculating the scattering process Δt by Δt. However, I knew that in the classical system, the microwave cavity loss, or wave propagation in a lossy medium may be represented by a complex dielectric function and so on, the solution usually involves a complex wave vector or a complex frequency. I have read quite a few accounts about the difference between classical systems and quantum mechanical systems. I think most arguments are missing the target. The substitution of classical canonical variables by operators into commutation brackets leads to quantum mechanics. As soon as nonlinear terms are included, the Hamiltonian is non-Hermitian. The real issue is not a fundamental one, rather a practical one. If the loss is so great that $Q < 1$, it is certainly futile to describe states and a propagation constant, and so on. However, for a rather meager quantum device that has a $Q \sim 2-5$, there are $2-5$ cycles before disappearance of the excitation; certainly characterizing the system with a wave equation is useful and correct. Before I leave this subject, I should point out that the scattering matrix

S shares the same flexibility as the Green's function. Since the impedance function Z is only a special case of the S matrix, engineers are on safer ground, because damping is usually introduced in the impedance function Z.

Something of fundamental interest is discussed near the end of the first volume of Morse and Feshbach (1956), pointed out to me by N. Horing. Let us take a non-Hermitian operator n and its adjoint n^+ with their respective state vectors ψ and χ, or

$$\hbar\psi_\mathrm{m} = \lambda_\mathrm{m}\psi_\mathrm{m} \tag{2.21}$$

and

$$n^+\chi_\mathrm{n} = \mu_\mathrm{n}\chi_\mathrm{n}. \tag{2.22}$$

If $n \neq n^+$, we can neither make ψ_m and ψ_n orthogonal, nor can we make χ_m and χ_n orthogonal, but ψ_m and χ_n can be made bi-orthogonal. This is very serious. Not only do we give up eigenvalues being constant at all time, the nonorthogonality wave function means that the wave functions belonging to two different states, m and n are not independent. This is another way of saying that the different states are not uniquely defined. We must give up the notion that a broad resonant state may be viewed as an eigenstate. Actually, a good experimenter would have asserted that transition into or out of a very broad state is not very interesting. *One of the first concepts I learned from Leo Esaki was that one should not waste time on a very broad spectrum, because it is unlikely to exhibit any intriguing physical phenomena.* With all these discussions, I hope we are prepared to move on to what we can do with the introduction of a complex wave vector.

2.5.3 Resonant Tunneling with Damping

As early as December 1973, I started putting a damping term, $i\hbar\nu \equiv i\Gamma$, in the Schrödinger equation. With the wave function $\psi = \psi(\mathbf{r})\psi(t)$, where $\psi(t)$ is the usual harmonic time variation, the time-independent equation for $\psi(\mathbf{r})$ becomes

$$-\frac{\hbar^2}{2m}\nabla^2 + [(V(\mathbf{r}) - E) - i\Gamma]\psi(\mathbf{r}) = 0. \tag{2.23}$$

Taking the wave vector $\mathbf{k} = (\mathbf{k_t}, k_x + i\kappa_x)$, where $k_x + i\kappa_x$ is the solution of

$$(k_x + i\kappa_x)^2 = \frac{2m}{\hbar^2}(E_x - V(x) + i\Gamma_x). \tag{2.24}$$

For resonant tunneling structures (RTS), $V(x) = V(x_j)$, where j denotes the sections of the structure, the band edge offset of the structure plus the applied potential across the RTS. In the actual calculation, because the transverse component of a $\mathbf{k}$

is assumed to be conserved from a separable variable, we simply have an equation in the variable x with the usual boundary conditions. Figure 2.13 shows the DBRT structure. First we compute $k_x + i\kappa_x$ in each section from Eq. (2.24) and A_+ damps to the right and B_+ damps to the left. The rest of the computation is the same as before. I would like to point out the difference in practice between the 1970s and the twenty-first century. In the 1970s, my matrix inversion required that I separate the real and imaginary parts and arrange into doubling the rank of the matrix; 2×2 becomes 4×4. The calculation is shown in Figure 2.12.

It is commonly assumed that the use of $i\Gamma$ (we drop the subscript x) does not introduce a self-energy shift (Gupta and Ridley, 1988). However, our solution of Green's function with the same $i\Gamma$ shows a self-energy shift. The reason is that the self-energy term comes only from the imaginary part of Γ for purely real k. Because k is complex in the result shown in Figure 2.14, there is a shift in the peak position of the resonant current. In the next section, the use of Green's function will be fully treated by $H_1 = -i\hbar/\tau$, or $H_1 = 2q$ (d/dx) as a damping term added to the Hamiltonian H_0.

In the years 1983−1985, I was a Professor at the Institute of Physics and Chemistry of the Universidade de Sao Paulo, Brazil. I was told by Liderio Ioriatti that I should look at the book by Isihara (1971), referring to Langevin's equation.

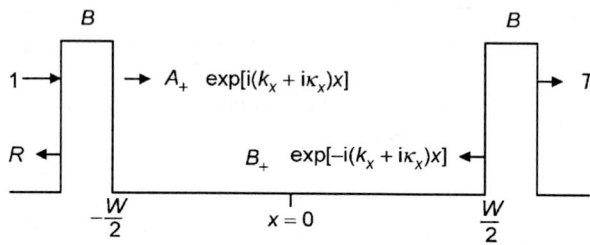

Figure 2.13 A DBRT structure showing the barrier width B and well width W. Values for $k_x + i\kappa_x$ in A_+ and B_+ are determined from Eq. (2.24), consistent with boundary conditions.

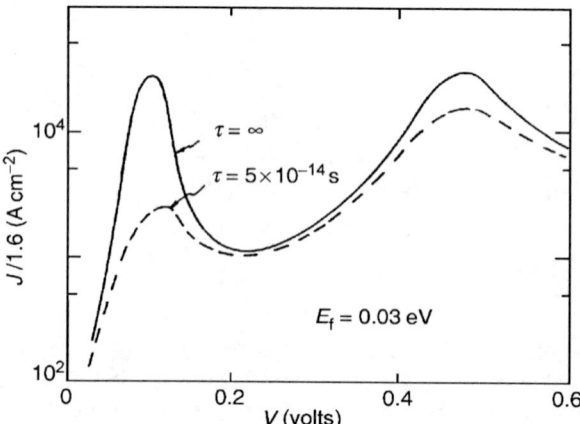

Figure 2.14 Resonant tunneling with time damping, $\tau = 5 \times 10^{-14}$ s for RTS with $B = 2.5$ nm and the well width $w = 6$ nm for GaAs/GaAlAs RTS. Note that there is a drastic reduction in the peak-to-valley ratio, particularly via the first resonant state. This calculation was repeated using the 35−65 rule for band-edge offset of the unpublished calculated in 1974 with the 20−80 rule. *Source*: After Tsu and Zypman (1990), with permission.

I seized on his suggestion and delved into the book. This resulted in my first serious attempt to use the Green's function approach (Tsu, 1985a,b), although I had partially finished work on Green's function using time damping even before 1974. Since Langevin's damping term gives damping in space, and my original approach was on time damping, I shall start with a later paper (Tsu, 1989), where I treated both cases. Although the primary target was the treatment of resonant tunneling involving amorphous materials, the approach may be applied to general cases of damping. Mathematically, time damping appears to be simpler and more stable without the need to switch the sign of the damping term every time the wave is reflected. However, transforming relaxation in time to spatial damping results in the imaginary part of k depending on the real part. Nevertheless, it is popular because it has been applied all along, as by Heitler (1954).

2.5.4 Green's Function in a Damped Free-Electron Schrödinger Equation

2.5.4.1 Relaxation Time

If we introduce $H_1 = -i\hbar/\tau$ into the free-particle Schrödinger equation, there are two possible cases. The first has complex energy and real momentum, which are useful for time damping such as cyclotron resonances and atomic transitions. The second has complex momentum and real energy, which is what we shall examine in detail. For real k and complex E,

$$\psi(x, t) = e^{\pm ikx}e^{-\Gamma t}e^{-i\omega t}, \tag{2.25}$$

where the energy $E \to \hbar\omega - i\Gamma$, is complex with $\Gamma = \hbar/\tau$. For complex k and real E, which is the case we want to apply to damping in space from relaxation time, we substitute $k = k_r + ik_i$ into $\psi_\pm(x) \sim \exp(\pm ikx)$ in Schrödinger equation,

$$(H + H_1)\psi_\pm = i\hbar\partial\psi_\pm/\partial t \quad \text{or} \quad (\partial^2/dx^2 + k_0^2 + iQ^2)\psi_\pm(x) = 0,$$

where

$$k_0^2 \equiv 2mE/\hbar^2, \quad k_r^2 = k_0^2 + k_i^2, \quad k_i = Q^2/2k_r, \quad \text{in which } Q^2 \equiv 2m/\hbar\tau. \tag{2.26}$$

The one-dimensional Green's function $G_\pm(x, x')$ satisfies

$$\left(\frac{\partial^2}{\partial x^2} + k_0^2 + iQ^2\right)G_\pm(x, x') = -4\pi\partial(x - x'), \tag{2.27}$$

with

$$G_\pm(x - x') = \frac{2\pi i}{k}e^{\pm ik(x-x')}. \tag{2.28}$$

Note that we must fix the factor associated with the term $\partial(x - x')$. Sometimes -4π, which I used, and sometimes -1 is used. What I usually do is to fix the factor consistent with $n(E)$, the density of states (DOS), obtained from Im $G(x = x')$. Therefore, the factor $-(2m/\hbar^2\pi)$ should be in front of $\partial(x - x')$ in Eq. (2.27). Why do most people simply take the factor as 1? It is because mathematically, we define Green's function as the solution of a general differential equation with a forcing term, an excitation term, replaced by a delta function. The forcing term depends on what is the variable for the differential equation. For simplicity, we set it to 1 and decide later to be consistent with the forcing term. We do not put this factor in but must realize that it should be there. For $Q \to 0$, $k_i = 0$, the Hamiltonian operator in Eq. (2.27) is Hermitian and the Fourier transform of $G_\pm(x, x')$ reduces to Eq. (2.18). However, for $Q \neq 0$ in Eq. (2.26), $G_\pm(x, x')$ is still given by Eq. (2.28), except k is now complex, so that the DOS,

$$n(E) = \text{Im } G_\pm(x',x') = \frac{2\pi i}{k} = \frac{2\pi k_r}{k_r^2 + k_i^2} \tag{2.29a}$$

and

$$\text{Re } G_\pm(x',x') = \frac{2\pi k_i}{k_r^2 + k_i^2}. \tag{2.29b}$$

Maximizing $n(E)$ in Eq. (2.29a), by setting the derivative to zero, we obtain a peak for $n(E)$ located at $k_r^2 = \sqrt{3/4}Q^2$, corresponding to $k_0^2 = Q^2/\sqrt{3}$. For $k_i = 0$, $n(E) = 2\pi/k$. Similarly, for the Re $G_\pm(x', x')$, there is also a very broad maximum located at $k_r^2 = Q^2/2\sqrt{3}$, or $-k_0^2/\sqrt{3}$. We see that between the maxima of Im $G_\pm(x', x')$ and that of Re $G_\pm(x', x')$, $\Delta k_0^2 = 2Q^2/\sqrt{3}$. For $Q^2 = 1$ in the example shown in Figure 2.15, $\Delta k_0^2 = 2/\sqrt{3}$. What happens to Re $G_\pm(x', x')$ for $Q^2 \equiv 0$? Re $G_\pm(x', x')$ is a delta function, $\partial(k_r) = \partial(k)$, at $k = 0$ or $E = 0$. Unlike for discrete states such as with oscillators, whenever the imaginary part goes through a peak, the real parts go through zero. We shall see later in Green's function for QWs that the real part does indeed go through a zero when the imaginary part goes through a peak.

Figure 2.15 shows k_r, k_i, and $n(E)$ versus E, for $Q = 0$ (no damping) and $Q = 1$ (with damping). Now $n(E)$ of the localized states has a tail in the normal forbidden gap. And the peak, the maximum of $n(E)$, is upshifted due to the presence of damping, from $k_0^2 = 0$ to $k_0^2 = Q^2/\sqrt{3}$. Setting $\exp(-k_i\ell) = 1/e$, defines $k_i \equiv \ell^{-1}$, where ℓ is obviously the mean free path. With Q^2 in Eq. (2.26), we see that $\ell = v\tau$, where the velocity $v = \hbar k_r/m$. Let us further note that the demarcation of the local and nonlocal set by $k_0^2 \geq Q^2/\sqrt{3}$ is equivalent to $k_r\ell \geq 1$. Therefore, we can state that $k_r\ell \approx 1$ demarcates the local and nonlocal, or evanescent and propagating regimes. In a QW with $k_r = n\pi/w$, $k_r\ell \geq 1$ means $\ell > w/n\pi$ is the criterion for the existence of quantum states. For amorphous solids, $\ell \approx 0.5-1$ nm, dictating the well width of a QW, $w \leq 1.5-3$ nm. The maximum of the DOS appears at an upshifted energy, which is therefore, not only qualitatively but quantitatively consistent with

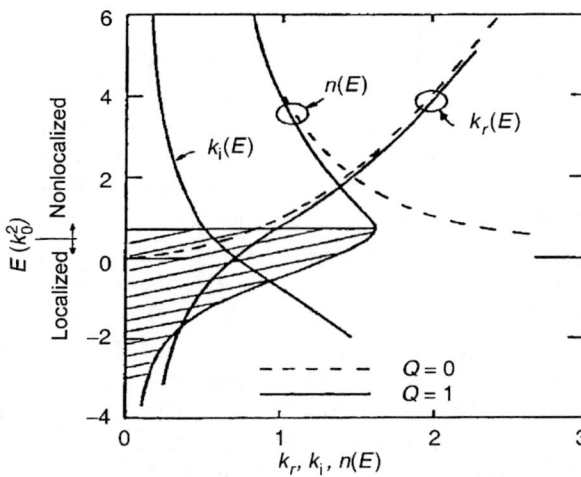

Figure 2.15 k_r, k_i, and $n(E)$ versus E. Now $n(E)$ of the localized states resembles an exponential tail. The peak demarcates the local (shaded) from nonlocal states.
Source: After Tsu (1989), with permission.

the concept of the mobility gap in amorphous solids going from localized states to nonlocal states as shown in Figure 2.15 (Mott and Davis, 1979).

In summary, above the mobility edge, the dispersion approaches that of a free particle. The imaginary part of the wave vector decreases rapidly compared to the real part as energy is increased. The nonlocal states are pushed up beyond $k = 0$. Qualitatively, the consequence of damping is to broaden and shift up the quasi-states in energy. What happens is because the self-energy term is responsible for broadening and level shift. Since localized states are already confined, confinement serves further to move up the extended states with respect to the localized states. In the process, trapping levels originally at resonance with band states are now left in the expanded localized region of energy, which may be a blessing for amorphous materials. The widening of optical transition has been attributed to quantum confinement in amorphous QW structures. For $\ell \sim 1$ nm, one should see quantum effects for a well width $w \leq 3$ nm. Therefore, the observed quantum confinement effects reported by Abeles and Tiedje (1983) and Hirose and Miyazaki (1984), and resonant tunneling reported by Pereyra et al. (1987), are indeed consistent with the present theory. We have shown that damping can also significantly push up the energy states involved in optical transitions. Therefore, one must be careful to distinguish between what is due to confinement and what is due to damping as presented. Furthermore, with time damping, the localized states form continuous tail states as shown in Figure 2.15. This simple theory gives tail states approaching zero at $E = -\infty$, reminding us of the exponential tail due to randomness.

Let us start with Eq. (2.26),

$$k_r^2 = \frac{k_0^2}{2} \pm \frac{1}{2}[k_0^4 + Q^4]^{1/2}.$$

We used the $(+)$ sign for Figure 2.15. However, if we use the $(-)$ sign, the real and imaginary parts of k are interchanged, but no new solutions are generated.

We know that the effective band gap in amorphous silicon is increased from 1.1 eV for c-Si to 1.7 eV for a-Si. Suppose we use this value 1.7 eV in Eq. (2.26) in the peak of the DOS at $k_0^2 \geq Q^2/\sqrt{3}$, we obtain a mean free path $\ell = 0.7$ nm, which is quite close to what we expect. In a way, this may be due to some remarkable coincidence. We normally represent amorphous materials in terms of randomness in bond angle distribution, where a model of "frozen-in" phonons can describe many properties of a-Si, including the explanation of optical absorption without any adjustable parameters (Tsu, 1985a,b). What we present here is a dissipative mechanism represented by a non-Hermitian operator. And yet, everything seems to fit together for a-solids (Mott, 1970). We may now ponder what is meant by random inelastic scattering. I showed these results to Sir Neville Mott. He said that he wanted to include them in his new book. Alas, he passed away not long after. One of the first books I carried back and forth from my home to my office at BTL was by Heitler (1954). In those days, I thought it presented the most inclusive treatment on damping. I noticed then that the so-called Lorentzian line is nothing but our Green's function for free particles [Eq. (2.18)]. I also knew that the isolated damped oscillator is Lorentzian and that an infinite number of coupled oscillators result in a Gaussian lineshape, the so-called Central Limit Theorem. This early introduction to fundamental ideas may have provided me with such conviction that damping is a better way to deal with inelastic losses affecting coherent interferences and the effects of finite mean free path on man-made quantum systems as a whole. Because these man-made quantum systems are most likely not to have as high Q quantum states as those in atoms and molecules, damping must be introduced even if for engineering design purposes alone.

As in the use of the Laplace transform, Green's function $G(\mathbf{r}, \mathbf{r}', p)$ has a singularity in three dimensions, which drastically complicates the matter in comparison with our one-dimensional treatment. This can be seen simply with the free space Green's function, $\exp(ikr)/r$ with $n(E) \sim k_r$. Applying such a function to the three-dimensional structure represents a formidable task.

2.5.4.2 Spatial Damping with Langevin's Term

The quantum mechanical analog of Langevin's braking term in classical systems is to include a term

$$H_1 = \mp \frac{2\hbar^2 \mathbf{q}}{2m} \cdot \nabla$$

in H_0, which is

$$\mp \frac{2\hbar^2 q}{2m} \frac{\partial}{\partial x}$$

in one dimension, with $|q| = 1/\ell$, where ℓ is the mean free path. As usual, $\psi_\pm$ representing a wave to the right with a $(+)$ sign, and to the left with a $(-)$ sign, satisfies

$$(H + H_1 - E)\psi_\pm(x) = 0. \tag{2.30}$$

Multiplying Eq. (2.30) by $2m/\hbar^2$, we obtain

$$\frac{d^2\psi_\pm}{dx^2} + k_0^2\psi_\pm + 2q\frac{d\psi_\pm}{dx} = 0, \tag{2.31}$$

where $k_0^2 = 2mE/\hbar^2$ and $q = [2\theta(x - x') - 1]|q|$. Note that when using the Heaviside function for q, there is no need to put in $\mp$. The reason we need this $\mp$ in H_1 is that without it, $\psi_\pm$ diverges at $x = \pm\infty$. However, it is obvious that the differential equations to the right and left are not the same. For a free particle without reflections, once a ∂-function is placed at x', waves to the right and left emergent from $x = x'$ do not overlap, so that we still can define Green's function. There are discontinuities at the point of the delta function, as well as at each point where reflection of the wave takes place, requiring the sign of the damping factor q to be switched, with complicated procedures for several points, $x_1, x_2, \ldots$, in general. From Eq. (2.31)

$$\left(\frac{d^2}{dx^2} + k_0^2 + 2q\frac{d}{dx}\right)G_\pm(x, x') = -4\pi\partial(x - x'). \tag{2.32}$$

Now, we have an extra discontinuity in addition to the usual jump in the derivative of Green's function at $x = x'$, a discontinuity arising from the change of sign of q at $x = x'$. To find G, we first perform $\int_{x'-\varepsilon}^{x'+\varepsilon}$ [Eq. (2.32)]dx. The first term gives the discontinuity in dG/dx, the second term vanishes because $G_\pm(x, x')$ is continuous, however, the third term does not vanish because of the Heaviside function for q. Before we find $G_\pm(x, x')$, let us substitute $\psi_\pm \sim \exp(\pm ikx)$ into Eq. (2.31) with $k = k_r + k_i$. Equating the real and imaginary parts, we obtain $k_0^2 = k_r^2 + q^2$, $k_i = +|q|$ for propagation to the right, $x > x'$, and $k_i = -|q|$ for propagation to the left. Taking Green's function $G_\pm(x, x') = A\exp(\pm ik(x - x'))$, gives $A = 2\pi i/(k - i2|q|)$. From $k_i = +|q|$, we obtain

$$G_\pm(x, x') = \frac{2\pi i}{k_r - i|q|}\exp\left[\pm i(k_r + i)|q|x\right]. \tag{2.33}$$

Note that if q does not change sign, the denominator in Eq. (2.33) would be same as k in the numerator for Green's function for $q = 0$. We see now in this form that we do not need to worry about the $\pm q$. Note that Eq. (2.33) is an improved version of what I published before (Tsu, 1985a,b). Surprisingly, the DOS is same as before,

$$n(E) = \operatorname{Im} G_\pm(x', x') = \frac{1}{\pi}\left(\frac{2m}{\hbar^2}\right)\frac{k_r}{k_0^2}, \tag{2.34a}$$

and

$$\text{Re } G_{\pm}(x',x') = \frac{1}{\pi}\left(\frac{2m}{\hbar^2}\right)\frac{-q}{k_0^2}. \tag{2.34b}$$

except that the prefactor Eq. (2.34a) is same as $n(E)$ given by Tsu (1989), or in the transformed form by Tsu (1985a,b). At $q = 0$, Im $G_{\pm}(x', x') \propto 1/k$ and Re $G_{\pm}(x', x') \propto \partial(k)$.

As shown in Figure 2.16, Langevin's spatial damping gives rise to results that are very similar to those from the time-damping case. However, $n(E)$ peaks at $k_0^2 = 2q^2$ or $k_r^2 = q^2$ without tailing into the forbidden gap and Re $G_{\pm}(x', x')$ has a broad maximum at $k_0^2 = q^2$. Except for the lack of tailing, all physical parameters obtained, such as the quality factor Q, the self-energy shift and the linewidth are basically the same. In a way this is significant, because essentially our intuitive back of the envelope estimate is remarkably good. I am not saying that we do not need to examine individual interactions such as impurity scattering, inelastic phonon scattering and more exalted ones such as surface roughness and structural defects. What I do want to emphasize is the fact that man-made quantum systems are far from atomic states, certainly far from states giving rise to superconductivity. The states we deal with perhaps have no more than $Q \sim 10$ and frequently more like $2-5$. Presently, the electronic world is moving toward quantum dots (QDs), by virtue of the fact that the smaller size will have more atoms on the surface which are subject to unwanted reactions and need complicated reconstruction to gain stability. I have pointed out on several occasions that the most important quantum mechanical effects we recognize in electronic devices are no more than the simple factor of two for the two spinning electrons occupying a given state. Most cases, electrons act almost classically. Therefore, we must view in the damping as representing an overall benefit quantum system in devices, because it is a powerful measure of the state of coherence. We shall come to acknowledge the usefulness of the

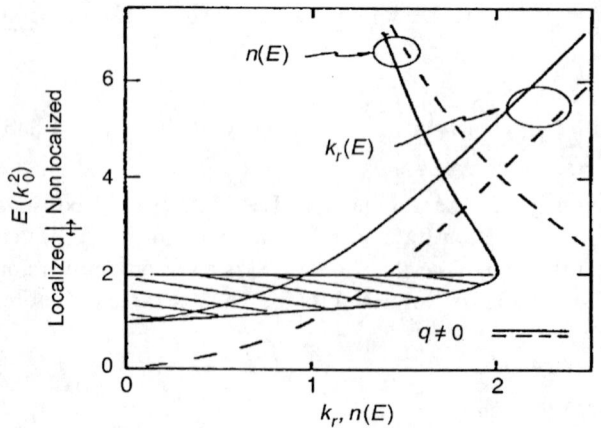

Figure 2.16 k_r and $n(E)$ versus E for spatial damping. As a result of damping, $q \neq 0$ the localized states (shaded) are separated from the nonlocal states.
Source: After Tsu (1989), with permission.

quick ways of calculating linewidth, shifts, and the quality factor Q drawing on what we have presented, particularly using Green's function for QWs.

2.5.5 Green's Function for Damped QW Structures

Green's function used in conjunction with the Laplace transform for studying tunneling time using the time-dependent Schrödinger equation is presented in Section 2.4. In this section, the exact Green's function with damping is presented. Here, we are in a position to examine some of those techniques based on physical concepts. First, we take an isolated QW, generalized to a finite barrier width useful for resonant tunneling. The steps needed to determine approximate values involve finding the phase change for an electron reflected from two barriers. We are all familiar with the fact that Im $G(x', x')$ gives the DOS. However, it is not so well known that the zeros of Re $G(x', x')$ give the eigenstates for an isolated QW without damping. We want to establish in what way with damping, the zeros of Re G are related to the use of the method of phase of constructive interference to find the energy state, which is exact without damping. We shall see how valuable these quick and simple ways are in determining linewidth and level shift. Moreover, with a finite barrier width and damping, the resonant state is further broadened by the combined effects of tunneling out of the well and damping in the well. Last, for tunneling through a DB structure, the energy spectrum of the transmitted electrons gives a linewidth. A mathematical model is used to calculate the transmission through the structure. For an isolated QW between two very thick barriers, there is no way to place the energy detector for the transmitted electrons. Even worse is the fact that we do not even know how to set up the problem in order to calculate the eigenstates. This is where Green's function comes in. Placing a delta function inside a QW, the Im $G(x'|x')$ gives the spectrum of the DOS, which exactly describes the position of the eigenstates as well as the linewidth when losses are present. Jumping ahead, we shall use $G(x'|x')$ to examine how well these approximate methods work involving the quality factor Q, and others.

2.5.5.1 Another Look at the Free Particle Green's Function

The Fourier transform of Green's function

$$G_\pm(E) = [(E - H) \mp \Sigma]^{-1}, \quad \Sigma = \Sigma_r + i\Sigma_i. \tag{2.35}$$

For $\Sigma_r = 0$ and as $\Sigma_i \to 0$, it is obvious that the Im $G \sim \partial(E - E_k)$ and the zero of Re G gives the eigenstate, $E = E_n$. However, for $\Sigma \neq 0$, the zeros of the Re G give the shifted resonant state and the self-energy shift, and the Im G gives the broadened DOS. To my mind, we should not be bound by the Hermitian assumption of the Hamiltonian. After all, we are dealing with the zeros and poles in the complex E plane. In this regard, Σ can be anywhere in the complex plane. The usual Green's function of Hermitian operators corresponds to $\Sigma = -i\Sigma_i$ and $\Sigma_i \to 0^+$ that apply to a closed system. For an open system or semi-open system, $\Sigma = \Sigma_r + i\Sigma_i$ lying off the real axis in the complex plane. In complex variable theories, whenever a pole moves off the real axis, damping follows. Back in 1967 when I was working on the theory of electron–phonon interactions in polar solids, I learned from Ted Schultz, who had just published a book on many body theory (Schultz, 1964), that the simplest way to include real-time damping is to move the pole from ω_0 to $\omega_0 + i\nu$, with a finite ν (Tsu, 1967). I was still overly concerned with criticisms received for incorporating non-Hermitian operators in the 1970s. When my manuscripts on damping (Tsu, 1985a,b; Tsu and Zypman, 1990) were rejected by the reviewers, I quoted Breit and Wigner (1936), to no avail. Perhaps, I should have simply stated my case without further ado as in Mehan (2000), that with damping, Σ is complex! Now that these points have been noted, I shall first derive Green's function for the case of an isolated QW with damping extending to the DB structure. But first, we shall derive the reflection coefficients with damping to see in what way the reflection coefficients enter various methods of approximation.

2.5.5.2 Reflection from a Barrier

Although reflection from a barrier is found in most elementary quantum mechanics books, for the convenience of the readers, we shall include a brief account here particularly when damping is present. We shall need these to treat the problem of damping using the concept of Q, the quality factor of a resonating system.

On the left side of Figure 2.17, with time damping, $\psi_\pm$ represent the wave to the right and to the left, respectively. Then

$$\psi_1 = \exp[i(k_r + ik_i)x] = R \exp[-i(k_r + ik_i)x], \quad \psi_2 = T \exp(-\alpha x), \tag{2.36a}$$

with $\alpha^2 = \alpha_0^2 - k_0^2$, in which $\alpha_0^2 = 2mV/\hbar^2$. Equating ψ and ψ', assuming equal masses at $x = w/2$, we obtain

$$R = \frac{ik_r + (\alpha - k_i)}{ik_r - (\alpha + k_i)} e^{-2k_i w} e^{-i2k_r w} = \left| \frac{k_r^2 + (\alpha - k_i)^2}{k_r^2 + (\alpha + k_i)^2} \right|^{1/2} e^{-2k_i w} e^{i(\theta - 2k_r w)}$$

$$= \left| \frac{\alpha_0^2 + 2k_i^2 - 2k_r \alpha}{\alpha_0^2 + 2k_i^2 + 2k_r \alpha} \right|^{1/2} e^{-2k_i w} e^{i(\theta - 2k_r w)}, \tag{2.36b}$$

$$\tan \theta = (2k_r \alpha)/(\alpha^2 - k_i^2 - k_r^2). \tag{2.36c}$$

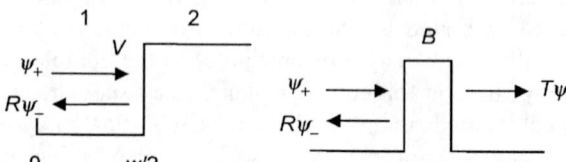

Figure 2.17 Left, an electron reflected from an infinite barrier; right, a finite barrier.

The phase factor $e^{-2k_i w}$ appears here because the barrier is placed at $x = w$ rather than at $x = 0$. Note that the imaginary part, $k_i = Q^2/2k_r$, which complicates a great deal. In the past I assumed a couple of numerical iterations by putting $k_r \approx \pi/w$ for k_i to find θ which could be used in the method of phase to calculate the QW state with damping. For $k_i = 0$, the reflection coefficient has a unity modulus with only a phase shift. What is interesting is when $\alpha = 0$, $k_i \neq 0$, $|R| = e^{-2k_i w}$, which tends to zero at large w, meaning that at large w, reflection disappears. This result is correct, because in a lossy system at large distance, nothing is left to be reflected back. Therefore, all coherence effects disappear.

Similarly, for the case shown on the right side of Figure 2.17, which allows tunneling to the right for finite barrier width B and barrier height V such that $\alpha^2 = \alpha_0^2 - k_0^2$, with $k_0^2 = 2mE/\hbar^2$ and $\alpha_0^2 = 2mV/\hbar^2$; assuming damping in the region to the left of the barrier, and no damping to the right of the barrier, then

$$R = |R|e^{i\theta}$$

$$|R|^2 = \left| \frac{(k_r - \alpha \sin \beta)^2 + (k_i - \alpha \cos \beta)^2}{(k_r + \alpha \sin \beta)^2 + (k_i + \alpha \cos \beta)^2} \right| \tag{2.37a}$$

and

$$\tan \theta = \frac{2\alpha(k_r \cos \beta - k_i \sin \beta)}{\alpha^2 - k_i^2 - k_r^2}, \tag{2.37b}$$

where

$$\tan \beta = \frac{-2 \exp(-2\alpha B)\cos \vartheta}{1 - \exp(-4\alpha B)} \tag{2.37c}$$

and

$$\tan \vartheta = \frac{2k_0 \alpha}{\alpha^2 - k_0^2}. \tag{2.37d}$$

As $B \to \infty$, $\beta \to 0$, and Eq. (2.37b) reduces to Eq. (2.36c).

2.5.5.3 The Method of Phase

We shall see how we can use the phase angle [Eq. (2.36c)] for the calculation of the energy state of an isolated QW shown on the left of Figure 2.17, and Eq. (2.37b) for the energy in a QW with a barrier of finite thickness shown on the right of Figure 2.17. If all we need is an approximate value, we can simply use the method of phase with $k_r w + \theta = n\pi$. For exact values we must use the solution Re $G(x'|x') = 0$. Let us take a simple example to obtain an understanding of the resonant tunneling through a DB structure by a simple fundamental concept, perhaps even with reasonable quantitative results. This aspect was first presented by Tsu (1985a,b), and will be duplicated here.

Let us use the method of phase, involving adding the phase $k_r w$ of a wave in a path w and the phase change upon reflection θ for the condition of constructive interference. The resonant energy is obtained by this simple process. Specifically, we equate the total phase change after the wave has experienced two bounces from the two walls to $2n\pi$ or $2k_r w + 2\theta = 2n\pi$, then

$$k_r = (n\pi - \theta)/w, \qquad\qquad\qquad\qquad (2.38)$$

where θ is the phase angle of the reflection coefficient given by Eq. (2.37b). From Eqs. (2.37) and (2.38), we can find the energy at resonance. As the barrier height is lowered, θ goes up, resulting in a lowering of k_r, and lowering the energy state with a finite barrier height. Let us take the case of a single well for illustration, with $k_i = 0$ and $B = \infty$, the solution for the eigenvalue is given by Schiff (1955),

$$k \tan kw/2 = \alpha \quad \text{and} \quad \tan \theta = 2 \tan(\theta/2)/[1 - \tan^2(\theta/2)].$$

Together with Eq. (2.38), we have $\tan \theta = (2k\alpha)/(\alpha^2 - k^2)$, which is identical to Eq. (2.37b) for $k_i = 0$ and $B = \infty$. We shall see how the method of phase can be used in Eq. (2.37b) to find the eigenvalues when $k_i \neq 0$, how the method of phase is related to the peak in energy corresponding to the maxima in the DOS given by Im $G_+(k)$ and the zeros of Re G_+, and furthermore, how well it performs.

2.5.5.4 Green's Function for an Isolated QW

We shall first work out Green's function for an isolated QW in detail, generalized to the case of a well with a finite barrier width on each side, allowing tunneling from the structure. Our method is to place $\partial(x - x')$ inside the well for $G(x - x')$, satisfying all the usual boundary conditions as well as the change of sign in q, in the case of spatial damping. For planar structures, variable separable allows the use of the one-dimensional Green's function. The problem is really straightforward. Figure 2.24 shows two cases: an isolated QW on the left and one that allows tunneling out on the right. The regions I and II are as shown. To reduce the complexity, we place $\partial(x - x')$ in the middle, or as $\partial(x)$. This geometry reduces the work by half because of symmetry, without loss of generality.

Let us first derive Green's function with time damping for the isolated QW on the left of Figure 2.18.

$$\text{I}: x > 0 \quad A_+ \exp(ikx) + B_+ \exp(-ikx), \quad \text{II}: C_+ \exp(-\alpha(x - w/2)) \qquad (2.39a)$$

$$\text{I}: x < 0 \quad A_- \exp(ikx) + B_- \exp(-ikx), \quad \text{II}: C_- \exp(\alpha(x + w/2)) \qquad (2.39b)$$

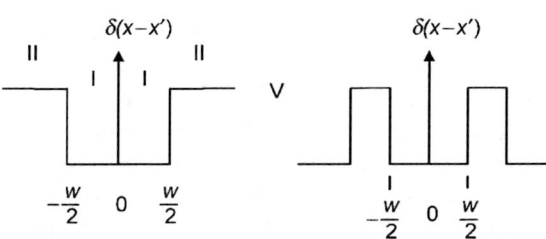

Figure 2.18 Left, barriers on each side have infinite extent; right, barrier widths are finite so that tunneling into, and out of, are involved.

At $x = 0$, $A_+ + B_+ = A_- + B_-$ and $(A_+ - B_+) = (A_- - B_-) + 4\pi i/k$ from the jump in the derivative due to the delta function at $x = 0$.

$$\text{II: } x = \pm w/2, \quad A_\pm \exp(\pm ikw/2) + B_\pm \exp(\mp ikw/2) = C_\pm \tag{2.39c}$$

Because of the symmetry with the delta function in the middle of the QW, $A_+ = B_-$, $A_- = B_+$, $C_+ = C_-$. At $x = 0$, $G'_+(0^+) - G'_-(0^-) = 2ik(A_+ - B_+) = -4\pi$, then

$$G(0|x) = (4\pi i/k)[y \exp(ikx) + (y - 1)\exp(-ikx)] \tag{2.40}$$

where $y \equiv [|R|\exp(-k_i w + i(k_r w + \theta)) + 1]^{-1}$, in which $|R|$ and $\tan\theta$ are same as the magnitude of R in Eqs. (2.36b) and (2.36c), respectively. And

$$G(0|0) = (4\pi i/k)(2y - 1). \tag{2.41}$$

The notations are getting too complicated, so let us set

$$R\exp(ikw) = |R|\exp(-k_i w)\exp[i(k_r w + \theta)] \equiv \rho\exp(i\Phi) = \rho(\cos\Phi + i\sin\Phi). \tag{2.42}$$

$$\text{Re } G(0|0) = \frac{2k_r\rho\sin\Phi + k_i(1 - \rho^2)}{(k_r^2 + k_i^2)(1 + \rho^2 + 2\rho\cos\Phi)} \tag{2.43a}$$

$$\text{Im } G(0|0) = \frac{(1 - \rho^2)k_r - 2k_i\rho\sin\Phi}{(k_r^2 + k_i^2)(1 + \rho^2 + 2\rho\cos\Phi)} \tag{2.43b}$$

Re $G(0|0) = 0$ gives $2k_r\rho\sin\Phi + k_i(1 - \rho^2) = 0$. For $Q = 0$, so that $k_i = Q^2/2k_r = 0$, it becomes $\Phi = kw - \theta = n\pi$, with $n = 1, 3, 5$. (This is because the even integers do not satisfy the symmetrical structure we are treating.) Thus, we see that without damping, the energy eigenstates are obtained with Re $G(0|0) = 0$, which is identical to our method of phase.

The expression for the numerator in Eq. (2.43a) when equal to zero involves k_i as a function of k_r, however, it can be computed simply with a pocket calculator. Once $\sin\Phi$ is determined, it is entered into Eq. (2.43b) with $\cos\Phi$ in the denominator for Im $G(0|0)$, we can then calculate Im $G(0|0)$ and hence $n(E)$. Let us see what happens when $k_i \neq 0$. The method of phase involves only the term with sin Φ. Although sin Φ itself involves k_i, it would have appeared as though the method of phase can be extended to $k_i \neq 0$. However, if the term with k_i is left out, the upshift of the energy of the resonant state would not have been present. Because $k_i(1 - \rho^2)$ in Eq. (2.43a) is always positive, then $\Phi \geq \pi$, which means that the resonant energy is always upshifted from the eigenstate when $Q = 0$ and cannot be obtained without the Green's function. This is what we are looking for; the resonant energy is upshifted due to damping, similar to the physical picture of the free electron Green's function (Tsu, 1985a,b). And for large k_i, the use of the method of phase is definitely not correct.

In Figure 2.19, the Re $G(0|0)$ and Im $G(0|0)$ calculated from Eqs. (2.43a) and (2.43b) for an isolated QW width $w = 6$ nm and a barrier height of $V = 0.4$ eV, are shown using a relaxation time $\tau = 10^{-14}$ s corresponding to $\ell \sim 8$ nm. The peak of the Im part corresponds to the zero of the Re part representing the resonant energy. What is satisfying is the fact that much of the understanding derived by intuitive reasoning is basically proved by the detailed Green's function. Various estimates using the quality factor and linewidth at the LWHM, give values in good agreement with the computations using Green's function. Furthermore, if the relaxation time is taken as the scattering time for the mobility $\mu = e\tau/m^*$, this chosen example corresponds to a μ of 1300 cm^2 V^{-1} s^{-1}, which is within expected range.

Next, let us see for $Q = 0$, what happens to Im $G(0|0)$ at the eigenstates. At Re $G(0|0) = 0$, Im $G(0|0) \equiv N/D$, where

$$N = (1 - \rho^2)k_r \quad \text{and} \quad D = k_r^2(1 + \rho^2 + 2\rho \cos \Phi).$$

As $\rho \to 1$, using L'Hospital's rule, Im $G(0|0) \to \eth(k_n)$, so that $n(E_n)$ are discrete values at the eigenstate E_n of the quantum well.

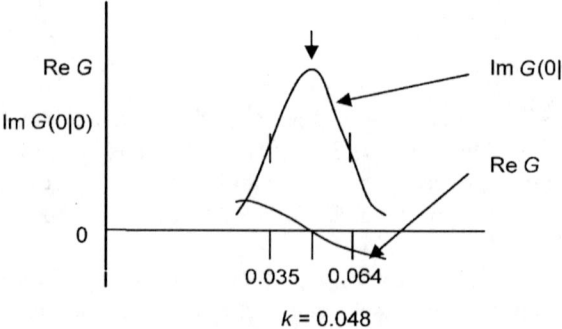

Figure 2.19 A sketch of Re- and Im-parts of $G(0|0)$ versus k_r calculated from Eqs. (2.43a) and (2.43b) for an isolated QW with well width 6 nm and barrier height of $V = 0.4$ eV, using relaxation time of $\tau = 10^{-14}$ s corresponding to $\ell \sim 8$ nm. The unit of k is in reciprocal angstroms. The zero of Re G is at $k_r = 0.048$, or $E = 0.1$ eV. For $\tau \to \infty$, $E_0 = 0.076$ eV. There is an upshift of 0.024 eV due to damping.

Green's function for the QW case with Langevin's damping is more complicated because of $\pm q$ for the waves going to the right and going to the left. For a free space Green's function, a Heaviside step function is used to describe $\pm q$, resulting in the wave equation itself being a function of x and x', or $\psi(x, x')$, even without the $\partial(x - x')$ term. What I have done with the free space Langevin's damping is to treat this term together with the delta function term as represented by a generalized Green's function and converting into integral equations. With barriers located at $x = w$, or in general at $x = x_1, x_2, \ldots$, the procedure needed to keep track of the relationship of the imaginary part of k, with the sign of q, is quite complicated. For a simple isolated QW, Green's function for Langevin's damping is almost identical to time damping, except that the imaginary part of k is a constant, as described in Eq. (2.37) for the free space case. I think the validity of this procedure may rest with whether we can satisfy all the boundary conditions with two waves from opposite directions in two different differential equations. Certainly, we can no longer use the eigenvalue problems with completeness and closure. However, we still have the principle of superposition. Before I leave this subject, I want to show another possible differential equation capable of describing damping in space.

2.5.5.5 Another Possible Equation for Spatial Damping

Let us take the nonlinear equation

$$\frac{d^2\psi}{dx^2} + 4\gamma x \frac{d\psi}{dx} + 4\gamma^2 x^2 \psi + k_0^2 \psi = 0, \tag{2.44a}$$

with the solution

$$\psi = e^{\pm ikx} e^{-\gamma x^2}, \tag{2.44b}$$

where $k_0^2 = k^2 + 2\gamma$, such that the wave damps in both $\pm k$. It can be shown that the boundary conditions are still represented by the usual continuity of ψ and ψ'/m. Moreover, the reflection coefficient is $R = (ik + \gamma)/(ik - \gamma)$. In fact, this equation leads to a good description of the effects of finite mean free path with one big problem that is shared by all others with damping, i.e., the differential equation is not the usual Schrödinger equation. The extra terms are not Hermitian and are non-linear. I leave these ideas to the amusement of the reader.

Now for the DB problem shown on the right of Figure 2.18, repeating the above procedures is really too complicated and generally require computers. Let us remember that even for the part of the Green's function with the Laplace transform, damping is not considered. However, one can use an age-old method for the calculation of losses in resonant systems, to be carried out next. We first calculate the quality factor Q, knowing the position of the resonant states, and the linewidth is obtained. In this way, we avoid solving the complete Green's function. This approach was published by Tsu (1985a,b), buried in the book in a form generally not accessible to most people. Thus, it is reproduced here.

2.5.6 Approximation Method for Linewidth and Level Shift Using Q

To find Q, we let an electron bounce between two barriers having a magnitude of reflection denoted by $|R|$. It is important to use only the magnitude to prevent any phase coherence effects. After n bounces, the density of the electron $\psi^*\psi$ is given by $|R|^{2n}\exp(-2nqw)$, with a well width w. The distance covered is $\ell = nw$, so that

$$|R|^{2n}\exp(-2nqw) = |R|^{2\ell/w}\exp(-2q\ell) = 1/e.$$

Taking the natural logarithm ℓn on both sides gives

$$\ell = \frac{1}{2q}\left[(1 + 1/qw)\ell n(1/R)\right]^{-1}. \tag{2.45}$$

Next, we use Matthiessen's rule for uncorrelated scattering events, $\ell^{-1} = \ell_0^{-1} + \ell_B^{-1}$, where $\ell_0 = 1/2q$ and $\ell_B = w/2\ \ell n|1/R|$, where ℓ_0 is a mean free path in the absence of confinement and ℓ_B is the mean free path when damping is zero so that it only takes into account the tunneling out of the barriers. In order to find the shift ∂E for the DB, we use the method of phase, which we earlier proved to be exact. We simply equate the total phase change after the wave has experienced two bounces from the two walls to $2n\pi$, i.e. $2k_r w + 2\phi = 2n\pi$, giving $k_r = (n\pi - \phi)/w$, in which ϕ is the phase change upon reflection from the barrier.

As an added example of our approach, we shall compare the method of calculating ℓ from $|R|$ with a physical approach. Using the third definition of the quality factor Q, we may write $Q = 2\pi J/JT$, where J and T are the current and period, respectively, given by

$$J = |A_0|^2\hbar k/m \quad \text{and} \quad T = 2\pi\hbar/E.$$

Taking a wave function in the well to be $\psi_W = (2/W)^{1/2}\sin kx$, together with results from Schiff (1955, p. 95),

$$|A_0|^2 = \frac{2}{w}\exp(-2\alpha B)\frac{4\alpha^2 k^2}{\alpha_0^2},$$

we arrive at the quality factor for tunneling through the barrier of width B,

$$Q_B = \frac{kw\alpha_0^4}{16\alpha^2 k^2}\exp(2\alpha B). \tag{2.46}$$

Using R from Eqs. (2.36b) and (2.36c) in Eq. (2.45) with $q = 0$ for ℓ_B, $\ell_0 = 1/2k_i$, with Matthiessen's rule, $\ell^{-1} = \ell_0^{-1} + \ell_B^{-1}$, we obtain the same expression as in Eq. (2.46) for Q_B, the quality factor due to tunneling out of the DB. From that the linewidth ΔE is obtained from (1) or (2) for Q defined in Section 2.5.1. This expression may be compared using the Green's function or by straightforward numerical computation of the linewidth of the transmission peak in resonant tunneling.

As presented in Tsu (1985a,b), from Im $G(x = x')$, the linewidth $\Delta E = 2E/k_r\ell = 2E/Q$ and the level shift $\partial E = E/4Q^2$, where Q is the quality factor. For small Q, the level shift is overshadowed by the broadening of the line. For large Q, the level shift dominates over the broadening factor. For typical crystalline materials, $k \sim 10^7$ cm^{-1} and $\ell_0 \sim 30$ nm, giving $Q \sim 30$ and $\partial E/\Delta E = 1/8Q$; for a linewidth of 200 meV, the shift is only ~ 1 meV, which is really difficult to observe. However, with poor quantum confinement, $Q \sim 1-3$, the shift is considerable. From an experimental point of view, it is easier to measure a 0.1 V shift than 0.8 V linewidth. A linewidth of a few tenths of an electron volt simply does not stand out! At this stage, I want to emphasize that we can estimate a relaxation time from the mobility. With this, we can obtain a line broadening and a level shift. One may ask why we want to go through the complexity of using Green's function. It is true that every engineer can have a good design based on a few important concepts, often from intuition, or simply some rule of thumb. A detailed mathematical model is extremely important for the overall understanding in establishing the rule-of-thumb, as well as, quite often, leading to new frontiers. Let us give an example of how good an approximate estimate can be made using the procedure just described by comparing with the use of Im $G(0|0)$ for the case of an isolated GaAs QW width $w = 6$ nm and a barrier height $V = 0.4$ eV. The spectra for various ℓ_0 are shown in Figure 2.20. The quality factor Q used is taken as $k_n\ell_0$, which is very close to the linewidth $2E_n/\Delta E_n$. The cases marked 1−5 with peak positions shown by arrows are summarized in Table 2.1.

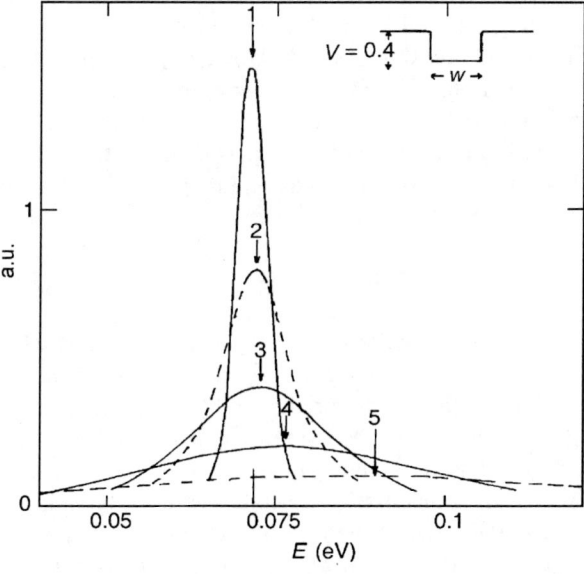

Figure 2.20 Energy spectra of an isolated GaAs QW with width 6 nm and a barrier height of 0.4 eV obtained from the zeros of Green's function, for five values of τ, corresponding to the mean free path ℓ_0. The arrows indicate the peak positions of the resonant energies.

Table 2.1 The Linewidth Calculated from Im $G(0|0)$ for the Case of an Isolated GaAs Quantum Well with Width $w = 6$ nm and a Barrier Height $V = 0.4$ eV

Case	ℓ_0 (nm)	Q	E_1 (eV)
1	96	12.8	0.0712
2	48	6.4	0.0715
3	24	3.2	0.072
4	12	1.6	0.076
5	6	1.0	0.089

Note that the shift is rather small except in the last two cases where the quality factor falls below 2. Therefore, as we intuitively guessed, as long as $Q > 2$, we should not worry too much about anything else. After all, the rule of thumb is no more than Heisenberg's principle. In a way, I am saying that all these theoretical criticisms about damping creating a non-Hermitian Hamiltonian, and supposedly violating quantum mechanics, are really groundless. I shall close here by pointing out that when $Q < 1$, the spectrum is really not discrete and assigning Pauli's exclusion principle, or even the Fermi distribution function, is really far-fetched for a state that is not even well defined.

To summarize, the approximation involves using damped electron wave functions. Next, we will find the amplitude and phase of the reflection coefficient $|R|$ and ϕ of a wave incident on a barrier. Without damping, the method of phase gives us exactly the energy corresponding to the resonant tunneling which is identical to the maximum in the DOS. With the method of incoherent multiple reflections, we arrive at Q, which is identical to the usual definition when allowing a wave to leak out of a storage system, via tunneling out and internal damping. All we need to do is to apply Eqs. (2.38) and (2.45) for Q and thus for ΔE. However, the self-energy shift δE must come from Green's function, $k_r^2 = k_0^2 - k_i^2$, with k_i from the mean free path ℓ_0 or the scattering time τ. For $k\ell \gg 1$, ΔE should be used and for $k\ell \ll 1$, and since the linewidth is so broad, it is more convenient experimentally to find δE. Therefore, δE should be used for most amorphous cases. From Eq. (2.43a), we have

$$\sin \Phi = -k_i(1 - \rho^2)/2k_r\rho, \quad \text{instead of } \sin \Phi = 0, \tag{2.47}$$

as with the method of phase. The difference is the factor $-k_i(1 - \rho^2)/2k_r\rho$, which is really quite small for $k_i \ll k_r$. Nevertheless, this term gives rise to an additional increase in k_r, upshift of the peak, a fact even predicted by the free electron Green's function with damping. What happens if we simply substitute k by $k_r + ik_i$ and find the transmission as well as the resonant tunneling as shown in Figure 2.20? We can either solve the transmission problem with a complex k, which is correct but lacks explicit elucidation of the role of the self-energy shift, or use Green's function, for an expression of the shift δE. The total shift comes from damping by the reflection coefficient, which involves k_i^2 even with the method of phase, but misses a larger shift from Eq. (2.47), involving k_i. For $k_i \ll k_r$, the shift is dominated by k_i in Eq. (2.47), in which it is erroneously stated that there is no shift (Gupta and Ridley, 1988). Their calculated result is correct, because the shift is so small, however, their conclusion is off the mark. As long as we recognize that ΔE is correctly approximated by $\Delta E = 2E/Q$ and the shift $\delta E = E/4Q^2$ from Green's function (Tsu, 1985a,b), the extremely simple approximation leading to the line broadening and the shift is now complete, giving a remarkably accurate estimate that includes the mobility involved. Last, the all-important peak-to-valley ratio in resonant tunneling is, therefore, prominently reduced by damping. I claim that the most important "figure of merit" in quantum devices, whether for an isolated QW, where the states with damping manifest in the broadening of the photons involved, or the peak-to-valley ratio of the transmitted current in resonant tunneling, is the mean free path, which was used in the very first introduction to man-made quantum systems (Esaki and Tsu, 1970).

The use of Green's function is more powerful than looking for the transmission peak. For an isolated QW, we can calculate Green's function, but we cannot set up the transmission case.

2.6   Very Short ℓ and w for an Amorphous QW

The problem with modeling an amorphous QW is when the mean free path ℓ, or the mean free time τ, are so short that even without quantum confinement, the energy is pushed up due to the self-energy shift. Therefore, modeling quantum confinement for the amorphous case requires a complicated procedure. Before we can use the measured data, let us summarize what we need to do. We use the results of the Green's function and the definition of $Q \equiv k\ell_0$ or $Q \equiv \omega\tau$, with $\tau = \ell_0/(\hbar k_r/m^*)$, for the energy linewidth ΔE and the level shift δE. To illustrate the procedure, we list in Table 2.2, important parameters obtained from our approximate procedures.

In Table 2.2, the appropriate parameters are consistent with the following summary using the free particle Green's function with loss:

$G(k)$ has a pole at $\sqrt{k_0^2 - k_i^2} + ik_i$
$n(k) = \text{Im } G(k)$ has a maximum at $k^2 = k_0^2 - k_i^2[1 - (k_i^2/3k_0^2)]$

Another modification is when the masses are not equal. Taking the mass in the barrier as m_1 and that of the well as m_2, because of the current continuity, α in

Table 2.2 Typical Parameters Characterizing a Damped System
Particularly Appropriate for Amorphous Quantum Systems

ℓ_0 (nm)	τ (ps)	Q	E (meV)	δE (meV)
100	0.06	40	2.5	0.3
10	0.006	4.0	25	3.0

Eq. (2.36) should be replaced by $\alpha' = \alpha\sqrt{m_2/m_1}$, effectively reducing the barrier height with respect to the QW. The consequence is to reduce the energy E.

For each ℓ_0, the energy E is calculated for a given well width w. From the plot of $E(w)$ versus ℓ_0, we can immediately determine the energy as a function of the mean free path. However, the matter is complicated by the further reduction of ℓ_0, as the width w is reduced due to various deleterious effects, such as induced interface strains and defects, and so on. We plotted the successive difference in $E_n(w_2) - E_n(w_1)$, $E_n(w_3) - E_n(w_2)$, ... , with the guide that $E_n(w \geq 6 \text{ nm}) = E_n(w \to \infty)$. With this procedure, the mean free path ℓ_0 versus w is obtained as shown in Figure 2.21, taken from Tsu (1985a,b, figure 2) with data from a-Si:H/a-SiN$_x$:H (Abeles and Tiedje, 1983). The important result is the fact that the mean free path is less than 0.5 nm, almost a factor of 2 below what is usual for a-Si:H, possibly due to the fairly complicated process of fabrication of the multilayer structure. As quantum confinement is a reality in amorphous solids, one wonders why no one is pursuing this route for them and in polycrystalline solids? The inverse Q is plotted versus the mobility μ in Figure 2.22, and, the P/V, the peak-to-valley ratio is plotted against τ in Figure 2.23, in both figures for the first two resonant states for two GaAs RT structures, where (40, 0.3) and (20, 1.0) denote $w = 4$ nm and a barrier height of 0.3 eV. Although the first resonant peak usually has a higher transmission, the second peak dominates because of a much wider width for the second transmission peak.

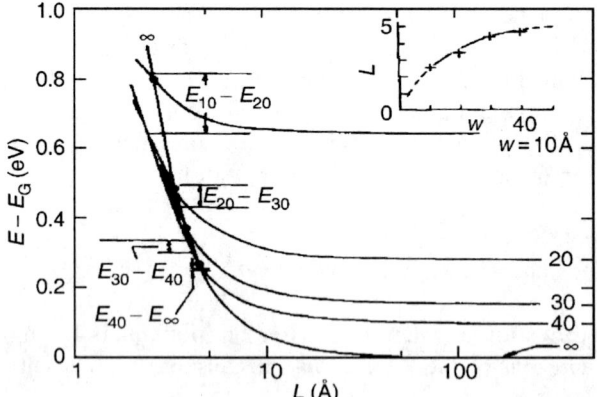

Figure 2.21 The energy states versus the mean free path L (same as ℓ used throughout this chapter). The notation E_{10}, E_{20}, and so on. refers to QW states with the width of the well, 1, 2, 3, 4 nm. A DB structure (a-Si:H/a-SiN$_x$:H) has dimensions $(B, w, B) = (3.5 \text{ nm}, w, 3.5 \text{ nm})$ with $w = \infty$, 4, 3, 2, 1 nm. The conduction band mass is taken as $m_c^* = 0.2\, m_e$. Inset shows the fit of data for $L = 0.26-0.5$ nm.
Source: After Tsu (1985a,b), with permission, using data taken from Abeles and Tiedje (1983).

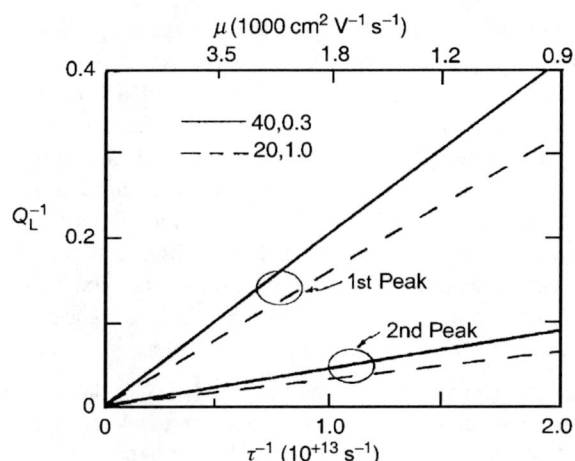

Figure 2.22 Inverse Q versus the first and second resonant peaks versus μ or inverse τ.
Source: Zypman et al. (1988, unpublished).

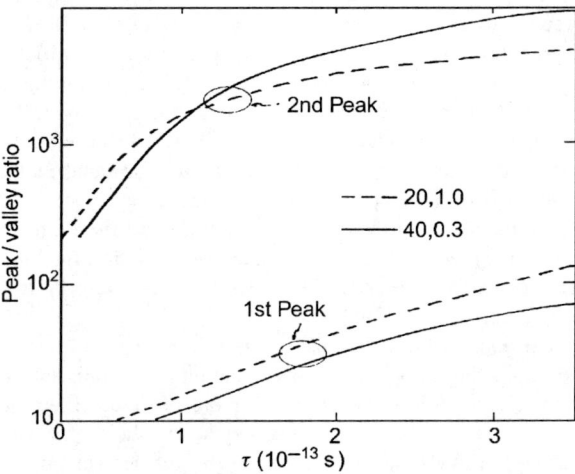

Figure 2.23 Peak-to-valley ratio versus τ for the same two structures as in Figure 2.22.

2.7   Self-Consistent Potential Correction of DBRT

Before we present some working rules for self-consistent potential calculations, we shall first cover some fundamental rules that govern junctions, such as the pn-junction and heterojunctions in general. The rules are in fact simple. I often wonder why most technical books on junctions still follow the "old fashioned," half-century-old rules, relying on equations to explain what happens in a junction. What we need to remember is that the Fermi level is similar to a water level between two containers with a connection between the two so that a common level is established at equilibrium. We plot the electron energy, with negative charge for electrons, so that the positive potential lowers the potential energy profile. Therefore, electrons fall down and holes climb up. If the two sides have different Fermi energies, at a distance far away from the contact, a common Fermi level exists. However, near the contact, the Fermi level is not definable because both sides near the contact are in depletion mode. Whenever a voltage is applied, we introduce quasi-Fermi levels that have a separation equal to the applied voltage. The higher the carrier density, the faster is the falling and climbing. Another principle I use is the fact that we should always assume a rule and follow that rule. For example, an applied voltage causes the electrons to move. A simple rule is based on preventing the universe from blowing up. If the motion results in increasing the potential that causes the motion in the first place, then the universe will blow up. Therefore, the motion must be such to cancel out the applied potential. Let us summarize the following important rules:

(1) Potential cannot change without charge and charge cannot exit without a volume, surface, or thickness. Take two conduction bands separated by a junction band-edge difference of $\Delta V = V_2 - V_1$ (the conduction band-2 is above the conduction band-1) and this difference ΔV is fixed at the junction. The electrons on the right fall toward the common Fermi level at $\pm\infty$, resulting in a depletion region to the right of the junction, exposing the $(+)$ dopant sites, and an accumulation of $(-)$ charges, the electrons to the left of the junction. On the right side, the potential falls until aligned with the common Fermi level. On the left, electrons accumulate. How fast the potential changes depends on doping. In this case, the slope of falling electrons depends on the doping density on the right. The slope of rising holes depends on how many electrons are available. Because of the boundary condition that the normal displacement vector is continuous, for equal dielectric constants on both sides, the downward slope on the right must be same as the upward slope on the left at the boundary. And the higher the doping and the more the electron density is available, the faster the change, or the thinner the depletion thickness on the right and accumulation thickness on the left. Therefore, with n-doping on both sides, the side with higher doping determines the outcome.

(2) For a pn-heterojunction with equal doping, the electrons fall into the side with holes, or we say the holes move up. The larger the n-doping, the faster the electrons fall, and the larger the p-doping, the faster the holes climb, and both sides are depleted. Electrons from the n-side leave the dopant $(+)$ sites in the depletion layer, and holes from the p-side climb and leave exposed $(-)$ sites. What happens when the p-side is negligibly doped? Then as in the n–n heterojunction, the higher doping side dominates. For example, if the n-side on the right is highly doped, many electrons from the right can fall into the accumulation layer on the left and the potential profile is now determined by the highly doped n-side.

(3) A superlattice and a QW have a quite different scenario. John Bardeen asked me during the Industrial Affiliates Conference at the University of Illinois in the late 1970s whether the Fermi levels are aligned in superlattices. I pointed out that the Fermi level is mostly in the middle because the superlattice is almost completely within the depletion region. This is because the layers are so thin that all the available carriers are used up trying to line up the Fermi level, but never quite make it. Of course, this situation arises because the band-edge alignment for the creation of the periodic potential is generally many times greater than the binding energy of the dopants. One can dope as much as one wants; the superlattice remains intrinsic. Since doping creates a scattering center and other deleterious effects, why should one dope the superlattice?

Just as important as the basic understanding is that the ultimate results are always based on mathematical modeling using a self-consistent calculation. In such a model, one often starts with no potential other than the applied voltage. The electron transport based on the wave-function solution of the Schrödinger equation is then solved together with the Poisson equation. The new potential is now added to the original zero-order solution. The process usually converges to the final satisfactory results as a self-consistent calculation. In some cases, if one needs faster convergence, the variational minimization technique is often used so that one iteration is often sufficient for satisfactory modeling. At this point I would like to remind the reader that the solution of a lossy wave-guide cavity utilizes the same procedure. First, one solves for the fields assuming perfect conductivity. The fields for the zero-order solution are now used to calculate the current and thus the power loss.

I mentioned before that the self-consistent calculation of Cahay et al. (1987) seemed quite straightforward to me. They divided the DBRT structure into three basic regions: electrons originate from the left-side contact, tunnel through the QW region, and are transmitted to the right. Even without assuming relaxation processes, the electrons in the QW create a space charge, raising the potential and requiring a higher applied voltage for the alignment of the discrete quantum states with the electron source at the left contact. In the actual model, there is a further complication, because two undoped sections were placed between the doped contacts and the barriers that form the QW structure in order to minimize Coulomb scattering in the quantum structure. Figure 2.24 shows their calculated results compared with the case where the self-consistent potential is not included. Note that the position of the resonant peak shifts to higher applied voltage and the peak height is lowered, as well, the linewidth increases. In short, the self-consistent results predict poorer performance. Simple inclusion of the space charge presented as follows shows essentially these findings.

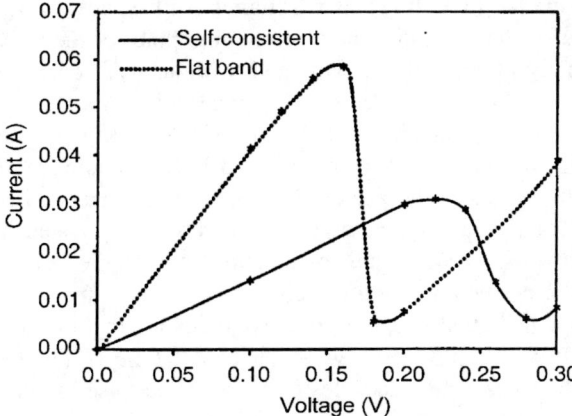

Figure 2.24 Current versus voltage, self-consistent and flat-band results for a DBRT GaAs with $Al_{0.45}Ga_{0.55}As$ barriers and two undoped GaAs spacers, with 2×10^{18} cm^{-3} Te n$^+$-doped contacts, all 5 nm thick. *Source*: After Cahay et al. (1987), with permission.

To calculate the space charge, we start with the two-dimensional DOS

$$n(E) = m^*/\pi\hbar^2. \tag{2.48}$$

In the QW of width w, assuming perfect confinement, the charge density in the lowest level E_1 is

$$e\psi^*\psi = e(2/w)\sin^2(\pi x/w),$$

so that the charge density inside the QW is

$$\rho = e(m^*/\pi\hbar^2)\Delta E_t(2/w)\sin^2(\pi x/w). \tag{2.49}$$

Solving the Poisson's equation

$$\nabla^2\phi_{SC} = \frac{d^2\phi_{SC}}{dx^2} = -\rho/\varepsilon,$$

we arrive at the potential energy of the space charge,

$$\phi_{SC} = (\rho_0/4\varepsilon)\{(w/\pi)^2\sin^2(\pi x/w) + wx - x^2\}, \quad 0 \le x \le w, \tag{2.50}$$

where $\rho_0 \equiv e(m^*/\pi\hbar^2)\Delta E_t(2/w)$. Note that the inclusion of the applied potential does not change the value of V_{SC}, because potentials may be superimposed.

The maximum value is at $x = 0.5w$, or the potential energy in eV units. (Note that in eV units, voltage V is numerically same as energy measured in eV.) $V_{SC}(w/2) = 0.25(\rho_0/\varepsilon)w^2(\pi^{-2} + 0.25)$. Taking $w = 5$ nm, $m^* = 0.07m_0$ $\varepsilon = 11\ \varepsilon_0$ for GaAs, the value of $V_{SC}(w/2) = 0.42\Delta E_t$. For $\Delta E_t \sim 80$ meV, corresponding to $n \sim 2 \times 10^{18}$ cm^{-3}, $V_{SC}(w/2) = 0.035$ eV. Near resonance the energy, $E_1 = eV/2$, then the increase in voltage at resonance with space charge correction $\sim 0.035 \times 2 = 0.07$ eV. The separation between the midpoint of the NDC in Figure 2.24, taken from Cahay et al. (1987), with and without self-consistent results, is $0.25 - 0.17 = 0.08$ V, indicating that our simple space charge increase of 0.07 V is really quite good. How do we explain the lowering of the peak value and broadening of the linewidth? As we know, higher voltage translates to more asymmetry, reducing $|T|^2$, which in turn reduces the peak value. The wave function used assumes complete confinement, or the quality factor Q of the resonating system is infinite. In reality, even without loss, $Q \ne \infty$ due to tunneling out of the barrier. However, even for a Q of 20, the correction using the eigenstate is very small. Our simple estimate indeed represents the actual situation, thus we conclude that the self-consistent correction in the left and right leads, where accumulation at the left contact and depletion at the right contact are much less important, because an electron confined in a resonating system has a overwhelming probability of being inside the resonating system.

2.8 Experimental Confirmation of Resonant Tunneling

I started on the theory a full 9 months before the publication of Tsu and Esaki (1973). After careful evaluation of the possible experimental verification, we at IBM, under the direction of Esaki, thought it too optimistic to expect that anyone could fabricate a tunneling structure without "pinholes" with an overall thickness of no more than 10 nm. One must understand the consequence of failure while the whole concept of the man-made superlattice was at stake. In late autumn of 1973, Esaki was notified about his Nobel Prize, which triggered a sense of confidence in all of us. Chang was instructed to start on the growth of the DBRT structure using the best computed results that I had obtained. I was instructed to prepare a mask for the structure that was suitable for $I-V$ measurements. My first try in late November 1973 used 50 μm dots for contacts, without having a single one showing NDC, which we had decided was the best evidence of resonant tunneling. We enlisted the best facilities at the Special Technique Group at IBM Research Center, with masks for dots of several micrometers. We decided on a conservative size of 6 μm dots for contacts. At the end of 1973 and the beginning of 1974, we were able to find NDC in more than 30% of all the dots, resulting in the now-well-known Chang et al. (1974) experimental confirmation of DBRT. Figure 2.25 shows the measured conductance and current at 77 K. Note that the barriers were much too thick by today's standards. The reason is that we simply could not believe that anyone could make a 2-nm barrier. We used $x = 0.5$ in $Ga_{0.5}Al_{0.5}As$. Later, we learned from the work at BTL (Dingle et al., 1974) about the alloy $Al_xGa_{1-x}As$, where $x \leq 0.3$, which is good for lattice matching. The peak positions are fairly close to the calculated values.

Figure 2.26 shows another design with a 4-nm QW, leading to a larger measured voltage than calculated voltage, which we believed was partially due to the use of exponential functions instead of the Airy functions needed for large fields. I would like to mention some of our computational problems. At the time, I was using the newly introduced personal terminal using the APL program at IBM instead of the mainframe computer with FORTRAN and punched cards. One paid a dear price for convenience, with only a 32-kB memory. This meant that a highly efficient program had to be written to run without allowing an approximation of the linear voltage by many steps, but permitting the use of exponential functions. Therefore, we were only taking the average values of the potential at the midpoints of the barriers and the well, which seriously limited the accuracy at larger voltage. There was not even a subroutine for Airy functions available with the APL, so I had to use the asymptotic expressions and create my own subroutine! In short, the agreement for the 4-nm well is not as good as for the 5-nm well, possibly because the larger voltage resulted in a poor approximation by the exponential functions. Nevertheless, as pointed out by Chang et al. (1974), the overall comparison between the calculated and measured values was satisfactory.

Resonant tunneling in DB structures was begun in 1973–1974, and has developed into a major field in quantum devices leading the way (and metamorphous) to

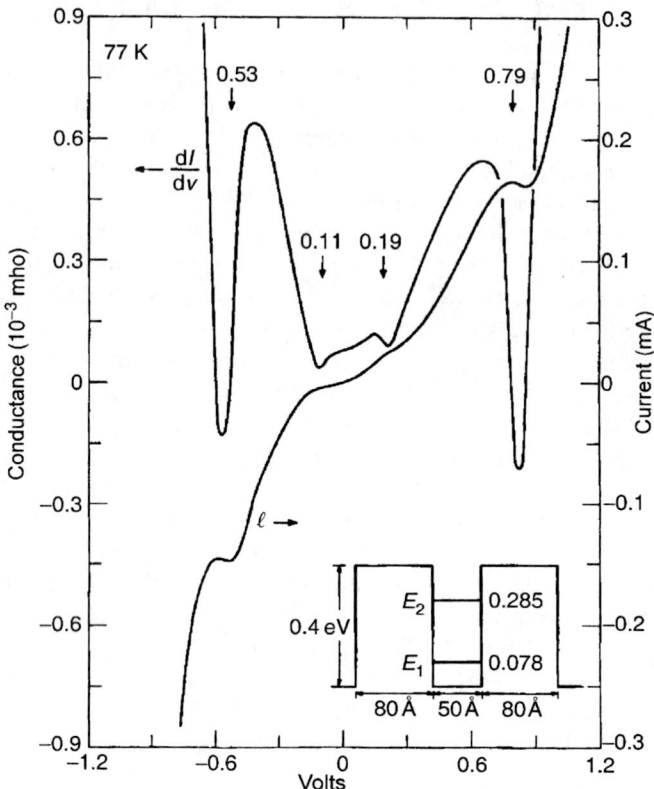

Figure 2.25 Current and conductance versus the applied voltage of a DBRT structure using 6 μm Au dots as contacts.
Source: After Chang et al. (1974), reprinted with permission.

the nanoelectronics of today. Understanding of the subject is still evolving. My personal view is that damping mechanisms need further attention. On the other hand, unlike QD devices, there is no input/output problem. A terahertz oscillator has been realized with RTD, a subject we shall treat later. Even if we allow that full adoption of RTD devices in everyday electronics may still be some years away, what is the future for QD electronics?

2.9 Instability in RTD

2.9.1 The Goldman–Tsui–Cunningham Instability

Instability was first reported by Goldman et al. (1987), who asserted that j plotted against V with the inclusion of self-consistent potential results in instability and hysteresis near the resonant peak. However, Sollner (1987) explained that oscillation is due to external circuits and oscillation manifests itself in hysteresis. Before

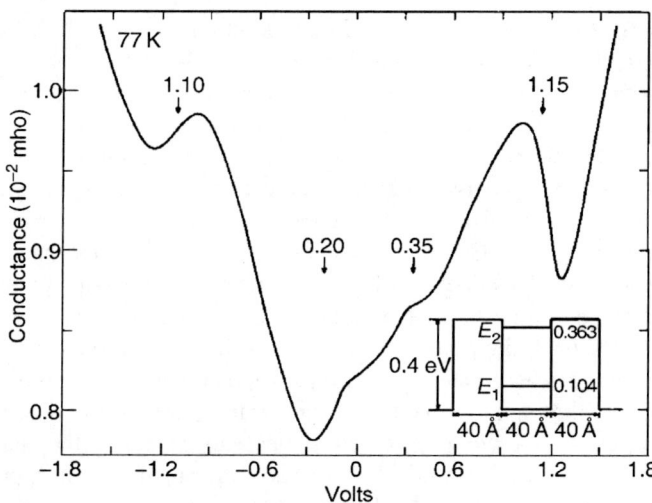

Figure 2.26 Conductance versus the applied voltage. Note that the measured peak positions do not agree with the calculated QW, for the reasons given. Positions aside, there is no NDC. At the time, we believed the problem was with the scattering; the higher the energy, the lower the mean free path.
Source: After Chang et al. (1974), reprinted with permission.

I try to resolve this important issue, I want to go back to explain oscillation and the appearance of hysteresis in terms of fundamental physics. Oscillation in almost all Radio Frequency (RF) devices, including microwave oscillators and lasers, is caused by the presence of gain, whether it is due to population inversion leading to negative resistance, as in gas and injection lasers, or simply gain, as in negative differential resistance (NDR), including parametric oscillators, with the frequency fixed by an external resonating system known as the cavity or laser mirrors. These external resonators are usually, but not always, tuned to a frequency near the maximum gain. In second harmonic oscillators, the external cavity is tuned to the second harmonic where the gain is usually much lower. In Gunn (1963, 1976) oscillators, a typical device with NDC, field-induced domain is caused by uneven partitioning of the applied voltage over a region, as a consequence of inhomogeneity. Fundamentally, domain formation is due to the system seeking a minimum total energy for the system; in this case, almost all of the energy is electrical. Therefore, space charge does not induce instability by itself.

Next, let us look at the typical hysteresis of the magnetization B−H curve. When I was at IBM Research, I asked H. Chang, who was in charge of the Magnetic Memory at the IBM Research Center, to explain to me whether magnetic domains, usually described in terms of a domain delineated by surface energy, involve some kind of phase transition. He said that the domains are formed as the system tries to minimize the total energy, and therefore, at least some sort of local minimization of the total energy is involved. The Curie temperature is just the energy needed for the configuration to return to a more random orientation. This process was very much on my mind when (Nicollian and Tsu, 1993) first observed switching and hysteresis in tunneling through 2- to 3-nm QDs of Si. We observed high frequency instability of greater than megahertz values, as well as low frequency switching of less than 10 Hz. Owing to the slowness of the switching, we knew we were dealing with trapping, where electrons are trapped onto a defect site resulting in negatively charged clusters, defects or dislocations, and so on. The structure relaxes to minimize the strain energy resulting in hysteresis. More will be devoted to this subject in Chapter 7. For example, why is trapping more prevalent in QDs than in QWs? I hope that I have at least set the stage for the discussion about why I think Sollner's explanation is close to being correct. I acquired this sort of wisdom during the couple of years I worked on porous Si (Feng and Tsu, 1994), where there was great debate about whether the visible light emitted from PSi is due to quantum confinement or surface crud. I came to realize that whenever two groups of fairly knowledgeable people heatedly disagree, the chances are they are both partially correct. We now know that both quantum confinement and surface crud together play important roles in the observed visible luminescence in PSi. As long as the two key points are clear in our minds, we can indeed resolve the issue at hand. Basically, our present debate is very much like the example I give here involving PSi.

1. No instability-induced self-oscillation is possible with a single QW—no domains, no coupled oscillators, and thus no self-oscillation.
2. No trapping follows by reconfiguration and there is no hysteresis.

Why then did the self-consistent calculations sometimes show stability (Cahay et al., 1987) while sometimes instability appears, as shown in Figure 2.27 (Berkowitz and Lux, 1987; Goldman et al., 1987; Alves et al., 1988), and Zhao et al. (2003)? The answer lies in the fact that two or more QWs are involved. Zhao was the first to point out that in addition to the DB−QW, an additional triangular-shaped QW, "EQW" (Zhao's terminology), is created by the applied voltage in the undoped section, called a buffer, used to separate the charged dopant sites from the structure. This "EQW," as in the inversion region of a metal oxide silicon field effect transistor (MOSFET), predates the man-made heterojunction DBRT structures. Figure 2.28 shows that the coupling between the two QWs introduces bistability, as in any ordinary flip-flop system. Let me comment on the need to use the time-dependent Schrödinger equation, or at least some time-dependent formulation such as the density matrix, or the Wigner function used by Zhao et al. (2003). I have worked out the response function for the superlattice (Esaki and Tsu, 1970; Tsu, 1990), using the simplified path integral time-dependent approach. The Wigner function is superior because it contains spatial boundary conditions. But it is also possible to include spatial boundary conditions in Pippard's integral. I can visualize that the result will be quite similar to the Wigner function in one dimension used by Zhao. I think the primary virtue of the Wigner function is that it is so simple, and quite likely, good enough; although the use of the density matrix and a three-dimensional calculation is probably better, it involves huge computational complexity.

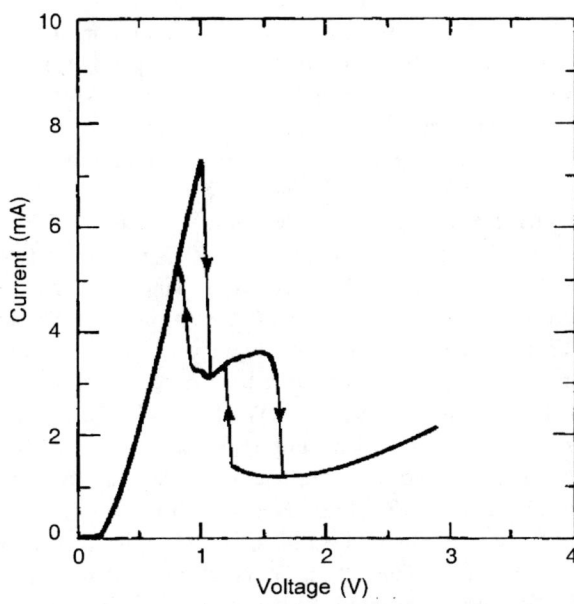

Figure 2.27 Current versus voltage of a DBRT with hysteresis.
Source: Sollner (1987), with permission.

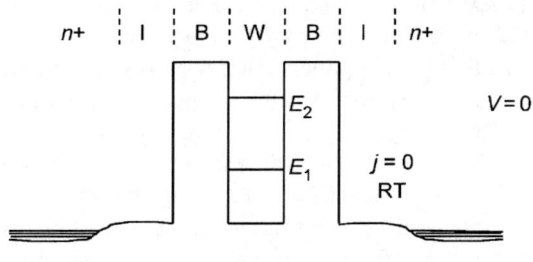

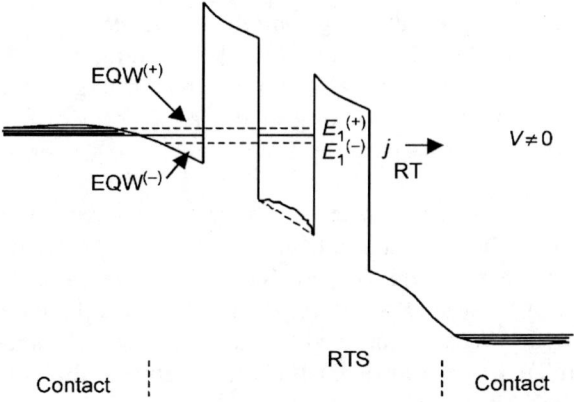

Figure 2.28 A single QW at $V = 0$, top, becomes a double QW system, bottom, with the additional EQW at $V \neq 0$. The coupling between the resonant state EQW and the QW state E_1 results in coupled modes, $EQW^{(\pm)}$ and $E_1^{(\pm)}$, which may explain the source of instability. The bowing of the potential profile is due to space charge.

Before Zhao et al. (2003), Woolard et al. (1996) introduced an inductor in their equivalent circuit that gave me a clue that internal oscillation can be present, since we know that the circuit representation of a resonating system with a cavity involves an inductor to store magnetic energy and a capacitor to store electrostatic energy. I have already concentrated on the role of the undoped buffer region transformed into a triangular QW by an applied voltage as shown in Figure 2.28. Interestingly, Woolard et al. (1996) did not even mention the second QW, EQW, until a later paper by Zhao et al. (2003). Figure 2.28 shows a RTS including two undoped sections on either side of the structure and the contacts. The notations are I for intrinsic, B for barriers, and W for the QW. The RTS portion is drawn to scale, but the contacts are not. This is because the linear dimension of the structure consists of 5 nm for each section, while the contacts have a long depletion layer for alignment of the Fermi level of the n^+ region. The two QW states, E_1 and E_2, are quite a way from the bottom of the conduction band edge so that at low applied voltage, there is no current. With an applied voltage, the I region forms into a triangular QW similar to the inversion layer of a MOSFET. Because this I region is so close to the heavily doped contacts, the mean free path is low resulting in fairly broad quantum states, the EQW states, following the nomenclature used by Zhao. No current appears until this EQW state is aligned with the source of electrons. There is coupling between the EQW state and the QW state E_1 as shown in Figure 2.28. The maximum coupling is at a voltage such that the EQW state and E_1 are aligned, resulting in coupled modes, denoted by $E_1^{\pm}$ and EQW $^{\pm}$.

In the original Tsu and Esaki (1973) paper, three peaks appeared in Figure 2.3, although in Figure 2.27 of Sollner only two peaks are evident. I think the reason is that the lower mobility in the EQW leads to much broader states. Also, we know that any quantitative agreement with measured $I-V$ requires a full self-consistent calculation. Thus, at least quantitatively we can account for the measured $I-V$, which is still more than we expected to do. What is important is that we can restore our confidence in our physical picture of the resonant tunneling mechanism. What has happened to the statement by Alves et al. (1988) that bistability results from the stored space charge in the QW? In their figure 1 for the structure that their calculation is based upon, there is a triangular well in the emitter at an appreciably high applied voltage, a similar situation that was reported by Zhao et al. (2003). In particular, reading between the lines of Sollner (1987), we must recognize what Sollner was trying to tell us: "Do not jump to a conclusion simply by looking at the $I-V$." However, I think a time-dependent calculation, like that used by Zhao et al. (2003), is probably more reliable.

There is another potentially very important issue I shall treat next—excitation inside of a QW that results in tunneling out through both barriers of the DBRT structure. This situation applies to cases where trapped electrons may re-enter the QW, affecting the space charge of the whole structure. If this is indeed the case, there will be a source of instability because this additional space charge comes from the traps which are not related to the resonant tunneling between the contacts. This possible source of instability requires no coupled wells. However, in reality, a coupled system of QW states and the states of the trapping sites are again involved.

2.9.2 Decoupling of the Two Barriers by an Excitation Within the Well

We will treat here the consequence of introducing an excitation function within the QW. The general formulation requires the use of the Laplace transform that will be presented in the next section. We will show that resonance, due to coherent interference of the electron wave function inside the well, gives way to exactly the same case as when the two barriers have been decoupled when a delta function excitation is introduced into the well. Let us first generalize the equations leading to Eq. (2.7). Let an electron wave function in region 1,

$$\psi_1 = A \exp(ik_1 x) + R \exp(-ik_1 x),$$

incident on a square barrier with width B and barrier height V in region 2, exit into region 3 with wave function

$$\psi_3 = T \exp(ik_3 x), \quad \text{with } k_2^2 = \frac{2m_2}{\hbar}(E - V).$$

Usually, one sets $|A| = 1$, however, we leave A to be arbitrary because later we need it when we introduce an excitation within the well. The incident current j_i, the reflected j_r, and the transmitted j_t are respectively

$$j_i = \frac{\hbar k_1}{m_1}|A|^2, \quad j_r = -\frac{\hbar k_1}{m_1}|R|^2 \quad \text{and} \quad j_t = \frac{\hbar k_3}{m_3}|T|^2.$$

Explicitly

$$\frac{j_t}{j_i} = \frac{k_3 m_1}{k_1 m_3}\left|\frac{T}{A}\right|^2 = \frac{k_3 m_1}{k_1 m_3}\left[\frac{4}{\alpha_1^2 \cosh^2 k_2 B + \beta_1^2 \sinh^2 k_2 B}\right], \tag{2.51}$$

where $\alpha_1 \equiv (1 + (m_1 k_3/m_3 k_1))$ and $\beta_1 \equiv ((k_2 m_1/k_1 m_2) - (k_3 m_2/k_2 m_3))$.

Since $j_i + j_r = j_t$, for $|A| = 1$ and equal masses, we recover Eq. (2.7).

Next, we take the case of two barriers with widths B_2 and B_1 as shown in Figure 2.26 and regions I, II, and III, with subscript "a," to the right of the ∂-function excitation, shown as an arrow pointing upward, and subscript "b," to the left of the excitation function, and with T_1 transmitted to the right and T_2 transmitted to the left (Figure 2.29).

The wave functions are:

$$\psi_1 = A_1 \exp ik_1(x - x') + B_1 \exp - ik_1(x - x') \quad \text{Region I}_a,$$
$$\psi_2 = A_2 \exp ik_1(x - x') + B_2 \exp - ik_1(x - x') \quad \text{Region I}_b,$$

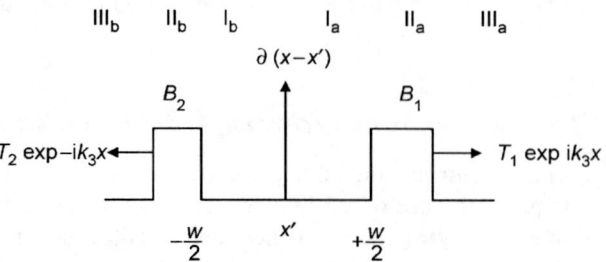

Figure 2.29 A DB structure with regions, I, II, III. Subscripts "a" and "b" are for $x > x'$ and $x < x'$, respectively. The excitation tunnels in both directions, T_1 and T_2.

with the boundary conditions: $\psi_1 = \psi_2$, and $\psi_1' - \psi_2' = 1$. Then, the normalized current transmission $J(I_a \to III_a) \equiv J_R$ becomes

$$J_R = \frac{m_1 k_3}{m_3 k_1} \left| \frac{T_1}{A_1} \right|^2 = \frac{4(k_3 m_1 / k_1 m_3)}{\alpha_1^2 \cosh^2 k_2 B_1 + \beta_1^2 \sinh^2 k_2 B_1}. \tag{2.52}$$

Similarly, the normalized current transmission $J(I_b \to III_b) \equiv J_L$ becomes

$$J_L = \frac{m_1 k_3}{m_3 k_1} \left| \frac{T_2}{B_2} \right|^2 = \frac{4(k_3 m_1 / k_1 m_3)}{\alpha_2^2 \cosh^2 k_2 B_2 + \beta_2^2 \sinh^2 k_2 B_2}, \tag{2.53}$$

with $\alpha_2 \equiv (1 + (m_3 k_1 / m_1 k_3))$ and $\beta_2 \equiv ((k_2 m_1 / k_1 m_2) - (k_1 m_3 / k_3 m_1))$.

Comparing both Eqs. (2.52) and (2.53) with Eq. (2.51), it is obvious that the current transmissions are governed by tunneling through a single barrier. Therefore, the presence of a source decouples the DBs into two single barriers. For $B_1 = B_2$, the total current to the right I_a,

$$j(I_a) = \frac{\hbar k_1}{m_1} \left[|A_1|^2 - |B_1|^2 \right], \quad \text{so that} \quad \frac{J(III_a)}{J(I_a)} = \frac{k_3 m_3}{k_1 m_1}.$$

The total current to the left in I_b,

$$j(I_b) = \frac{\hbar k_1}{m_1} \left[|B_2|^2 - |A_2|^2 \right], \quad \text{so that} \quad \frac{J(III_b)}{J(III_b)} = \frac{k_3 m_3}{k_1 m_1}.$$

both involve the same factor of $k_3 m_3 / k_1 m_1$. Now we can apply our results to a general case of m barriers to the left of a well with an excitation and n barriers to the right of the same well with an excitation, the $m + n$ barrier system breaks down into two simple systems, one with m barriers and the other with n barriers. Our results bring some very important physics into the understanding of excitations, even having an impact on the self-consistent potential calculation. Charges entering and leaving the QW act in a way similar to the excitation function described here. The result is to decouple the two barriers in phase, switching to a mode where only tunneling in a single barrier seems to matter. This is another manifestation of the reason why the sequential model seems to apply. However, without any sort of phase randomizing relaxation, instability due to charging cannot occur. And since the trapped charges re-enter the QW randomly, it is a source of instability, the famous $1/f$ noise (Kogan, 1996).

Some of the disagreements in the computed results have been clarified and at least two sources of possible instabilities have been identified and discussed in a qualitative way with some quantitative backing: coupled QW to "EQW" formed by an applied voltage and coupling of the QW to traps, which simply constitutes another system. Thus, instability, if it occurs, is not an intrinsic property of the NDC in an RTD. In fact, I cannot agree with Woolard et al. (2003), who call the existence of oscillation in their time-dependent computation an intrinsic instability in fact not much different from the solution of the Laplace transforms.

2.10 Summary

In summary, I knew that a given model had limitations, particularly one that involves a fair degree of complexity. The approach started by Stevens using Laplace transform represents a step beyond time-independent treatments. Sen first called my attention to Stevens's work. The tunneling time of a typical RTD is a couple of hundred of femtoseconds, confirming that the ultimate device speed is $\sim$10 THz. I went to great lengths to show that it is important to account for the losses inside the QW, if nothing else, to provide the answer why the P/V ratio of RTD is much reduced from the computed value even when the space charge effects are included. I want to go into some conceptual points. Quantum mechanics was designed to take care of atoms and molecules. The mathematical description of an eigenstate and an eigenvalue is just the right tool because atoms and molecules do last indefinitely, or very long. Therefore, the use of the Hermitian Hamiltonian is natural. Most quantum devices do not last more than several periods. We see that even to account for losses in a laser cavity, we solve for the modes without losses. Then we calculate losses using these state functions. Simply put, we use perturbation. Similarly, we calculate the wave function for H_0, then we use the eigenfunction to calculate the losses. We can even go one step further to balance the detail. In the density matrix formulation and Boltzmann equation, the collision terms are calculated in terms of the eigenfunctions of the Hamiltonian H_0. The reason why these accepted approaches are not adequate is because the losses are usually very high, so that the Q of the system is typically no more than $5-10$. The self-energy shift may be more than line broadening. This is why we must resort to the use of Green's function with complex energy or complex momentum and why I introduced the idea that we could perhaps cut the system into parts, representing open systems, each having non-Hermitian terms. During our debates, I joked that we could move the poles along the imaginary axis to account for losses and still insist that the Hamiltonian is Hermitian. My theorist friends reminded me that the end result on the surface is the same as using Green's function, so I should simply accept the formalism of Green's function. I reminded them that Green's function formalism does not cover the totality of non-Hermitian operators. For example, nonlinear terms may not be represented by a complex energy or momentum! In fact, all we need is to include two different loss mechanisms; Green's function does not have the simple form represented by Eq. (2.18). Using the concept of the quality factor Q, the linewidth is $\propto Q^{-1}$, but the level shift is $\propto Q^{-2}$, giving a simple method of estimating the performance of RTD. We conclude that even the reported resonant tunneling in amorphous materials is basically sound. However, we need to recognize that the upshift of states is partially due to the short mean free path, in addition to quantum confinement.

Now I have discovered that the use of Wigner function may even be more suitable because scattering may be included. However, time-independent models appear to have pushed beyond their validity in situations where interactions with other systems are involved. Since no modeling is precise enough, as far as I am concerned, fundamental physics and understanding based on even the best models should not be more than guiding the thinking. Nevertheless, it is satisfying to know that these simple models of resonant tunneling are basically sound. I was quite alarmed when intrinsic instability was supposedly discovered in RTD. I am happy to state that I am much relieved that the "second" QW, in addition to possible traps, provides the missing coupling, which may lead to instability. The NDC in RTD is no different from any other oscillators and amplifiers possessing NDC and, therefore, RTD does not involve intrinsic instability. However, one wonders whether the use of an intrinsic buffer is the best way to produce a terahertz (THz) oscillator. Would a gradual grading of doping, eliminating the possibility of an additional QW under applied voltage, offer a better solution? Last, with all the recent emphasis on nanoelectronics, the RTD has no implementation problems. Quantum nature is dictated by the dimension along the device structure, which being a planar device, does not have problems with contacts for input/output.

The amount of literature related to RTDs is enormous and a simple search can be started the Internet. Nevertheless, in spite of the laboratory successes in THz RTD oscillators, there is still no RTD oscillator that has a respectable output in the milliwatt range. I think the villain is still posed by defects resulting in traps. As the number of electrons participating in the active process decreases in a device, trapping becomes more and more pronounced. This tendency is particularly great in QDs where the wave functions of the intended device and defects become almost indistinguishable, a subject I shall go into in some detail in Chapter 7. The RTD, manifests in the fundamental understanding of solids, as well as serves as testing grounds for devices involving or dominated by the wave nature of electrons, naturally leads the frontiers of science into nanoscience.

References

Abeles, B., Tiedje, T., 1983. Phys. Rev. Lett. 51, 2003.
Alves, E.S., Eaves, L., Henini, M., Hughes, O.H., Leadbeater, M.L., Sheard, F.W., et al., 1988. Electron. Lett. 24, 1190.
Bandara, K.M.S.V., Coon, D.D., 1989. J. Appl. Phys. 66, 693−696.
Berkowitz, H.L., Lux, R., 1987. J. Vac. Sci. Technol. B 5, 987.
Breit, G., Wigner, E., 1936. Phys. Rev. 49, 519.
Cahay, M., McLennan, S., Datta, S., Lundstrom, M.S., 1987. Appl. Phys. Lett. 50, 612.

Carslaw, H.S., Jaeger, J.C., 1948. Operational Methods in Applied Mathematics, Oxford University Press, Oxford, chapter IX.

Carslaw, H.S., Jaeger, J.C., 1959. Conduction of Heat in Solids, Oxford University Press, Oxford.

Chang, L.L., Esaki, L., Tsu, R., 1974. Appl. Phys. Lett. 24, 593.

Coon, D.D., Liu, H.C., 1985. Appl. Phys. Lett. 47, 172−174.

Datta, S., 1995. Electronic Transport in Mesoscopic Systems, Cambridge University Press, Cambridge.

Dingle, R., Wiegmann, W., Henry, C.H., 1974. Phys. Rev. Lett. 33, 827−830.

Duke, C.B., 1969. Tunneling in Solids, Academic Press, New York.

Esaki, L., Tsu, R., 1970. IBM J. Res. Dev. 14, 61.

Feng, Z.C., Tsu, R., 1994. Porous Silicon, World Scientific, Singapore.

Goldman, V.J., Tsui, D.C., Cunningham, J.E., 1987. Phys. Rev. Lett. 58, 1256−1259.

Gunn, J.B., 1963. Solid State Commun. 1, 88.

Gunn, J.B., 1976. IEEE Trans. Electron. Dev. ED 23, 705.

Gupta, R., Ridley, B.K., 1988. J. Appl. Phys. 64, 3089.

Heitler, W., 1954. Quantum Theory of Radiation, Oxford University Press, Oxford.

Hirose, M., Miyazaki, S., 1984. J. Non-Cryst. Solids 66, 327.

Isihara, A., 1971. Statistical Physics, Academic Press, New York.

Kogan, Sh., 1996. Electronic Noise and Fluctuations in Solids, Cambridge University Press, Cambridge.

Landauer, R., 1957. IBM J. Res. Dev. 1, 223.

Landauer, R., 1970. Phil. Mag. 21, 863.

Landauer, R., 1992. Phys. Scripta T 42, 110.

Luryi, S., 1985. Appl. Phys. Lett. 47, 490.

Matthiessen, A., 1857. Phil. Mag. 13, 81.

Mehan, G.D., 2000. Many Particle Physics, third ed. Plenum Press, New York.

Mitin, V.V., Kochelap, V.A., Stroscio, M.A., 1999. Quantum Heterostructures—Microelectronics and Optoelectronics, Cambridge University Press, Cambridge.

Morse, P., Feshbach, H., 1956. Methods of Theoretical Physics, McGraw-Hill, New York.

Mott, N.F., 1970. Phil. Mag. 22, 7.

Mott, N.F., Davis, E.A., 1979. Electronic Processes in Non-Crystalline Materials, Clarendon Press, Oxford.

Nicollian, E.H., Tsu, R., 1993. Appl. Physics 74, 4020.

Noteborn, H., 1993. Quantum Tunneling Transport of Electrons in Double Barrier Heterostructures, Thesis, Physics Department, Eindhoven University of Technology, pp. 54−59.

Payne, M.C., 1986. J. Phys. C: Solid State Phys. 19, 1145.

Pereyra, I., Carreno, M., Onmori, R.K., Sassaki, C.A., Andrade, A.M., Alrarez, F., 1987. J. Non-Cryst. Solids 97/98, 871.

Schiff, L.I., 1955. Quantum Mechanics, McGraw-Hill, New York.

Schultz, T.D., 1964. Quantum Field Theory and Many Body Problem, Gordon & Breach, New York.

Sen, S., 1989. Time Dependent Solution in Double Barrier Resonant Tunneling. MS Thesis, Department of Electrical Engineering, A&T State University.

Sollner, T.C.L.G., 1987. Phys. Rev. Lett. 59, 1622.

Stevens, K.W.H., 1983. J. Phys. C 16, 3649.

Tsu, R., 1967. Phys. Rev. 164, 380.

Tsu, R., 1985a. In: Adler, D., Fritzsche, H. (Eds.), Tetrahedrally Bonded Amorphous Semiconductors. Plenum Press, New York, p. 433.

Tsu, R., 1985b. J. Non-Cryst. Solids 75, 463−468.

Tsu, R., 1989. J. Non-Cryst. Solids 114, 708−710.

Tsu, R., 1990. Proc. SPIE 1361, 231.

Tsu, R., 2001. Challenges Nanoelectron. Nanotechnol. 12, 625.

Tsu, R., 2003. Microelectron. J. 34, 329.

Tsu, R., 2007. Microelectron. J. 39 (2008), 335−343.

Tsu, R., Esaki, L., 1971. Appl. Phys. Lett. 19, 246.

Tsu, R., Esaki, L., 1973. Appl. Phys. Lett. 22, 562.

Tsu, R., Zypman, F., 1990. Surf. Sci. 228, 418.

Tsu, R., Li, X.-L., Nicollian, E.H., 1994. Appl. Phys. Lett. 65, 842.

Vassell, M.O., Johnson, L., Lockwood, H.F., 1983. J. Appl. Phys. 54, 5206−5213.

Weil, T., Vinter, B., 1987. Appl. Phys. Lett. 50, 1281.

Whittaker, E.T., 1928. Bull. Cal. Math. Soc. 20, 199.

Woolard, D.L., Buot, F.A., Rhodes, D.L., Lu, X.J., Lux, R.A., Perlman, B.S., 1996. J. Appl. Phys. 79, 1515.

Woolard, D.L., Cui, H.L., Gelmont, B.L., Buot, B.L., Zhao, P., 2003. IJHSPES 13, 1149.

Zhao, P., Woolard, D.L., Cui, H.L., 2003. Phys. Rev. B 67, 085312−085321.

3 Optical Properties and Raman Scattering in Man-Made Quantum Systems

3.1   Optical Absorption in a Superlattice

The optical absorption of a superlattice where the minibands are fairly distinct may be modeled simply as shown in Figure 3.1, which depicts the band gap shifted up due to confinement effects in (a), $E - \mathbf{k}_t$ in (b), and the two-dimensional density of states (DOS) in (c), where the states in the longitudinal direction are taken as sharp discrete levels. Optical absorption is proportional to the product of the |transition matrix|^2 and the joint DOS. It is not a subject involving new concepts. However, experimentally, the superlattice structure, or multiple quantum well (QW) structures, consists of a thin layer not much thicker than 10 nm on top of a substrate at least several tens of times as thick. For this reason a challenge is presented if the substrate is not removed. I have in the past had occasion to use reflectivity at various angles of incidence. Reduction of such data is not well known. For this reason, I want to present the bulk of results of Tsu et al. (1975b).

Optical transitions involve transitions between the $1-2$, $1'-2'$, $1'-1$ states, etc. The first two processes involving intra-bands have been theoretically studied by Kazarinov and Shmartsev (1971). The present treatment involves a $1-1'$ transition without taking excitons into account. The wave function of the nth band, ψ_n is given by

$$\psi_n = f_n(x)U(\mathbf{k}_t, \rho)\exp(\mathrm{i}\mathbf{k}\cdot\mathbf{r}). \tag{3.1}$$

The matrix element for transition

$$H_{nn'} = \frac{\mathrm{i}e\hbar A}{2mc}\int \psi_{n'}\exp(\mathrm{i}\mathbf{q}\cdot\mathbf{r})\hat{\alpha}_0\nabla\psi_n\mathrm{d}\mathbf{r},$$

with the vector potential $\mathbf{A} = \hat{\alpha}_0 A$, for vertical transition (direct)

$$H_{nn'} = -(eA/2mc)\alpha_0 P_{nn'},$$

where

$$P_{nn'} \equiv -\mathrm{i}\hbar\int f_n^*(x)U^*(\mathbf{k}_t, \rho)[k + \mathrm{i}\nabla f_{n'}(x)U(\mathbf{k}_t, \rho)]\mathrm{d}\mathbf{r}.$$

Superlattice to Nanoelectronics. DOI: 10.1016/B978-0-08-096813-1.00003-5

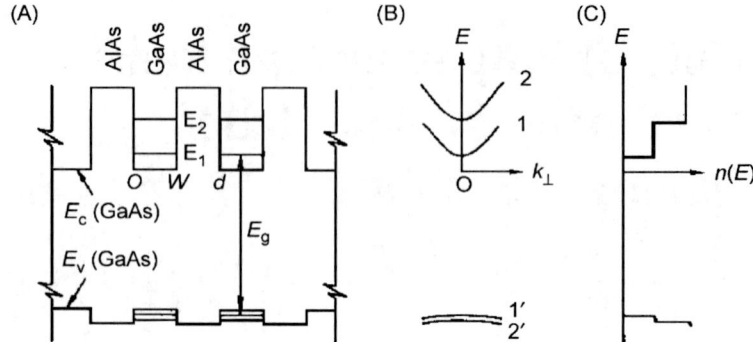

Figure 3.1 (A) Section of a superlattice potential profile, (B) $E - \mathbf{k}_t$, (C) density of states. *Source*: From Tsu et al. (1975b), with permission.

The energy difference between the initial and final states

$$E_c - E_v = \sum_{nn'}(E_n + E_{n'}) + \frac{\hbar^2 k_t^2}{2\mu}$$

in which μ is the transverse reduced mass. For the special case, where the valence band is either flat or even not quantized owing to poor mobility,

$$E_1 = E_g + (1 - \cos k_x d)\varepsilon_0, \tag{3.2}$$

where $2\varepsilon_0$ is the width of the miniband. The absorption coefficient α is given by

$$\alpha\hbar\omega = (e^2\mu/2\pi m^2 h)|\overline{P}_{nn'}|^2 k_{max}, \tag{3.3}$$

where k_{max} is given by $\hbar\omega - E_g = \varepsilon_0(1 - \cos k_{max}d)$ and $|\overline{P}_{nn'}|$ is the average over the solid angle, implying that we ignore any polarization effects. Figure 3.2 shows the qualitative absorption for constant matrix element. The dashed curve applies to zero bandwidth for large barrier height, generally for multiple QWs. Thus, absorption is essentially the joint DOS for the superlattice, or multiple QWs.

The sample under study consists of 100 periods of GaAs/Ga$_{0.5}$Al$_{0.5}$As with a period of 4.5 nm. For comparison, we use a Ga$_{0.5}$Al$_{0.5}$As alloy 6.5-nm-thick film. A freshly evaporated film of Ag was used for reference. We did try to remove the substrate with a hole varying in thickness but it was thought to be inadequate for serious measurement. In retrospect, taking note of the more direct results obtained by Dingle et al. (1974), which appeared after the manuscript of this work was submitted, we agreed that removing the substrate, even if the sample was not ideal, would have given more dramatic data. Nevertheless, useful characterization was possible owing to the use of a small incident angle that accentuated the reflectivity fringes. What follows demonstrates this point. Figures 3.3 and 3.4 show the reflectivity of the superlattice compared with the alloy. The number shows the order of the interference fringe.

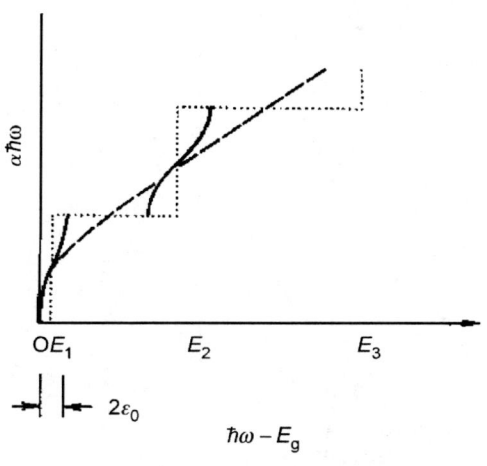

Figure 3.2 Qualitative absorption versus $\hbar\omega - E_g$ for constant matrix element; __ __ superlattice; . . . , multiple QWs, where the states are discrete.

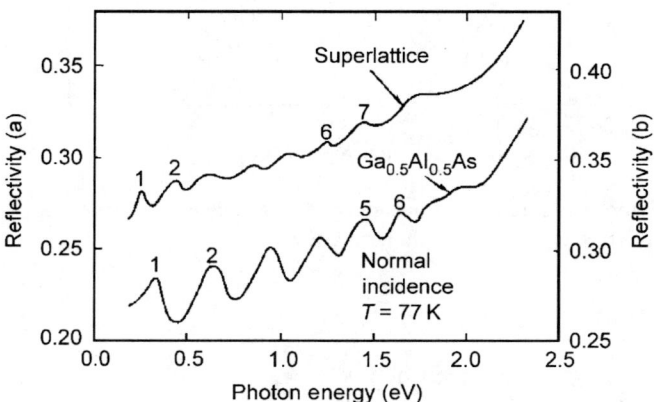

Figure 3.3 Reflectivity at normal incidence of the superlattice at 900 nm and alloy at 6.5 nm. *Source*: From Tsu et al. (1975b), with permission.

Note that the fringes are much more visible and the positions are different allowing a cross-check for fitting. In the analysis of data with a substrate attached, because of the complexity of the exact expression for the reflectivity, the data rarely fit to the whole expression for the determination of the optical constants. Usually one obtains the refractive index from the fringes. The procedure described here is very powerful, because both the refractive index and the absorption constants may be obtained without the use of Kramer−Krönig relations, and without the use of either the reflectivity or the transmission. Let k_1, k_2, and k_3 be the propagation constants for the vacuum, film, and substrate, respectively, for light incident at an angle θ_1 from the vacuum and θ_2 and θ_3, the propagation directions in the film and substrate, all measured from the surface normal.

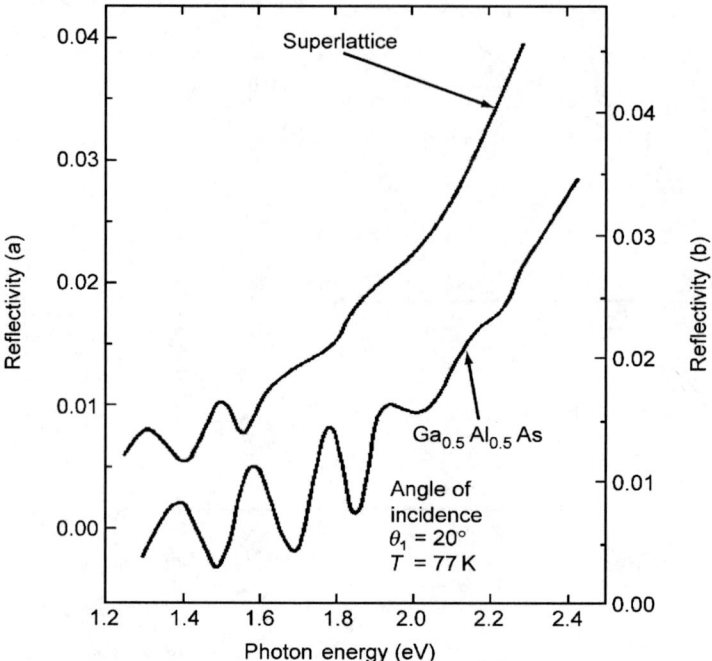

Figure 3.4 Same as Figure 3.3 but the angle of incidence is 70°, almost at grazing.
Source: From Tsu et al. (1975b), with permission.

Snell's law gives

$$\frac{\sin\theta_1}{\sin\theta_2} = \frac{k_2}{k_1} = n + iK, \qquad \frac{\sin\theta_1}{\sin\theta_3} = \frac{k_3}{k_1} = n',$$

where the refractive index is n for the film and n' for the substrate and d is the thickness of the film, then

$$R^2 = \left[(1-a)/(1+a)\right]^2, \qquad a = \frac{1-\exp(i2k_2d\cos\theta_2)Q_{23}}{1+\exp(i2k_2d\cos\theta_2)Q_{23}}P_{12}$$

$$Q_{23} = \frac{k_3\cos\theta_2 - k_2\cos\theta_3}{k_3\cos\theta_2 + k_2\cos\theta_3}, \qquad P_{12} = \frac{k_1\cos\theta_2}{k_2\cos\theta_1}. \qquad (3.4)$$

For relatively low absorption, $n \gg K$, the extinction coefficient, we neglect K except in the exponential terms. Then some surprising results follow. The maxima and minima occur at wavelengths determined by

$$(4\pi d/\lambda)(n^2 - \sin^2\theta_1)^{1/2} \; = 2\pi p, \qquad \text{for } R_+ \text{(maxima)}$$
$$= 2\pi(p + \tfrac{1}{2}), \quad \text{for } R_- \text{(minima)} \qquad (3.5)$$

where p is any integer. And for normal incidence,

$$R_+ = \frac{n'-1}{n'+1} \quad \text{and} \quad R_- = \frac{n^2-n'}{n^2+n'}. \tag{3.6}$$

We see that the maxima, R_+, are determined only by the refractive index of the substrate.

The procedure led to an n for the alloy within 3% of the value for GaAs. Since the visibility of the fringes is proportional to $a_+ - a_-$, and inversely proportional to $\cos\theta_1$, the grazing angle gives more accurate values for n and K. Once n and n' have been obtained, remember that $k_2/k_1 = n + iK$, therefore K, or the absorption coefficient $\alpha = 4\pi K/\lambda$, can be determined for each wavelength.

Figure 3.5 shows the refractive index for the substrate, the superlattice, and the alloy. Note that, far from resonances, at long wavelengths, the alloy is close to the average value for GaAs and AlAs obtained by Onton (1970).

At the time that we first obtained the result, we were exuberant because we had found a way to obtain the absorption coefficient without removing the substrate (Figure 3.6). However, a few months later when the absorption of a multiple QW appeared (Dingle et al., 1974), we recognized that more impressive data could be obtained with the substrate removed and wedged to suppress interference fringes. Wedge is used to avoid the interaction of interference fringes and discrete energy states of the QW, creating a very complex spectrum. Figure 3.7, taken from Dingle et al. (1974), shows the absorption spectra of GaAs/Al$_{0.2}$Ga$_{0.8}$As multiple QW structures, with well width L_z at 2 K. The value of $\alpha \sim 2.5 \times 10^4$ cm^{-4} for $L_z = 400$ nm at the exciton peak. For $L_z = 21$ and 14 nm, the first excitons as well as subsequent excitons for transitions involving $n = 2, 3$, also show up clearly. They indicate that quantum effects at

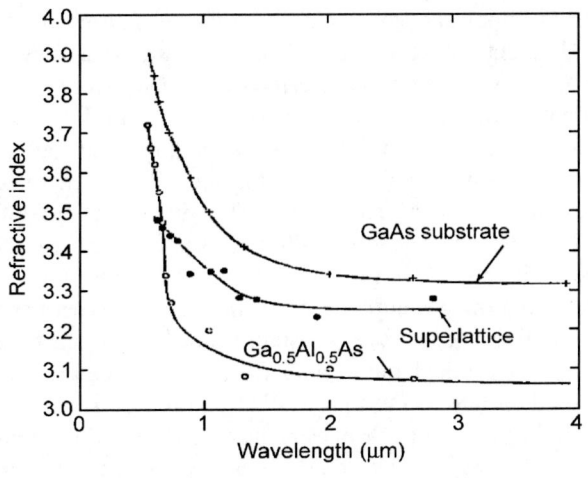

Figure 3.5 Refractive indices for GaAs, the superlattice, and the alloy at 77 K.
Source: From Tsu et al. (1974), with permission.

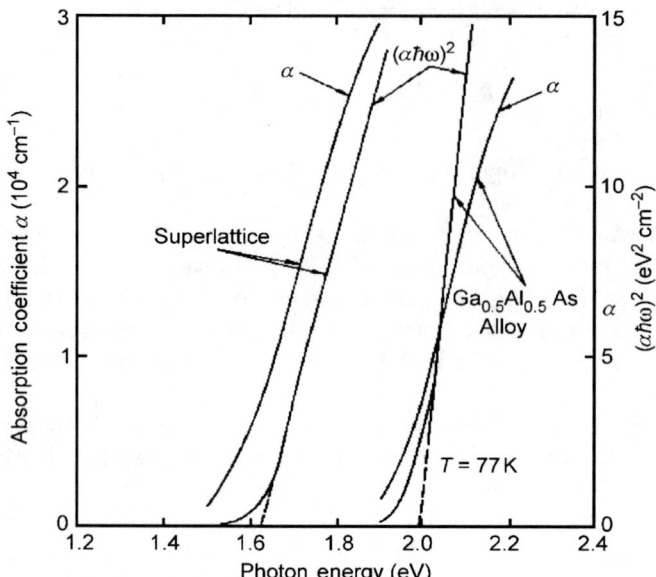

Figure 3.6 Absorption coefficient α (10^4 cm^{-1}) versus photon energy in eV at 77 K. The best straight line fit for $(\alpha\hbar\omega)^2$ versus $\hbar\omega$ gives the position of the absorption edge, 1.63 eV for the superlattice and 2 eV for the alloy.
Source: From Tsu et al. (1975b), with permission.

2 K start at $L_z \sim 50$ nm. The finding is consistent with the bound states of the well given by

$$E = E_n + \frac{\hbar^2}{2m^*}\mathbf{k}_t^2.$$

I want to point out a couple of important points regarding the results of Dingle et al. (1974). First, I have already mentioned the removal of the substrate, and even more importantly, how wedging is used to suppress interference fringe. Second, we should not be too dogmatic about fitting theory to experimental data. Initially, we at IBM used a barrier height of 80% of the difference between the band gap of AlAs and GaAs for band-edge alignment. Dingle et al. using their data seemed to have presented a convincing argument that the band-edge offset in the conduction band should be 88% of the difference in the band gap. Some 10 years later, we all agreed on 60% of the difference. Therefore, the fit would not have been very good if the correct value for the band-edge offset had been used. In fact, my experience convinced me that detailed qualitative consistency is usually more important than quantitative agreement between a theoretical model and fit to experimental data. Why we did not seriously pursue the route of multiple QWs? The reason is simple. We were too wrapped up in proving our point. We were having a hard time convincing anyone, particularly those at IBM, that we indeed had something to offer in

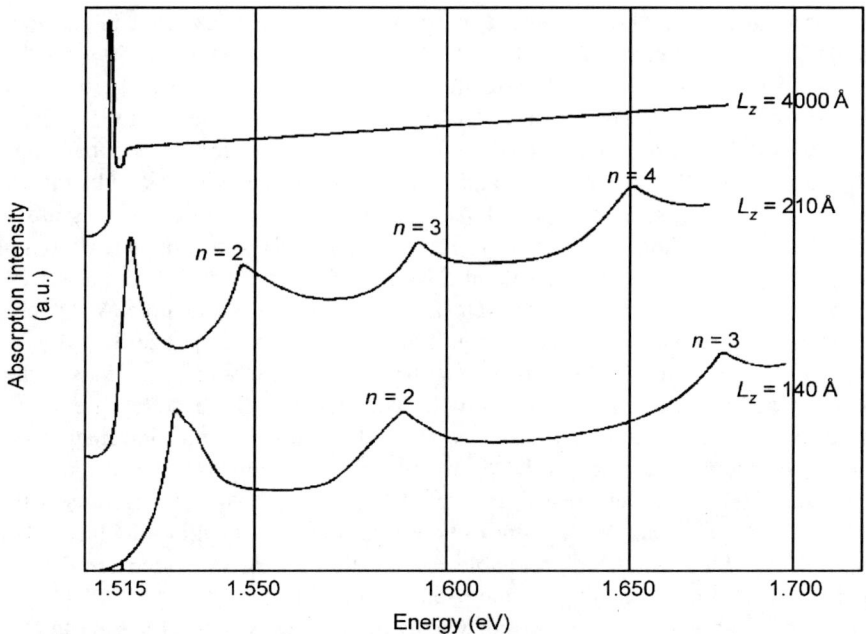

Figure 3.7 Absorption spectra of excitons in the multiple QW GaAs/Al$_{0.2}$Ga$_{0.8}$As, with well width L_z at 2 K. $\alpha \sim 2.5 \times 10^4$ cm^{-4} for L_z = 400 nm at the exciton peak. For L_z = 21 and 14 nm, excitons associated with the nth bound states are evident.
Source: After Dingle et al. (1974), with permission.

the form of man-made quantum states to broaden the class of electronic materials and devices utilizing quantum states. Finally, we simply did not understand well enough the importance of keeping $x < 0.3$ in Al$_x$Ga$_{1-x}$As before 1974–1975. The vast experience in working on heterojunction lasers gave BTL an edge over us at IBM. By 1975, I knew that the QW structures at BTL were better than those we had at IBM, because by then we had removed and wedged the sample–substrate and repeated the absorption measurement. Our exciton peaks were only barely visible at 4.2 K.

3.2 Photoconductivity in a Superlattice

Photon conductivity serves as a powerful technique for characterizing the optical and transport properties of a superlattice and QW. Because this technique is not as widely used, I would like to give an example, from a fairly early stage of the development of superlattices and QWs for resonant tunneling, of how the photoconductivity experiment reinforced our confidence that the research we had started was indeed heading in the right direction—although, as we shall see, many details do not fit together very well.

Superlattices with three different configurations have been used in the experiments (Tsu et al., 1975a): A—(100, 35, 35 − x = 0.8), B—(80, 50, 50 − x = 0.78), and C—(50, 110, 100 − x = 0.55), using the notation: sample—(number of periods, thickness of GaAs, thickness of GaAlAs—% of alloy composition of Ga). The total thickness of the superlattice (SL) is ∼1 μm in order to provide adequate photon absorption. The SL is undoped and fabricated on an n-GaAs substrate with $n \sim 10^{18}$ cm^{-3}. The semi-transparent Au 0.25×0.025 cm^2 top electrodes are used for both $I-V$ and photoexcitations. In Figure 3.8, the photo response is shown plotted against the photon energy in eV at 5 K, with E_1 and E_2 designating the two SL bands. The trend is there, showing E_2 broader than E_1, and sample A, which has narrower barriers shows a broader peak marked as $|E_1|$ or $|E_2|$. The Schottky junction between the Au−GaAs contact results in a built-in potential. In sample C, not only do the states move down with wider QWs, these states are more or less discrete, having smaller transitions from the valence band, dominated by the heavy hole band with a larger DOS. In the original analysis (Tsu et al., 1975a), it is estimated from the steady-state photocurrent that the $\mu\tau$ product will be $\mu\tau \sim 2-2.5 \times 10^{-10}$ cm^2 V^{-1}. Using $\mu \sim 10$ cm^2 V^{-1} s^{-1} (Esaki and Chang, 1974), $\tau \sim 10^{-11}$ s, which is not unreasonable for the lifetime of the electrons. Note that this lifetime is not the scattering time used for mobility.

Figure 3.9 shows the photocurrent versus the applied voltage for sample B at 5, 63, and 217 K at two photoenergies that correspond to the two peaks at E_1 and E_2. The positive voltage refers to $+V$ on the Au-electrode. The asymmetry is from the Schottky barrier. The most encouraging feature is the negative differential

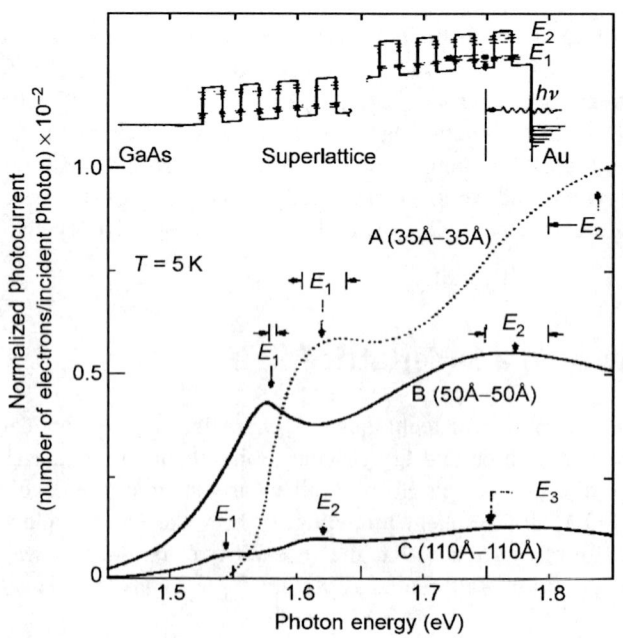

Figure 3.8 Number of electrons per incident photons $\times 10^{-2}$ versus photon energy in eV for the samples A, B, and C described in the text. Calculated energies of the bandwidths are marked with arrows showing fair agreement with the measured photocurrent peaks. The energy profile is shown in the top part. *Source*: After Tsu et al. (1975a), with permission.

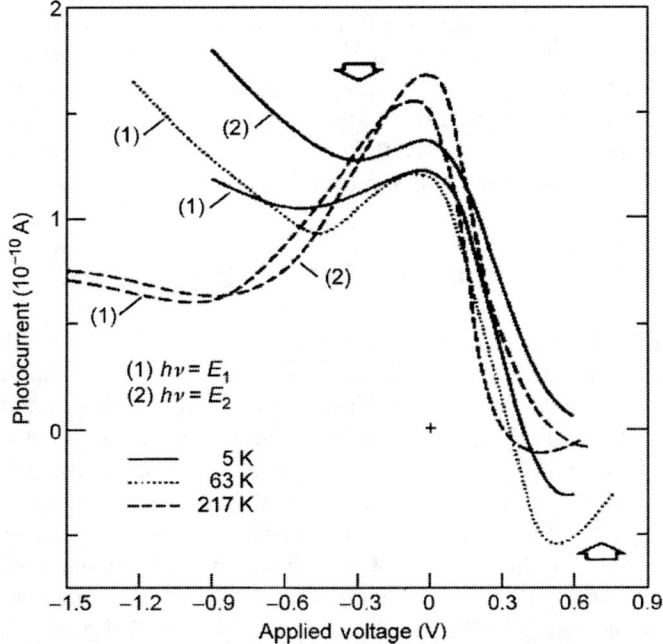

Figure 3.9 Photocurrent versus applied voltage for sample B at 5, 63, and 217 K with incident photoenergies corresponding to the two peaks at E_1 and E_2 of Figure 3.8. The position of the maximum NDR is marked by arrows.
Source: After Tsu et al. (1975a), with permission.

resistance (NDR) marked by arrows. This NDR is stronger at 217 K than at 5 K, indicating that the main transport is phonon-assisted hopping described in Section 1.6 of Chapter 1. As shown in Figure 1.11, the negative slopes at 77 and 300 K are nearly same, but the level of the tunneling current at 77 K is more than a factor of 3 lower than that at 300 K. If the lower limit of the tunneling current is set by the mean free path, as we know that mean free path has a maximum nearer to 77 K than 5 K, and since the level of the phonon-assisted hopping is much higher at higher temperatures, we think that the SL is basically behaving as Stark ladder, rather than using band conduction as described in Section 1.7. The voltage per period corresponds to $eFd \sim 10$ meV, which is greater than $\Delta E_1 \sim 7$ meV; therefore, it is within the phonon-assisted hopping regime. Therefore, we are quite certain that the observed NDR confirms the theoretical calculation of the NDR due to phonon-assisted hopping (Döhle et al., 1975; Tsu and Döhle, 1975). More recently (Capasso et al., 1985), in a $p^+ in^+$ junction that has an AlInAs/GaInAs superlattice in the *i*-region at 77 K used photoconductive measurements to exhibit NDR in the hopping regime, confirming earlier results (Tsu et al., 1975a).

Note that the band-edge offset used in this work was changed from 0.8 to $0.88\Delta E_g$. We know that a higher barrier gives rise to greater confinement that leads to a higher energy state. All our work at IBM, and later work at BTL, show good agreement. When later this number settled down to $0.6\Delta E_g$, the real agreement of measurements made by everyone should have been viewed as fortuitous. However, as I showed that damping upshifts these energies (Chapter 2) making it appear as though the barriers were higher, again I assert that we must not be too dogmatic about so-called very good fit. The truth of the matter is that we are making incremental but definite progress all the time and that process applies to all technical progress.

3.3   Raman Scattering in a Superlattice and QW

As shown in Chapter 1, the folding of the dispersion in the regular Brillouin zone (BZ) into minizones (MZs) for a superlattice describes most of the important features of the man-made superlattice. These features include the formation of minibands, localization with large gaps between minibands, Bragg reflection giving rise to Bloch oscillation, and the foundation of the appearance of NDR. The reciprocal space for phonons is exactly the same as for electrons because the real space and reciprocal space are determined by the structure. Therefore, we can expect most of these features that are due to zone folding to be applicable to phonons. A couple of acronyms are useful for the discussion: OPFA for optical phonons from folded acoustic phonons and OPFO for optical phonons from folded optical phonons. The idea that formation of the OPFA introduces photon—phonon interaction at low frequencies is very appealing. Before we explain the main reason why, as pointed out in the very first paper launching man-made quantum structures (Esaki and Tsu, 1970), the key is that the mean free path must exceed structure size, the period of the superlattice. This puts a stringent limitation on the phonons. Optical phonons do not propagate far, but low-frequency sound waves propagate almost unimpeded like in the crystal quartz, which has the highest Q (quality factor) resonator known to man, and was the time standard used by WWV radio signal before being replaced by the Ce-laser. Therefore, folded acoustic phonons should exhibit pronounced SL effects. The quasi particle of a photon—phonon coupled system is a polariton, with its manifestation in the dielectric function. In a normal solid, polaritons do not exist below the optical phonon bands because the dispersion $\omega - q$ for photons does not cross that of the acoustic phonons. OPFA allows the crossing of the two dispersions, creating polaritons at frequencies ranging from the acoustic phonons at the BZ all the way down to zero. We should realize that at such low frequencies, the wavelength is so long that we need a very long period SL to have sufficient coupling. Therefore, the real device is still limited near the terahertz scale so that an SL of practical thickness may be fabricated. Another aspect in which we study Raman scattering in SL is in characterization, e.g., determination of alloy compositions and thicknesses as discussed in Chapter 1, localization of excitons, as well as the strains at the heterojunction interface.

Not long after launching a man-made superlattice (Tsu and Jha, 1972) using a linear chain model, calculated the dispersion of phonons and polaritons in an SL showing that the frequencies of these modes may be prescribed by proper design. The occurrence of fairly strong OPFA in the far infrared should be useful for coherent frequency generators, Restrahlen filters, acoustic gratings, and so on. It was recognized that folding the BZ into MZ provides a means of mapping the phonon dispersion, which is particularly useful when samples are too thin for neutron-scattering experiments. Raman scattering was used routinely for the optimization and characterization of the superlattice systems, alloy compositions, alloy compositions. (Tsu et al., 1972), and a fairly long-period superlattice (Mayer et al., 1973; Tsu, 1981). The folding of the acoustic phonons was first experimentally confirmed by Colvard et al. (1980). However, folding of optical phonons is weaker because optical phonons simply do not last long enough because the group velocity of optical phonons is at least two orders of magnitude below that of acoustic phonons. The situation may be different if the superlattice were constructed with alternate layers of single-unit calls. The Raman spectra of optical excitation between the valence band and the first two QW states were first calculated by Tsu and Esaki (1975). In this work, optical phonons are in the extended zone but the electronic wave functions are quantized in the reduced zone. The spectra show a series of peaks below the main peak of the GaAs, thus clearly indicating the effects of confinement in the form of zone folding. Sai-Halasz et al. (1978) obtained Raman spectra suggestive of optical confinement; however, Merlin et al. (1980) explained their results in terms of anisotropy of polar phonons. However, I can simply dismiss Sai-Halasz's experiment by the fact that the two samples, 10 nm/10 nm and 5.2 nm/ 6.6 nm, give identical results. Barker et al. (1978) turned their attempt to identify folding optical phonons into a study of the interdiffusion process at the heterojunction interface. The first clear-cut evidence of the OPFO came from Jusserand et al. (1984, 1985). The OPFOs for both longitudinal-optical (LO)- and transverse-optical (TO)-phonons in GaAs were obtained by Sood et al. (1985). Raman scattering has also been used to determine deformation in GeSi/Si strain-layer superlattice (SLS) by Cerdeira et al. (1984). Their results indicate that the entire lattice mismatch is accommodated by a homogeneous tetragonal strain in the GeSi alloy layers only. It is important that the mechanisms of Raman scattering in man-made quantum structures are understood. Of all the subjects involving solids, I consider that a good working knowledge in Raman scattering is the most difficult because the subject is heavily involved with group theory, even with the use of double group.

3.3.1 Some Fundamentals

Starting with a simple system of acoustic and optical branches, many additional modes arise. What we should recognize is that these additional modes are weak. Figure 3.10 shows a simple sketch of displacement of various atoms indicated by arrows, its length referring to the magnitude of the displacements. This is the kind of model I used to gain physical insight. For example, in a diatomic system, even if we take the radical assumption that the force constants are all equal, then the zone center optical phonons have the highest frequency because the opposing motion of the two atoms has a reduced mass in the denominator to determine frequency. Next, the optical frequency at the BZ boundary is determined by the mass of the lighter atom of the two, while the acoustic phonon frequency at the BZ is

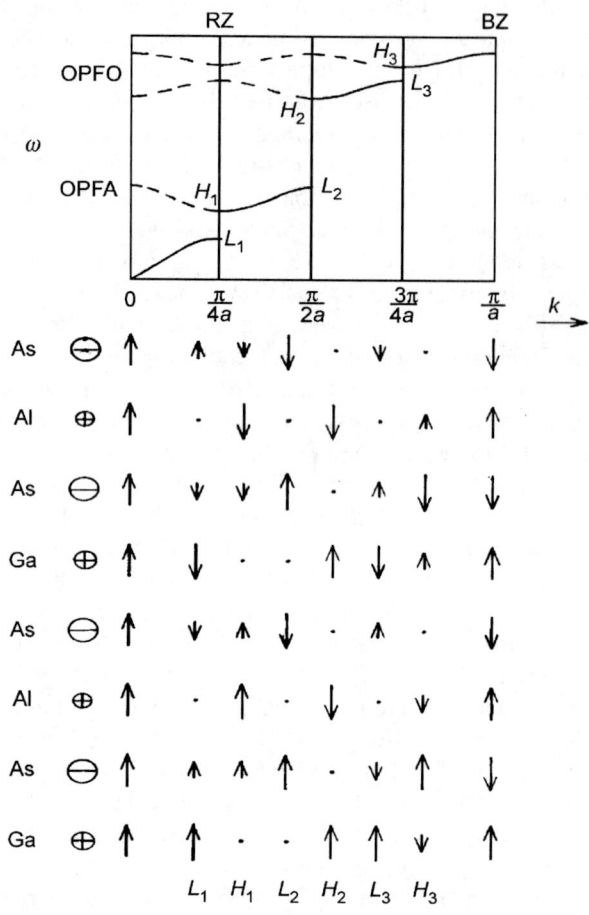

Figure 3.10 Displacements of various atoms and dispersion relations based on the linear chain model. OPFO and OPFA stand for optical phonons from folded optical or acoustic phonons. The modes are designated as L and H, and arrows show displacements.

determined by the mass of the heavier of the two atoms. For equal masses, the reduced mass is halved and the two frequencies at the BZ boundary are degenerate. This picture allows me to predict many eventualities. We can add complexity to this simple picture, for example, the distinction between longitudinal and transverse frequencies, and even include anisotropy.

Figure 3.10 shows the dispersion of the phonon branch in the original BZ folded twice with π/a of the BZ into $\pi/4a$ of the reduced zone (RZ). There are three kinds of atoms, Ga, As, and Al. Actually, we should differentiate between As in GaAs and As in AlAs, because the force constants, as well as the effective charges, are different in the two cases. Nevertheless, we take them to be equal. Because of the splittings, at each k there is a higher and a lower branch. For example, at $k = \pi/2a$, H_2 denotes the lowest OPFO and L_2 denotes the highest OPFA. When folded into the RZ, for $k = 0$ to $\pi/4a$, L_2 is folded into the zone center, shown as a dashed curve. At $k = \pi/2a$, at the very bottom, these mode locations are shown as $L_1, \ldots, H_3$. Let us focus on L_2. At the top next to As there is a large arrow pointing downward, but a dot for

Al indicating that the Al atom does not move. The arrow points upward for the next As atom, third from the top, followed by a dot for Ga. And this sequence repeats. This means that the two As atoms move to cancel each other, since they have the same charge sign and this large opposing motion does not generate a dipole moment. Next to L_2 is H_2, the As atoms do not move, all being shown as dots, but the opposing arrows for Ga and Al do not cancel each other, resulting in much stronger Raman scattering as well as some dipole moments. In reality there is a small dipole moment even at L_2 because the effective charges and masses are different owing to the difference in the ionic radii. At BZ, all pairs of up-down arrows for Ga−As and Al−As enforce each other resulting in strong dipole moments. Whether polaritons are present depends on crossing of the dispersion $\omega - q$.

3.3.2 Phonons and Polariton Modes in a Superlattice

In 1971, Tsu and Jha (1972) considered that phonons and polaritons in a superlattice described by zone folding should form optical modes with the acoustical phonon. What happens to the folding of the optical phonon? A periodic potential results in a series of MZs in reciprocal space, forming the SL. However, alternate layers of materials with different masses, force constants, and effective ionic charges result in a superlattice that has interesting optical properties due to the interactions of photons and optical phonons. When acoustical phonons are folded into an optical branch they interact with low-frequency photons within the normal acoustic phonon frequencies. Because the frequencies of the modes may be prescribed by the correct choice of thicknesses as well as of materials, we have man-made optical materials that have periods in the regime of phonon wavelengths that are usually close to the dimensions of man-made electronic superlattices. Therefore, periods are much smaller than those in photonic crystals that have dimensions in the range of photon wavelengths. What is the major consequence? Photonic crystals utilize interference properties with far larger dimensions. Let me give an example. A typical superlattice is characterized by dimensions governed by the mean free path, in the order of tens of nanometers. Photonic crystals usually have periods greater than few micrometers. And phonon superlattices have periods in the range of tens of nanometers, similar to superlattices.

To work out the zone folding and the opening up of minibands as in the case of an electronic SL, we used the linear chain model. For N atoms per unit cell, with m_p, x_p, e_p, and F_p being the mass, equilibrium position, effective charge, and force constant for the pth atom, the equation of motion becomes

$$m_p \ddot{y}_{rp} + \frac{F_p}{x_p - x_{p-1}} (y_{rp} - y_{r,p-1}) + \frac{F_p}{x_{p+1} - x_p} (y_{rp} - y_{r,p+1}) = e_p E, (p = 1, \ldots, N)$$

(3.7)

where y_{rp} is the displacement of the pth atom in the rth call, transverse to the propagation vector, and E is the self-consistent electric field. We set up the boundary conditions $x_1 - x_0 = x_{N+1} - x_N$, $y_{r,0} - y_{r-1,N}$, and $y_{r,N+1} - y_{r+1,1}$. For simplicity, we let $x_p - x_{p-1} = a$ and $x_{N+1} - x_0 = Na = d$, the period. With $E = 0$ and $y_p = \exp(iqx_p - i\omega t)$, in Eq. (3.7) and writing in matrix form

$$(K - \omega^2 M)Y = 0, \tag{3.8}$$

where K and M are the force constant matrix and mass matrix, following Verleur and Barker (1966). Eq. (3.8) may be reduced to an eigenvalue problem by substituting $u = M^{1/2}Y$ and $B = M^{1/2}KM^{-1/2}$, resulting in the complex matrix equations

$$(B^2 - \omega^2)u = 0. \tag{3.9}$$

The solution of this frequency equation gives the transverse phonon frequencies $\omega_{ti}(q)$, with the phonon wave vector q. In the presence of an electric field E, the dielectric function

$$\varepsilon(\omega) = \varepsilon_\infty + \sum_i{}' \frac{S_i \omega_{ti}^2}{\omega_{ti}^2 - \omega^2}, \tag{3.10}$$

where $\omega_{ti}(q = 0)$ and $S_i \equiv AQ_i^2/\omega_{ti}^2$ is the oscillator strength (Figure 3.11). The vector Q_i is

$$(Q_1, \ldots, Q_P, \ldots, Q_N) = (e_1/m_1^{1/2}, \ldots, e_N/m_N^{1/2})U, \tag{3.11}$$

where the unitary matrix U diagonalizes the force constant matrix B. The constant A in S_i satisfies the sum rule

$$\varepsilon_0 = \varepsilon_\infty + A \sum_i{}' \frac{Q_i^2}{\omega_{ti}^2} \tag{3.12}$$

in which ε_∞ and ε_0 are the high- and low-frequency limits of the dielectric constants, respectively. The sum in Eq. (3.12) excludes the point zero of the acoustic frequency. The polariton dispersion is obtained by putting $\varepsilon(\omega) = q^2 c^2/\omega^2$ in Eq. (3.12). The detail values for the dynamic force constant B, the effective charges e and ε are to be found in Tsu and Jha (1972), with most values taken from Dolling and Waugh (1963), giving the dynamic force constant B_{As-As} in GaAs more than double $B_{As'-As'}$ in AlAs. However, $e_{As} = -2.12e$ and $e_{As'} = -2.2e$ are nearly the same. For simplicity, we took ε_0 and ε_∞ to be the average of the two.

3.3.3 Calculation of Raman Scattering in Superlattice and Surface Quantization

3.3.3.1 Superlattice

The spectra of LO phonons from Raman scattering associated with quantum states in a GaAs/GaAlAs superlattice and the surface quantization in highly doped GaAs were calculated (Tsu and Esaki, 1975). Since the results were

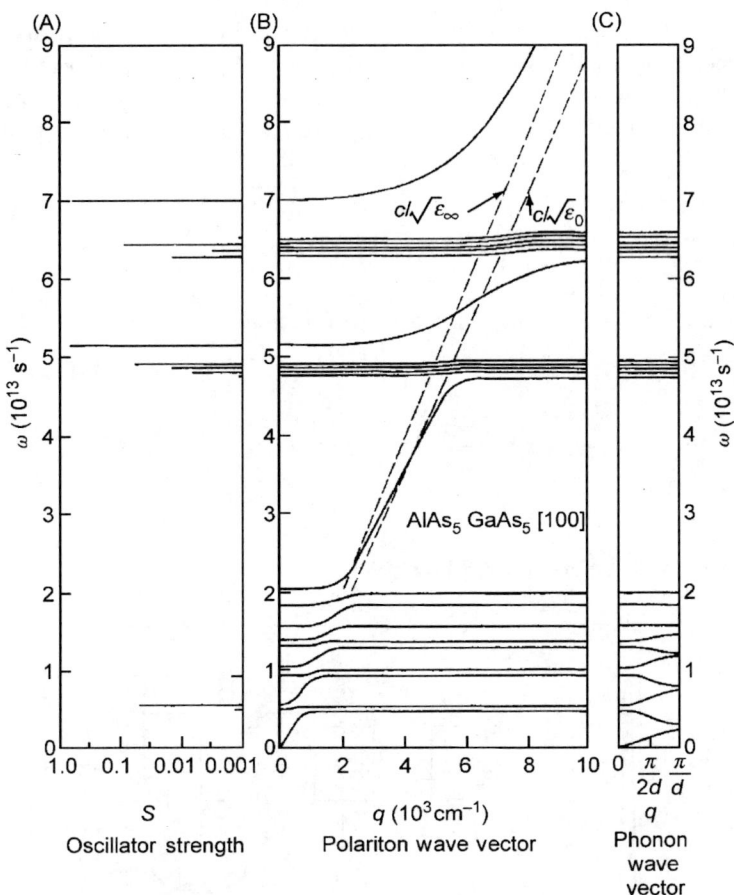

Figure 3.11 (A) Oscillator strength versus frequency ω, (B) $\omega-q$ for the polaritons, and (C) $\omega-q$ for the phonons in the minizone. The period of the superlattice consists of AlAs (five monolayers) and GaAs (five monolayers) along the (100) direction.
Source: After Tsu and Jha (1972), with permission.

published in a conference proceeding that is not generally available and because some details were cut from the original IBM Research Note, the bulk of the results based on the IBM Research Note are given in this section. Since the phonon branches are folded into the reduced MZ, these minibranches can now be excited near $q \sim 0$. If the oscillator strength of these folded branches is weak, i.e., ionic mass-, charge-, and force constant-differences are small, one may neglect the folding effects of the phonon branches and consider instead the *umklap process* of electrons in the reduced MZ, interacting with phonons in the extended MZs (regular BZs of the solid). The case being treated is a superlattice where the bandwidth of the quantum states is very narrow so that a double-well approximation is valid.

Figure 3.12 shows the potential profile in one direction. The energy states in the valence- and conduction-bands are labeled by E_v and E_α, respectively. Owing to coupling of wave functions in adjacent wells, the energy E_α split into an antisymmetric function ψ_- corresponding to $E_{\alpha-} = E_g + \varepsilon$, and a symmetric state ψ_+ having $E_{\alpha+} = E_g$. These wave functions

$$\psi_- = \frac{1}{\sqrt{2}}(|1\rangle - |2\rangle) \ \text{ and } \ \psi_+ \frac{1}{\sqrt{2}}(|1\rangle + |2\rangle) \tag{3.13}$$

are written in terms of the uncoupled wave functions

$$|1\rangle$$
$$\sim \sqrt{\frac{2}{W}} \sin\frac{n\pi}{W}\left(x \pm \frac{d-w}{2}\right)e^{ik_\perp\rho} \tag{3.14}$$
$$|2\rangle$$

for α level, $n = 1$, whereas for β level, $n = 2$.

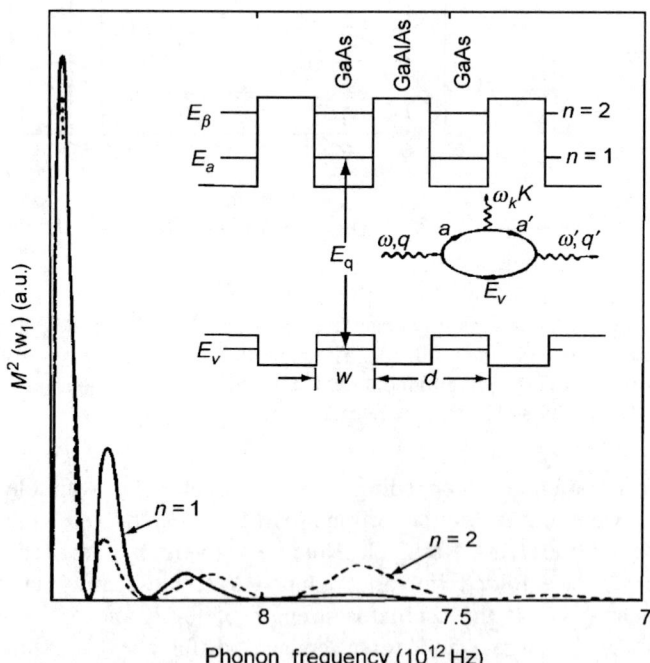

Figure 3.12 Calculated Raman spectra for a GaAs/GaAlAs superlattice with a period $d = 2w$ and $d/a = 5$, where a is the lattice constant. Inset shows the optical excitation for holes in the valence band to the electronic state in the conduction band.
Source: From Tsu and Esaki (1975), with permission.

The scattering process consists of excitation by the incoming photon, ω and $\mathbf{q}$, scattering an electron from E_v to $E_{\alpha+}$, emitting a phonon ω_K, $\mathbf{k}$ resulting in $E_{\alpha-}$ then returning to E_v while creating a photon at ω', $\mathbf{q}'$; in other words

$$\frac{d\sigma}{d\omega} \sim \sum_{\mathbf{K}} \sum_{v,\alpha,\alpha'} \left| \frac{\langle v|H_R|\alpha\rangle \langle\alpha|H_p|\alpha'\rangle \langle\alpha'|H_R|v\rangle}{(E_1-\hbar\omega)(E_2-\hbar\omega')} \right|^2 \delta(\omega_K - \omega + \omega')$$

$$\equiv \sum_{\mathbf{K}} |A + B|^2 \delta(\omega_K - \omega + \omega'). \tag{3.15}$$

Summing over the intermediate states involves integrations over the transverse momenta, $k_{\perp v'}$, $k_\perp$, and $k'_\perp$, and summations over the discrete states. In our case, the discrete states, α, β, etc., have only two states each, the symmetric and antisymmetric states. We are interested in resonance enhancement, therefore, we may treat α and β separately. The terms,

$$\langle v|H_R|\alpha\rangle \text{ has two terms}, A_q^- (\langle\psi_{+v}||e^{i\mathbf{q}\cdot\mathbf{r}}|\psi_\pm\rangle + \langle\psi_{-v}|e^{i\mathbf{q}\cdot\mathbf{r}}|\psi_\pm\rangle);$$
$$\langle\alpha|H_p|\alpha'\rangle \text{ has one term}, b_K^+ \langle\psi_\pm|e^{-i\mathbf{K}\cdot\mathbf{r}}|\psi_\mp\rangle \text{ and}$$
$$\langle\alpha'|H_R|v\rangle \text{ has two terms}, A_{q'}^+ (\langle\psi_\mp|e^{-i\mathbf{q}'\cdot\mathbf{r}}|\psi_{+v}\rangle + \langle\psi_\mp|e^{-i\mathbf{q}'\cdot\mathbf{r}}|\psi_{-v}\rangle).$$

Whenever there are two signs, the superior ones are used in A and the inferior ones in B. The energy denominators for A and B are different

$$A : (E_g + E_1(\perp) - \hbar\omega)(E_g + \varepsilon + \hbar\omega_K + E_2(\perp) - \hbar\omega)$$

and

$$B : (E_g + \varepsilon + E_1(\perp) - \hbar\omega)(E_g + \hbar\omega_K + E_2(\perp) - \hbar\omega),$$

in which

$$E_1(\perp) \equiv \frac{\hbar^2 k_\perp^2}{2m_e} + \frac{\hbar^2 k_{\perp v}^2}{2m_h} \quad \text{and} \quad E_2(\perp) \equiv \frac{\hbar^2 k_\perp'^2}{2m_e} + \frac{\hbar^2 k_{\perp v}^2}{2m_h}.$$

After summing over $k_{\perp v}$ and $k'_\perp$,

$$A \text{ or } B \sim \int_0^{k_{\perp max}} \frac{k_\perp \, dk_\perp}{(k_\perp^2 - a^2)(k_\perp^2 - b^2)} M^2(\mathbf{K}), \tag{3.16}$$

where

$$a^2 \equiv \frac{2\mu}{\hbar^2}(\hbar\omega - E_g) \text{ and } b^2 \equiv \frac{2\mu}{\hbar^2}(\hbar\omega - E_g - \varepsilon - \hbar\omega_K) \text{ for } A; \text{ and}$$

$$a^2 \equiv \frac{2\mu}{\hbar^2}(\hbar\omega - E_g - \varepsilon) \text{ and } b^2 \equiv \frac{2\mu}{\hbar^2}(\hbar\omega - E_g - \hbar\omega_K) \text{ for } B.$$

It may be shown that there is little resonance enhancement for $\hbar\omega_K \ll \varepsilon$ and $\hbar\omega_k \gg \varepsilon$: there is a logarithmic singularity represented by $(\hbar\omega_K)^{-1}\ell n(\hbar\omega_K/\delta)$, where $\delta \equiv \hbar\omega - E_g - \varepsilon$ for A; $\delta \equiv \hbar\omega - E_g$ for B. Therefore

$$\frac{d\sigma}{d\omega} \sim (\hbar\omega_K)^{-1}\ell n(\hbar\omega_K/\delta)D(\omega_K)\eta(\omega_K)M^2(\omega_K)|_{\omega_K=\omega-\omega'} \tag{3.17}$$

where $D(\omega_K)$ is the DOS, $\eta(\omega_K)$ is the Bose–Einstein function, and $M^2(\omega_K)$ is given by

$$M^2(\omega_K) = \int_0^{K_{\perp M}} \left(\frac{I^2(K_x)}{K^2}K_\perp dK_\perp\right) I^2(q)I^2(-q') \tag{3.18}$$

in which

$$I(q) \equiv \frac{2}{W}\int_0^W \sin^2\frac{n\pi x}{W}e^{iqx}dx,$$

and

$$I^2(K_x) = \frac{2\sin(K_xW/2)}{K_xW}\sin(K_xd/2)[1-(K_xW/2n\pi)^2]^{-1}.$$

For acoustic phonons, the factor K^2 in the denominator of Eq. (3.19) should be replaced by K in the numerator. The function $M^2(\omega_K)$ is plotted in Figure 3.12 for the case $d = 2W$. Note that there are various peaks corresponding to $K_x \sim (\pi/2W)$, $(3\pi/2W)$, etc. Since this resonance is due to the two-dimensional nature of the quantum states, it is possible to produce stronger resonance by applying a magnetic field along the superlattice direction creating true bound states. Our results indicate that it is possible to probe the acoustic and optical phonon branches for large momenta by sandwiching a given material between potential barriers. In fact, as far as Raman scattering is concerned, a perfect periodic structure is not necessary, as long as all well widths are kept reasonably constant. This is due to the relative broadness of the spectra. One might ask, what happens to the *umklap processes* for the regular crystal lattice? Since the reciprocal space for phonons and electrons coincide, all the *umklap processes* are redundant.

3.3.3.2 Surface Quantization

In an n-type semiconductor such as GaAs, a large DOS in the forbidden gap pins the Fermi level within the gap, causing the electronic energy bands to bend up toward the surface. The absorption length for 5145 Å in GaAs is only about a couple of tens of nanometers, so that the fraction of Raman scattering due to surface quantization from the usual coupling of plasmon-optical phonon is significant enough to be observable (Tsu et al., 1974). We shall calculate this effect by assuming that hole states are quantized at the surface with a ground state wave function given by Stern and Howard (1967),

$$|b\rangle \sim ze^{-\alpha z} u(k_{b\perp}, \rho) e^{ik_{b\perp\rho}} = u e^{ik_{b\perp\rho}} \int_{-\infty}^{\infty} \frac{e^{ik''z} dk''}{(k'' - i\alpha)^2}. \tag{3.19}$$

Many of the complicated steps were never published anywhere, so I shall present them here almost in full for those who want to see how it was done. From the first-order Raman cross section, $d\sigma/d\omega$ of Eq. (3.15), $\omega' = \omega - \omega_K$, where ω, ω', and ω_K refer to the incident and scattered photon and the phonon frequencies; and $E_1 = E_g + (\hbar^2/2m_e)k'^2 + (\hbar^2/2m_h)k^2$, $E_1 = E_g + E_b + (\hbar^2/2m_h)k^2$, where k and k' refer to electron in the valence band and conduction band, respectively. The H_R and H_p are

$$H_R = \frac{e}{mc} \left(\frac{2\pi\hbar}{\varepsilon \omega_q V} \right)^{1/2} \sum_j \left[a_q^- e^{i\mathbf{q}\cdot\mathbf{r}_j} e^{-i\omega_q t} + a_q^+ e^{-i\mathbf{q}\cdot\mathbf{r}_j} e^{+i\omega_q t} \right] \varepsilon_{\mathbf{q}} \cdot \mathbf{P_j}$$

where $\mathbf{P_j}$ is the momentum operator for the electrons and

$$H_p = \frac{e}{iK} \left(\frac{2\pi\hbar\omega_K}{V} \right)^{1/2} \left(\frac{1}{\varepsilon_\infty} - \frac{1}{\varepsilon_0} \right)^{1/2} \left[b_{\mathbf{K}} e^{i\mathbf{K}\cdot\mathbf{r}} + b_{\mathbf{K}}^+ e^{-i\mathbf{K}\cdot\mathbf{r}} \right].$$

In $\langle \mathbf{k}|H_R|\mathbf{k'}\rangle$, for sufficient layer thickness, $\sum_j e^{i(\mathbf{k}+\mathbf{q}-\mathbf{k'})\cdot\mathbf{R}_j} \to \partial(\mathbf{k} + \mathbf{q} - \mathbf{k'})$, and as usual for solids, the volume integral over the whole sample is converted to an integration in a unit cell and a delta function in k. In $\langle \mathbf{k'}|H_p|b\rangle$, momentum is only conserved in the transverse direction, the integration over the volume results in $\partial(\mathbf{k'_\perp} - \mathbf{K_\perp} - \mathbf{k_{b\perp}}) \int dz$, because bound state $E_b(z)$ is not a plane wave. The scattering via phonons from this ground state to a continuum state is represented by the matrix element, leaving out the factor $1/2\pi N$ for the moment,

$$\langle b|H_p|a\rangle \sim M(\mathbf{k}_b, \mathbf{k'}) \delta(\mathbf{k'_\perp} - \mathbf{K_\perp} - \mathbf{k_{b\perp}})/(k'_z - K_z - i\alpha)^2,$$

and similarly, the other two matrix elements in the Raman cross section are

$$\langle f|H_R|b\rangle \sim M(\mathbf{k'}, \mathbf{k}_b) \delta(\mathbf{k_{b\perp}} - \mathbf{k_\perp} - \mathbf{q'_\perp})/(\alpha + i(k_z + q'_z))^2$$

and

$$\langle a|H_R|i\rangle \sim M(\mathbf{k'}, \mathbf{k}) \delta(\mathbf{k} + \mathbf{q} - \mathbf{k'}).$$

The total sum is for a single bound state b,

$$\sum_{\mathbf{k},\mathbf{k'},\mathbf{k}_{b\perp},b} (\cdot) = \sum_{k_\perp, k_z, k_{b\perp}} (\cdot).$$

Note that the wave function in Eq. (3.19) destroys the momentum conservation in the first two matrix elements along the k_z direction. After summing over $\mathbf{k}'$ and $\mathbf{k}_{b\perp}$, taking only one bond state, neglecting the k-dependence of the M the scattering is proportional to

$$
\frac{d\sigma}{d\omega} \sim \int dk_z \int_0^{k_\perp M} \frac{k_\perp dk_\perp}{\left(k_\perp^2 - \frac{2\mu}{\hbar^2}(\hbar\omega - E_g) + k_z^2\right)\left(k_\perp^2 - \frac{2\mu}{\hbar^2}(\hbar\omega - E_g - E_b - \hbar\omega_K) + \frac{\mu}{m_h}k_z^2\right)}
$$
$$
\times (k_z + q_z - K_z + i\alpha)^{-2}(k_z + q_z' - i\alpha)^{-2},
$$

where $(k_\perp M)^2 = \frac{2\mu}{\hbar^2}(\hbar\omega - E_g - E_b - \hbar\omega_K)$, then

$$
\frac{d\sigma}{d\omega} = \int_{-\infty}^{\infty} \frac{\ell n\left(k_z^2 - \frac{2\mu}{\hbar^2}(\hbar\omega - E_g)\right) / \left(k_z^2 \frac{\mu}{m_h} - \frac{2\mu}{\hbar^2}(\hbar\omega - E_g - E_b - \hbar\omega_K)\right)}{(k_z - K_z - i\alpha)^2(k_z - i\alpha)^2\left(k_z^2 - \frac{2\mu}{\hbar^2}(E_b + \hbar\omega_K)(m_h/m_h - \mu)\right)} \tag{3.20}
$$

where E_b is the energy of the surface quantized state and $q_z \sim q_z' \sim 0$. In my worksheets, there are five pages of integration in the complex plane. Our society likes to refer to some people as bright and others as not so bright. Most often what we need is discipline to help us work through any complexities. Herring made this apt comment in the days when I was at BTL and he was looking into my work on the electron−phonon interaction in piezoelectric solids. No one but ourselves can check through complicated mathematical results. The poles in the third term in the denominator make no contribution. The double poles in the first and second terms make a small contribution and have a peak at $K_z = 0$, but the first-order pole in the first and second terms produces the spectrum that has a peak near $K_z = \alpha$. Since we are interested in highly doped cases where $\alpha \sim 2 \times 10^7$ cm^{-1} such that $\alpha^2 \gg (2\mu/\hbar^2)(E_b + \hbar\omega_K)$, there is no possibility of resonance. Then

$$
\frac{d\sigma}{d\omega} \sim \eta(\hbar\omega_K)I(\hbar\omega_K)D(\hbar\omega_K),
$$

in which $\hbar\omega_K = \omega - \omega'$, and

$$
I(\hbar\omega_K) \sim (K_z/\alpha)^2 / [(K_z/\alpha)^2 + 1]^3. \tag{3.21}
$$

We have used the dispersion relation for GaAs (Waugh and Dolling, 1963) to convert $I(K_Z)$ to $I(\hbar\omega_K)$ in order to compare the calculated Raman scattering involving a bound state with the experimentally observed Raman scattering in the depletion region of GaAs (Tsu et al., 1974). In the experiment, for doping of $n = 2.9 \times 10^{18}$ cm^{-3} and $\alpha = (12m^*e^2Nd/\varepsilon\hbar^2)^{1/3}$ from Stern and Howard (1967), $d = 20$ nm, and $\alpha = 2.1 \times 10^7$ cm^{-1}. In Figure 3.13, the measured data, shown as a dashed line, compares favorably with the calculated data with a peak at $K_z = \alpha$.

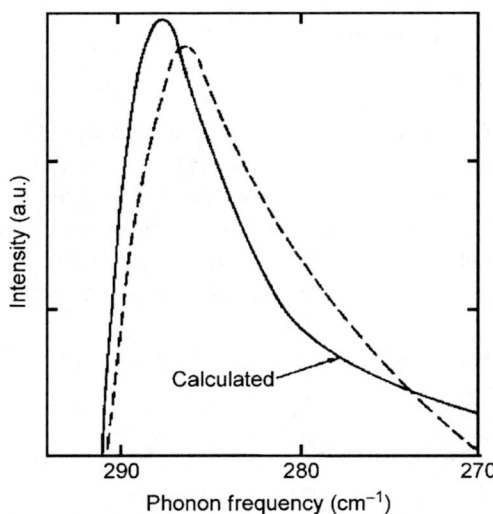

Figure 3.13 Raman scattering from highly doped $n = 2.9 \times 10^{18}$ cm^{-3} and (111) orientated GaAs. The dashed line shows the measured data. The calculated peak at $K_z = \alpha$ corresponds to phonon frequency of 286.5 cm^{-1}.

As we see that even localization is caused by a highly damped wave function of electrons—the Raman spectrum shares the same features as the localized state—only then the matrix element will be much reduced, resulting in smaller effects.

The two cases presented serve as a model for understanding Raman scattering where wave functions are localized. In normal bulk solids with continuous $E-k$ dispersion for the electrons, Raman scattering picks up the phonons at $q \sim 0$. However, with discrete electronic states, Raman spectrum picks up the q value of the phonons according to the discrete k value of the electrons. In the case of highly localized wave functions such as the surface quantized state as treated, or in the inversion region of the MOS capacitor, as well as in QWs and superlattices, the phonon involved is determined by the characteristic k. For the surface quantization treated, it is the parameter α. In QWs, it is the k of the discrete states. In a superlattice, it is determined by the zone folding of the phonon dispersion from the BZ into the MZ. As we have seen that folding transforms an acoustic phonon into an optical phonon, displaying the phonon dispersion in the MZ gives a series of peaks in the Raman spectrum as shown in Figure 3.12. One might ask, what happens to the *umklap processes* for the regular crystal lattice? Since the reciprocal space for phonons and electrons coincides, all the *umklap processes* are redundant. Note that in the theory we have completely ignored phonon localization because basically the difference in the elastic constants and the dielectric constants between GaAs and AlAs is simply not large enough to affect significant localization. Mathematically, we can predict that if the phonons are significantly localized, the Raman spectrum will exhibit even greater peaks owing to higher-order poles in the complex plane. But experimentally it is difficult to distinguish the cases with or without phonon localization. There have been several outstanding experiments proving the role of zone folding in Raman spectra, which we will take up next.

3.3.4 Experimental Confirmation of Zone Folding

3.3.4.1 Folded LA in Raman Scattering

The first experimental observation of OPFA modes was reported by Colvard et al. (1980). The samples consisted of 1720 periods of 1.36 nm GaAs/1.14 nm AlAs along the (100) direction. The acoustic phonons at $q = 2p/d$ at the center of the MZ were observed as the B_2 and A_1 modes of the D_{2d} group. These modes may be represented by the continuum elastic model first investigated by Rytov (1956). The dispersion relation for LA in a structure with thicknesses d_1 and d_2 is given by

$$\cos qd = \cos(\omega d_1/c_1)\cos(\omega d_2/c_2) - [(1+K^2)/2K]\sin(\omega d_1/c_1)\sin(\omega d_2/c_2),$$

$$(3.22)$$

in which $K = c_1\rho_1/c_2\rho_2$, where c_1, c_2 are the sound velocities for LA phonons along (100), and ρ_1, ρ_2 are the respective densities. In the case of thicker GaAs layers, $d_1 > d_2$, then $\omega(B_2) > \omega(A_1)$ and conversely, for $d_1 < d_2$, then $\omega(B_2) < \omega(A_1)$. Since the B_2 mode is odd under inversion, it is usually weaker than the symmetric A_1 mode. For $d_1 = d_2$, the Raman intensity tends to zero for even m, with m being an integer in the Bragg reflection, $m\lambda = d$. The second-order susceptibilities for the tetragonal D_{2d} symmetry applicable to the superlattice along (100), from Hayes and Loudon (1978), are

$$
\overset{A_1(Z)}{\begin{pmatrix} a & & \\ & a & \\ & & a \end{pmatrix}}
\overset{A_2}{\begin{pmatrix} & c & \\ -c & & \\ & & \end{pmatrix}}
\overset{B_1}{\begin{pmatrix} d & & \\ & -d & \\ & & \end{pmatrix}}
\overset{B_2(z)}{\begin{pmatrix} & e & \\ e & & \\ & & \end{pmatrix}}
\overset{E(y)}{\begin{pmatrix} & & f \\ & & \\ g & & \end{pmatrix}}
\overset{E(x)}{\begin{pmatrix} & & \\ & & f \\ & g & \end{pmatrix}}.
$$

It is simple to show that only the A_1 and B_2 modes are Raman active for back-scattering geometry, $z(x,x)\bar{z}$ for A_1 and $z(x,y)\bar{z}$ for B_2. In group theory, the irreducible representation of the polarizability tensor for the group D_{2d} must transform as xx or xy. Similarly, in infrared, they transform as x or y with two-dimensional E-modes which involve only transverse phonons. Figure 3.14 shows the measured Raman spectrum with sharp peaks near 65 cm^{-1}, the folded LA phonon modes. Dispersion of LA phonons is sketched in Figure 3.15, with an enlarged portion shown in (B) and the amplitude of the folded phonons shown in (C). AlAs layers are shown by shaded dots. Colvard et al. (1980) found that these phonon modes are quite localized in their respective layers using a simple calculation. This is not surprising, because optical phonons may be calculated with good results by two Keating force constants (Keating, 1966). What it means is that nearest neighbor coupling constants are all that is necessary. If that is the case, putting AlAs next to GaAs, which have different force constants without next nearest neighbor terms certainly will give localized modes in the respective layers. Therefore, I am not convinced by the arguments. Looking at the Raman spectrum that I calculated with zone folding, it would be very difficult to say anything at all about the degree of

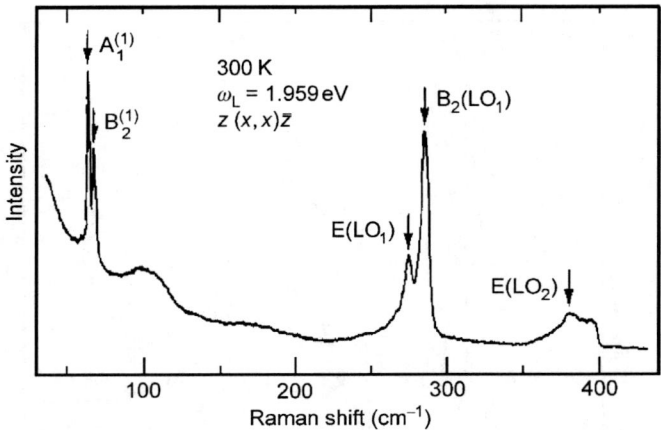

Figure 3.14 Raman spectra of a GaAs(1.36 nm)/AlAs(1.14 nm) SL.
Source: After Colvard et al. (1980), with permission.

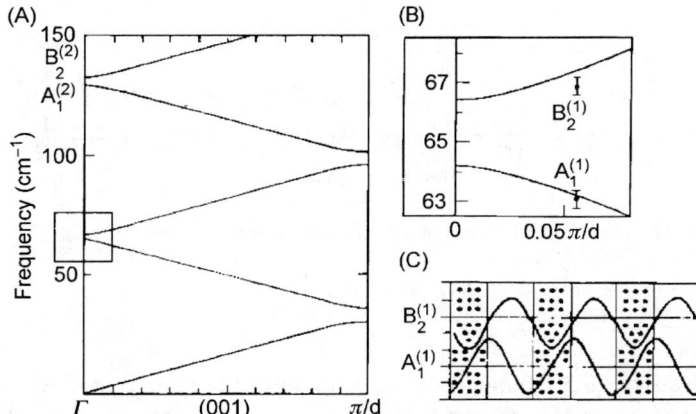

Figure 3.15 (A) Dispersion of LA phonons, (B) detail from (A), and (C) amplitude of the folded phonons. AlAs layers are shown by shaded dots.

localization of the phonons. It is simply due to the selection rule conserving the phonon q to the highly localized wave vector k of the electronic quantum structures.

3.3.4.2 Folded Optical Phonons in Raman Scattering

The first definite observation of OPFO in GaAs/GaAlAs superlattices was reported by Jusserand et al. (1984). For an Al molar fraction of 0.3, the Raman spectra of various samples in the $z(x, y)\bar{z}$ are shown in Figure 3.16. Using the notation (n, m) for $(GaAs)n-(GaAlAs)m$, the samples used in Figure 3.16 are characterized by the set: $S_2(6,4)$, $S_5(9,9)$, $S_7(12,7)$, and $S_9(17,12)$. The results agree with the calculation

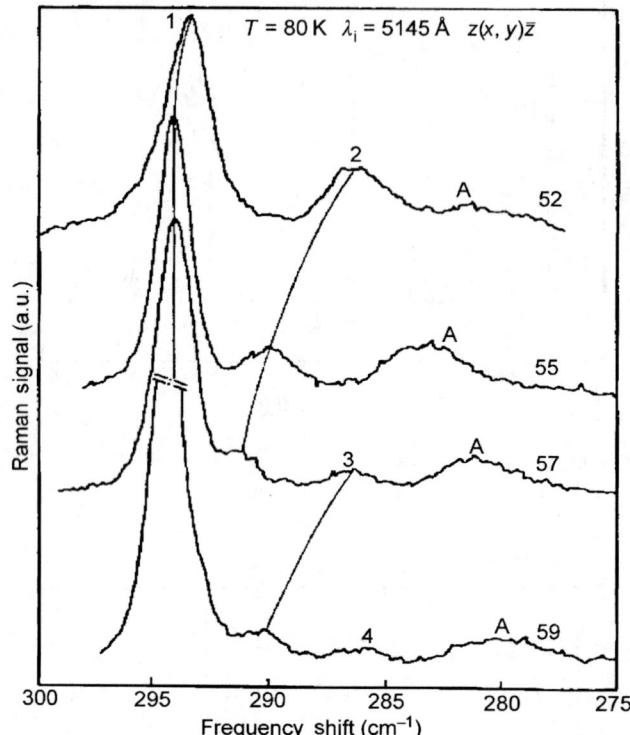

Figure 3.16 The folded LO for four samples S_1–S_4 clearly demonstrates the theoretical calculation in Section 3.3.3.
Source: From Jusserand et al. (1984), with permission.

given in Section 3.3.3, although the calculation does not employ the proper Raman tensor for D_{2d} symmetry so that there is no distinction between the polarization of the incoming and scattered light.

The Raman spectra for a 400 period GaAs(2 nm)/AlAs(6 nm) superlattice obtained by Sood et al. (1985), with both the folded LO and TO, are shown in Figure 3.17A and B. The series of peaks, labeled LO_m, corresponding to LO phonons of A_1 with even m, and B_2 with odd m for the D_{2d} point group, for two different laser frequencies, are the best experimental proof of the interaction of electrons in a superlattice with the folded phonons in the mini-BZ. In Figure 3.18, five TO peaks show up superimposed on a strong background, at an incident photon energy of ~1.9 eV. Since LO phonons with a long-range Coulomb interaction are more affected by Landau damping (Tsu, 1967), qualitatively, one should expect that TO at higher q will show up more easily than LO in Raman scattering. This explanation was not given by Sood et al. in their original publication. The strong background appears to be due to the polarization used. It is impressive that a mapping of both the LO and TO phonons of GaAs using only $\omega(q = m\pi/d)$, with d being the

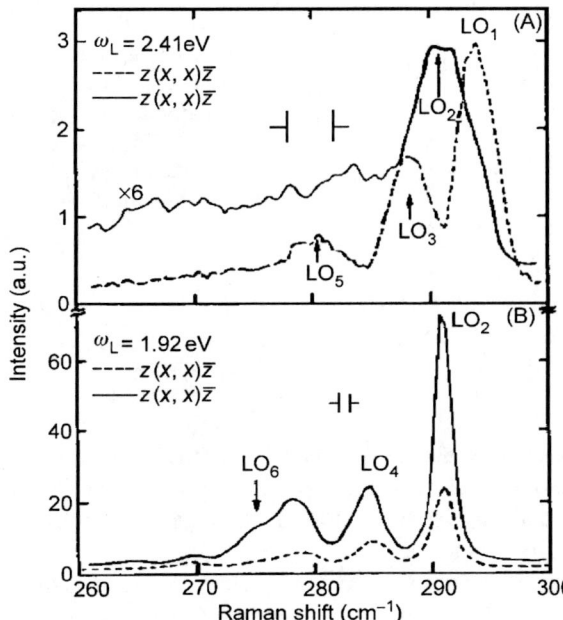

Figure 3.17 Off-resonance Raman spectra (A), and near resonance (B), showing folded LO phonons.
Source: After Sood et al. (1985), with permission.

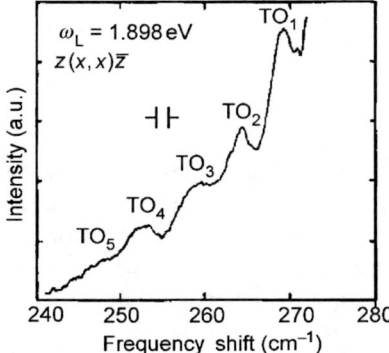

Figure 3.18 Raman spectra of folded TO phonon, showing five distinct peaks. The high background is due to the use of polarization $z(x, x)\bar{z}$.
Source: After Sood et al. (1985), with permission.

thickness of the GaAs layer, agrees so well with the data known from neutron scattering usually requiring a very thick sample.

3.3.5   Raman Scattering from a Strain-Layer Superlattice (SLS)

The use of Raman scattering to determine deformation in Si/GeSi (SLS) was reported by Cerdeira et al. (1984). According to Matthews and Blakeslee (1976), perfection may be enhanced by choosing film thicknesses below that at which misfit dislocations are formed when interfacial strain is high. These concepts are

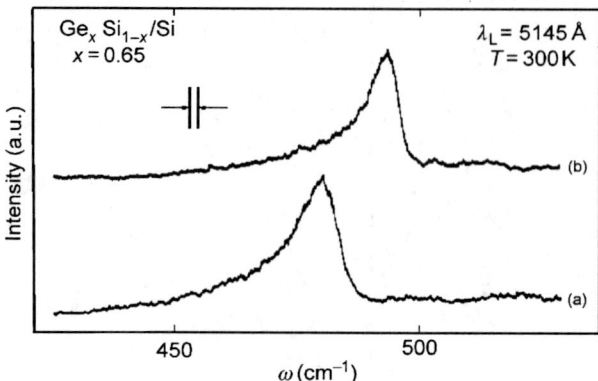

Figure 3.19 (a) $Ge_{0.65}Si_{0.35}$/ Si strain-layer superlattice curve (b) is upshifted from an alloy of the same composition, $Ge_{0.65}Si_{0.35}$. *Source*: After Jusserand et al. (1985), with permission.

incorporated in what is known as the strain-layer superlattice (SLS). In fact SLS has greatly broadened the choice of materials forming perfect epitaxial heterostructures for quantum confinement and superlattices. Figure 3.19 shows the Si–Ge peak in the alloy spectrum for (a) a single incoherent layer, thick enough that the stored strain energy from a lattice mismatch is much greater than what is required to create dislocations, and (b) an SLS. Note that the spectrum for (b) is narrower and strain shifted. Using the appropriate Gruneisen constants, Cerdeira et al. estimated the interlayer strain. With $Ge_{0.65}Si_{0.35}$, the strain is 2%. For GaSb/AlSb, a strain of 0.2% was measured by Jusserand et al. (1985).

3.4 Summary

To measure the optical absorption without removing the back substrate, we resort to a technique that takes the reflectivity at an angle in addition to the usual normal incidence, resulting in the determination of both refractive index and absorption coefficient. This technique is not new, but is rarely described in any books on optical measurements. This simple example again illustrates why one should think about the problem without rushing to look up the usual reported work. In fact the reviewer suggested to me that our method should constitute a separate paper for the *Journal of the Optical Society*. At the time we were too busy with new work, but more precisely, published work from BTL using the wedge shape to eliminate fringes clearly offers a superior measured exciton line for the QW structures, as shown in Figure 3.7. Again, because we did not remove the substrate, photoconductivity offers a good way to identify the miniband states of a superlattice experimentally as shown in Figure 3.8. In 1971, Jha and I calculated the zone-folding effects of a polariton and phonons in a superlattice using a linear chain model, again illustrating how a complicated problem may be simply modeled. Obviously, a more rigorous calculation would have been very complex. In 1975, Esaki and I published the first Raman spectrum of a superlattice as well as the spectrum of an isolated quantum state. The key to the success of the calculation involves the use of complex integration in terms of the poles, mostly double poles. This is a good illustration of how useful it is to have some working knowledge of a complex variable. I was very much influenced by my teacher, Walter Kohn, who was my advisor for little more than a couple of months in 1956. He told me that the single most important tool for a physicist is a working knowledge of complex variables. The first experimental confirmation of phonon zone folding was obtained by Colvard et al. (1980) and later by Sood et al. (1985). The Raman spectrum is capable of determining strain using the Gruneisen constant relating strain to frequency shift for Si/SiGe (Cerdeira et al., 1984; Jusserand et al., 1985). When I was learning about Raman scattering, I discovered that it is a very difficult subject because of the involvement of group theory and fairly complicated solid-state theory. However, experimentally one only needs to acquire some working knowledge, such as looking up the Raman tensors for a given crystal structure, and some understanding of crystal orientation with respect to the polarization of the incident and scattered photons. However, if spin flip is involved, some understanding of the double group, a much more complex subject, is needed. Raman scattering involving spins is so complex that it usually takes a year to master.

References

Barker, A.S., Merz, J.L., Gossard, A.C., 1978. Phys. Rev. B 17, 3181.

Capasso, F., Mohammed, K., Cho, A.Y., 1985. Physica. B & C 134B, 487.

Cerdeira, F., Pinczuk, A., Bean, J.C., Batlogy, B., Wilson, B.A., 1984. Appl. Phys. Lett. 45, 1138.

Colvard, C., Merlin, R., Klein, M.V., Gossard, A.C., 1980. Phys. Rev. Lett. 45, 298.

Dingle, R., Wiegmann, W., Henry, C.H., 1974. Phys. Rev. Lett. 33, 827.

Döhle, G.H., Tsu, R., Esaki, L., 1975. Solid State Commun. 17, 317.

Dolling, G., Waugh, J.L.T., 1963. In: Wallis, R.F. (Ed.), Lattice Dynamics. Pergamon, New York.

Esaki, L., Chang, L.L., 1974. Phys. Rev. Lett. 33, 495.

Esaki, L., Tsu, R., 1970. IBM Res. Dev. 14, 61.

Hayes, H., Loudon, R., 1978. Scattering of Light by Crystals, Wiley, New York.

Jusserand, B., Paquet, D., Regreny, A., 1984. Phys. Rev. B 30, 6245.

Jusserand, B., Voisin, P., Voos, M., Chang, L.L., Mendez, E.E., Esaki, L., 1985. Appl. Phys. Lett. 46, 678.

Kazarinov, R.F., Shmartsev, Yu.V., 1971. Sov. Phys. Semicond. 5, 710.

Keating, P.N., 1966. Phys. Rev. 145, 637.

Matthews, J.W., Blakeslee, A.E., 1976. J. Cryst. Growth 32, 265.

Mayer, J.W, Ziegler, J.F., Chang, L.L., Tsu, R., Esaki, L., 1973. J. Appl. Phys. 44, 2322.

Merlin, R., Colvard, C., Klein, M.V., Morkoc, H., Cho, A.Y., Gossard, A.C., 1980. Appl. Phys. Lett. 36, 43.

Onton, A., 1970. Proceedings of the 10th International Conference of Semiconductors Cambridge, U.S., Atomic Energy Comm., Washington, DC, p. 107.

Rytov, S.M., 1956. Sov. Phys. Acoust. 2, 68.

Sai-Halasz, G.A., Esaki, L., Harrison, W.A., 1978. Phys. Rev. B 18, 2812.

Sood, A.K., Menendez, J., Cardona, M., Ploog, K., 1985. Phys. Rev. Lett. 54, 2111.

Stern, F., Howard, W.E., 1967. Phys. Rev. 163, 816.

Tsu, R., 1967. Phys. Rev. 164, 380.

Tsu, R., 1981. Proc. SPIE 276, 78.

Tsu, R., Kawamura, H., Esaki, L., 1972. Int. Conf. Semi. Wlarsaw, 16, 1136.

Tsu, R., Döhle, G.H., 1975. Phys. Rev. B 12, 680.

Tsu, R., Esaki, L., 1975. In: Balkanski, M., Leite, R.C.C., Porto, S.P.S. (Eds.), Proceedings
    of the Third Light Scattering in Solids, Campinas. Wiley, New York, p. 533.

Tsu, R., Jha, S.S., 1972. Appl. Phys. Lett. 20, 1.

Tsu, R., Chang, L.L., Sai Halasz, G.A., Esaki, L., 1975a. Phys. Rev. Lett. 34, 1509.

Tsu, R., Kawamura, H., Esaki, L., 1974. Solid State Commun. 15, 321.

Tsu, R., Koma, A., Esaki, L., 1975b. J. Appl. Phys. Rev. Lett. 33, 827.

Verleur, H.W., Barker, A.S., 1966. Phys. Rev. 149, 715.

Waugh, J.L.T., Dolling, G., 1963. Phys. Rev. 132, 2416.

4 Dielectric Function and Doping of a Superlattice

The dielectric constant represents the screening of the applied electric field in a medium owing to the presence of charges and dipoles, and so on. Usually, it is considered in the random phase approximation in Lindhart's expression (Ziman, 1988) in terms of a sum over all the transitions due to an applied electric field, which is simply a statement that the total effect is a superposition of individual responses from individual transitions. What happens if two transitions are coupled? As long as it is possible to transform the coupled modes into the individual responses of the quasi particles, then we may simply sum all of the responses. Normally we separate the modes into various ranges in energy. In free molecules, the lowest term comes from the rotation of the molecule, which is not present in solids. In solids, the most important contribution comes from the optical response to an applied electric field that results from transitions between the valence band and the conduction band, involving the product of two terms, the matrix element of transition and the joint density of states. The matrix element has an energy denominator; the higher is the band gap of the solids and the lower is the dielectric function. Therefore, if the energy state is raised owing to quantum confinement, the dielectric function is reduced accordingly. As we have seen from Chapter 3, the dielectric constant enters into a multitude of physical situations, ranging from phonon dispersions to electron–phonon interactions. Often it is good enough to consider the bulk values in a situation, like the relation between the LO and TO phonons, in the Laddane–Sachs–Teller relation. However, in doping semiconductors, often the most important features in the operation of electronic devices, the binding energy of dopants is inversely proportional to the square of the dielectric constant, a major effect to be accounted for in three-dimensional quantum confinement which will be treated later. For a superlattice and a quantum well, quantum confinement is only along the layer structure. The major difference is not the overall screening, rather it is due to the joint density of states of the two-dimensional system. This aspect will be treated here.

4.1 Dielectric Function of a Superlattice and a Quantum Well

4.1.1 Longitudinal Dielectric Constants for Quantum Wells

The transverse dielectric constant of the $Ga_{0.5}Al_{0.5}As/GaAs$ superlattice has been measured by Tsu et al. (1975), showing that the refractive index lies between the

Superlattice to Nanoelectronics. DOI: 10.1016/B978-0-08-096813-1.00004-7

$Ga_{0.5}Al_{0.5}As$ alloy and GaAs. The dielectric constant is important for impurity levels, excitons, and carrier screening in general. For polar semiconductors like GaAs, the ionic contribution to the static dielectric constant represents a significant factor compared with the part contributed by its covalent nature. The present treatment follows closely the calculation by Tsu and Ioriatti (1985). Since the ionic part involves charge transfer between the Ga and As atoms, as treated by Ilegems and Pearson (1966), it is assumed that the quantum well plays a negligible role in the ionic part. For a typical quantum well of width $w = 5$ nm, due to the upshift of the energy, the dielectric constant is expected to be reduced. The matrix element between the initial and final states $\langle i|\mathbf{A} \cdot \mathbf{P}|f\rangle$ is almost a constant in comparison to the Van Hove singularity in the joint density of states at the Γ, X, and L points of the Brillouin zone (BZ), where the valence band runs almost parallel to the conduction band along the Λ and Δ directions. Assuming that the transitions near the three points of the BZ, Γ, X, and L, are additive, we calculate the difference between the unbound bulk and the bound case, the quantum well. Under the application of a perturbing electric field F in the direction of the quantum well, the wave function

$$\psi_{nk} = |nk\rangle + \sum_{nk} \frac{\langle nk|eFx|n'k'\rangle}{E_{nk} - E_{n'k'}} |n'k'\rangle, \tag{4.1}$$

where $\langle nk|x|n'k'\rangle = \langle nk|x|n'k\rangle \partial_{kk'}$, the Fermi function $f_{vk} = 1$, $f_{ck} = 0$, $n = v$, and $n' = c$, then

$$\varepsilon = \varepsilon_{\text{ionic}} + \frac{8\pi e^2 \hbar^2}{m_0^2} \sum_k \frac{P_{vc}^2}{[E_c(k) - E_v(k)^3]}. \tag{4.2}$$

We replaced the part in Eq. (4.2) that is due to the bulk by the corresponding quantum well part. Taking E_b, the energy at the barrier, as the upper limit of energy, the part to be subtracted is

$$\varepsilon_B(\Gamma) = \frac{8e^2 P^2}{\pi^2} (\mu_\Gamma/m_0)^{3/2} (2m_0 E_g/\hbar^2)^{1/2} (E_g)^{-2} \int_1^{E_b/E_g} \frac{(x-1)^{1/2}}{x^3} \, dx. \tag{4.3}$$

For a transition involving the light hole band to the conduction band at the Γ point, μ_Γ is the reduced mass for the light hole mass in the valence band and the conduction band. At the point L, μ_L is used; however, we shall see that P^2 in Eq. (4.3) are different at each point Γ, X, and L, as well as the two polarizations ($\parallel$, $\perp$). Similarly, for the bound states with wave functions for the quantum well:

$$\psi_{cv} = U_{cv}(r, k) \sin(m\pi/w) \exp(ik_t \rho), \tag{4.4}$$

with the matrix element $P_{mm'} = P_{mm}\partial_{mm'}\partial_{k\perp,k'\perp}$, we arrive at the term to replace $\varepsilon_B(\Gamma)$,

$$\varepsilon_{\parallel}^{\Gamma} = \sum_n \frac{2e^2}{w} \frac{|P_{mm}^{\parallel}|^2}{m_0} (\mu_{\Gamma}/m_0) \frac{(E_g/E_{mm})^2 - (E_g/E_b)^2}{E_g^2}, \qquad (4.5)$$

where $P_{mm}^{\parallel}$ is the momentum matrix with the electric field transverse to the well. For the electric field $\perp$ to the plane of the quantum well, $P_{mm}^{\parallel}$ is substituted by $P_{mm}^{\perp}$ and the total dielectric constant at Γ, X, and L becomes

$$\varepsilon_{\parallel,\perp} = \varepsilon - (\varepsilon_B^{\Gamma} + \varepsilon_B^{L} + \varepsilon_B^{x}) + (\varepsilon_{\parallel,\perp}^{\Gamma} + \varepsilon_{\parallel,\perp}^{L} + \varepsilon_{\parallel,\perp}^{X}). \qquad (4.6)$$

Let us go over what we did. Instead of calculating the dielectric constant we used the effective mass, which is usually only valid near these high symmetry points. Therefore, we subtracted the bulk contribution and replaced this with the quantum well expression, expressed by Eq. (4.6). Figure 4.1 shows the band structure of GaAs (solid) and $Al_{0.35}Ga_{0.65}As$ (dashed). We used 60% of the difference in the two band gaps E_g^{Γ} and E_b^{Γ}, and the band-edge offset at Γ is $0.6(3.31-3.3)$, according to Mendez et al. (1981). At point X, $E_g^X \sim E_b^X$, so that the offset is nearly zero as shown in Figure 4.1.

With the density of state mass at the point L $m_L^{3/2} = N_L \mu_t m_l^{1/2}$, where $\mu_t^{-1} = m_{ct}^{-1} + m_{vt}^{-1}$ in which $m_{ct} = 0.075 m_0$ and $m_{vt} = 0.2 m_0$; $N_L = 4$ and $m_{c\ell} = m_{v\ell} = m_{\ell} = 1.9 m_0$,

$$\varepsilon_B(L) = \frac{4e^2 |p|^2 m_l}{\pi m_0^2} (2m_L E_g/\hbar^2)^{1/2} \frac{(y+1)(y-1)^{3/2}}{y^2 E_g^2}, \qquad (4.7)$$

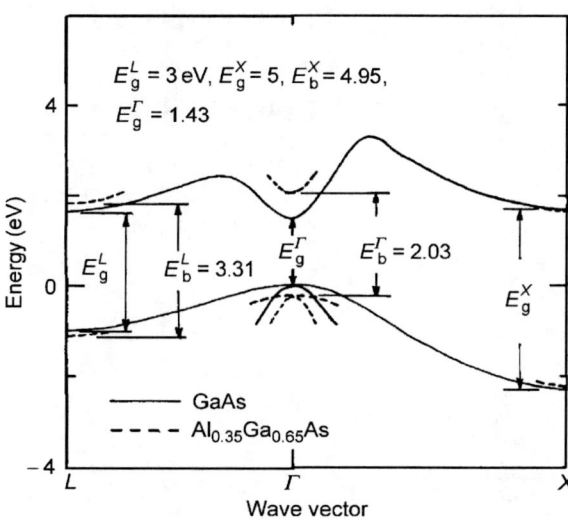

Figure 4.1 Energy band structure of bulk GaAs and $Al_{0.35}Ga_{0.65}As$ at Γ, L, and X. *Source*: After Tsu and Ioriatti (1985), with permission.

$E_g^L = 3\,\text{eV}, E_g^X = 5, E_b^X = 4.95,$
$E_g^{\Gamma} = 1.43$

E_g^L $E_b^L = 3.31$ E_g^{Γ} $E_b^{\Gamma} = 2.03$ E_g^X

Energy (eV)

—— GaAs
- - - - $Al_{0.35}Ga_{0.65}As$

L Γ X

Wave vector

where $y = E_b^L - E_g^L$. What are the values for the momentum matrix at Γ, X, and L? Let us first list the group designation of these points for the various bands in GaAs.

L	Γ	X	
L_6	Γ_6	X_7	Conduction band
$L_{4,5}$	Γ_8	X_6	Valence band
L_6	Γ_7	X_7	

According to Pollak and Cardona (1966), the momentum matrix for various cases listed as follows should be used.

At Γ: $\quad \varepsilon_\parallel^\Gamma(hh) - |P|^2 \qquad$ At L: $\quad \varepsilon_\parallel^L - |P|^2(4/3)$

$\qquad \varepsilon_\parallel^\Gamma(\ell h) - |P|^2(1/3)$

$\qquad \varepsilon_\perp^\Gamma(\ell h) - |P|^2(2/3) \qquad\qquad \varepsilon_\perp^L - |P|^2(2/3) \; |P|^2/m_0 = 7.5 \text{ eV}$

The calculated dielectric constant versus the well width is shown in Figure 4.2.

Below a well width of 2 nm, the quantum states are squeezed out of the well for the barrier height involved, giving a zero contribution. As the width is increased to beyond 7 nm, a transition involving the second quantum state gives rise to an additional contribution, resulting in a sudden rise. The states near the top of the barrier are neither two-dimensional nor three-dimensional in nature. They have not been taken into account in this calculation, contributing to the poor convergence toward the bulk value. As pointed out by Frank Stern, if a better procedure for incorporating the sum rule is used in the process of removal and insertion, a better convergence may result. Thus, this method does not apply to very high barriers. Nonetheless, our results should be applicable in a limited energy range near each quantum state. Let us summarize the general trend. The breakdown of symmetry results in an increase in the case of parallel polarization over perpendicular polarization. The

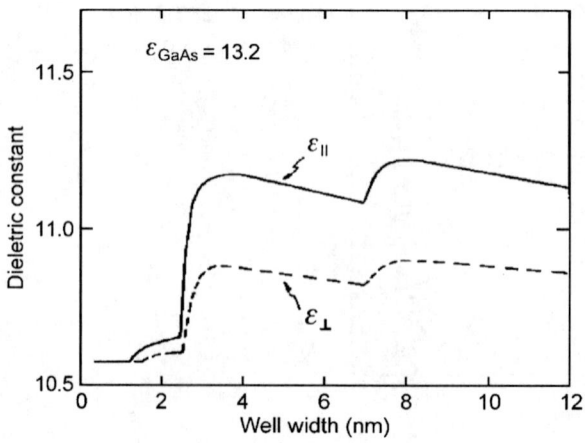

Figure 4.2 Calculated longitudinal dielectric constant $\varepsilon_\parallel$ and $\varepsilon_\perp$ versus well width. *Source*: After Tsu and Ioriatti (1985), with permission.

static dielectric constant approaches the value for the GaAs bulk for large well width and shows a 20% reduction for $w \leq 2$ nm. It is important to account for this reduction when dealing with impurity states and excitons.

4.1.2 Transverse Dielectric Constant of the GaAs/AlAs Superlattice

The calculation in this section is based on that of Kahen et al. (1985), where the individual contributions from the Γ, X, and L valleys are treated separately. Since confinement is in GaAs, an increase of the energy states lowers the dielectric constant, a general rule that applies to all cases.

The notation $e\ell h(1)$ refers to the transition from the light hole to the first superlattice band in the conduction band. The spin−orbit split off to the first superlattice band is marked as $eso(1)$. Note that for parallel polarization, the transverse dielectric constant is above that of the alloy. Since the electrons are basically confined in the GaAs well region, comparison to the alloy is not as meaningful as comparison to GaAs. It is clear that the dielectric constant is decreased owing to upshift of the energy state from confinement in the GaAs layer (see Fig. 4.3).

4.2 Doping a Superlattice

When the energy states in a quantum confined system are pushed up, the reduction in the dielectric constant increases the binding energy of the dopants. Since the impurity sites intermingle with the barriers, which confine the electrons, the binding

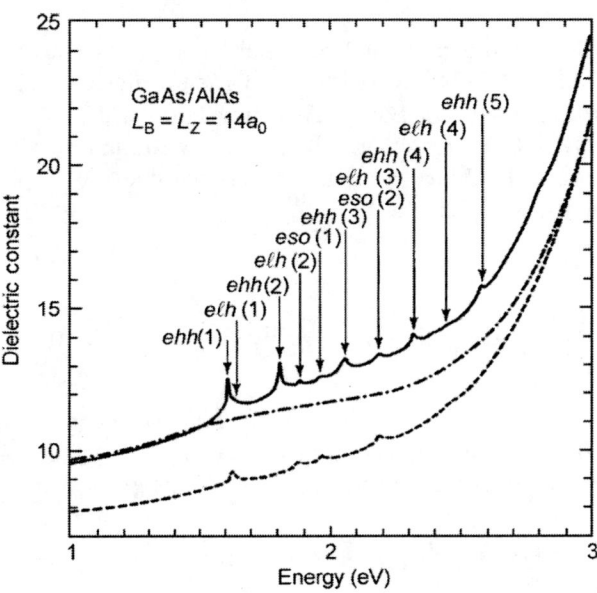

Figure 4.3 The calculated transverse dielectric constant for 158 periods of the GaAs/AlGaAs superlattice with a barrier width L_B and well width L_Z, with a_0 being the lattice constant. The solid and dashed curves depict polarization of the electric field parallel to and perpendicular to the superlattice layers, respectively. Arrows mark the Γ point transitions. The dash−dot curve represents the AlGaAs alloy.
Source: After Kahen et al. (1985), with permission.

energy is position dependent. This was first treated by Bastard (1981), using infinite barriers and was later extended to finite barriers by Mailhiot et al. (1982). Usually one starts with a trial wave function consisting of a bulk hydrogenic wave function modulated by a function depending only on the z coordinate, the axis along the potential variation, used in a variational procedure. Ioriatti and Tsu (1986) used a different approach, leaving the z-dependent function as an unknown. Upon integration over the transverse coordinates, the minimization of the total effective Hamiltonian leads to a well-known differential equation, Whittaker's equation, which is solved for any square-well potentials with arbitrary impurity locations. For an isolated quantum well, this approach leads to the same results found by Mailhiot et al. (1982) and Greene and Bajaj (1983). For a superlattice, where the width of the miniband is in the order of or larger than the binding energy, our results are substantially different from the isolated well result owing to the spread of the donor envelop function into neighboring wells. This effect tends to dominate to the point of reversing the trend, that is, there is an increase in binding energy with increasing well width. Basically, for narrow barriers, because of the lack of confinement, the mini-band states are more 3-D contrasting the 2-D nature of the impurity level in a quantum well.

Let the impurity be located at z_i in a GaAs/Al$_x$Ga$_{1-x}$As superlattice. The Hamiltonian may be written as

$$H = P^2/2m^* - (e^2/\varepsilon)[\rho^2 + (z - z_i)^2]^{-1/2} + V_{SL}(z), \tag{4.8}$$

where

$$V_{SL}(z) = \begin{array}{ll} 0, & |z - n(a+b)| < a/2 \\ V_0, & |z - (n + (1/2))(a+b)| < b/2 \end{array} \tag{4.9}$$

in which a and b are the widths of the wells and potential barriers, respectively, and V_0 is the barrier height. In Eq. (4.8) $\rho = (x^2 + y^2)^{1/2}$ and ε is the dielectric constant (if MKS units were used, $\varepsilon \to 4\pi\varepsilon_r\varepsilon_0$). For an Al composition with $x < 0.3$, we can simply use the same $m^* = 0.067m_0$ for both barrier and well, and for simplicity we further assume that $\varepsilon = 13$. Since a separable variable does not apply, the trial function for the ground state is taken as

$$\psi(\rho, z) = R(\rho, z)\phi(z), \tag{4.10}$$

$$R(\rho, z, \lambda) = \frac{2}{\lambda} \frac{\exp(|z - z_i|/\lambda)}{(1 + |z - z_i|/\lambda)^{1/2}} \exp\left\{-[(z - z_i)^2 + \rho^2]^{1/2}/\lambda\right\} \tag{4.11}$$

and $\phi(z)$ should be determined variationally. R is the 1s hydrogenic state containing a variational parameter λ, and is normalized. The ground state energy $E = \langle\psi|H\psi\rangle$ can be obtained by first integrating over the transverse variable. This gives

$$E = \int_{-\infty}^{\infty} dz\{\phi^*(-\hbar^2/2m^*)d^2\phi/dz^2 + \phi^*\phi[V_{SL}(z) + V_1(z)]\}, \tag{4.12}$$

in which

$$V_1(z) = -\frac{\hbar^2}{m^*\lambda^2}\left[\frac{(2\lambda/a_B^*) - 1}{1 + 2|z - z_i|/\lambda} + \frac{1/2}{(1 + 2|z - z_i|/\lambda)^2}\right] \tag{4.13}$$

is the average effective impurity potential after integrating out the transverse variable, and a_B^* is the effective Bohr radius, given in MKS unit by $a_B^* = 4\pi\hbar^2\varepsilon_r\varepsilon_0/m^*e^2$. Minimizing E with respect to ϕ for a given λ, Eq. (4.12) leads to a differential equation for ϕ. Making the following substitutions:

$$\beta^2 = 2m^*(V_0\zeta - E)/\hbar^2$$
$$K = [(2\lambda/a_B^*) - 1]/2\beta\lambda$$
$$x = \beta\lambda(1 + |z - z_i|/\lambda)$$

with $\zeta = 0$ or 1, for z in the well or barrier regions, respectively, we obtained the well-known Whittaker equation,

$$\frac{d^2\phi}{dx^2} + \left(\frac{-1}{4} + \frac{1}{4x^2} + \frac{K}{x}\right)\phi = 0. \tag{4.14}$$

The two linearly independent solutions are $M_{k,0}(x)$ and $W_{K,0}(x)$, the regular and irregular Whittaker functions (Whittaker, 1968). For $E > 0$, ϕ behaves like the exponential and trigonometric functions for $\zeta = 1$ or 0, respectively. The power law dependence in the asymptotic forms of the Whittaker functions reflects the long-range Coulomb term (Merzbacher, 1970). The boundary conditions, the continuity of ϕ and $d\phi/dx$ across the boundary and at the donor sites, determine E as a function of λ.

When $V_{SL} = 0$, the continuity of ϕ and ϕ' at the donor site and $\phi = 0$ at infinity leads to $\beta\lambda = 1$ and $K = 0.5$, giving $\lambda = a_B^*$ and $E = -\hbar^2/2m^*(a_B^*)^2$, the solution for the hydrogen atom. For a hydrogenic center located in the plane of an infinite potential and the boundary conditions $\phi = 0$ at $z = z_i$, ∞; then $K = 3/2$ and $\beta\lambda = 1$, giving $\lambda = 2a_B^*$, so that $E = -\hbar^2/8m^*(a_B^*)^2$, which is exact. Finally, situating an impurity between two hard walls separated by an infinitesimal distance, setting $z = z_i$ in Eq. (4.13), and minimizing the resulting expression with respect to λ, we obtain the exact solution for the two-dimensional hydrogen atom (Levine, 1965).

We will elaborate further for a superlattice. We have taken an effective range for the potential $V_1(z)$ as $2L$, defined by $|z - z_i|$, which is several a_B^*. Outside this effective range the potential is represented by the square-well Krönig–Penney (KP) model. Inside this range, we apply at each interface (the impurity site and the right and left termination points are also considered as interfaces) a two-component vector composed of the value of the wave function and its first derivative at the corresponding interface. The two linearly independent solutions of Eq. (4.14) are then used to relate vectors associated with two consecutive interfaces. A transfer matrix is thus associated unambiguously with each pair of consecutive interfaces. Continuous boundary conditions for both ϕ and ϕ' across all the interfaces allow us to represent the propagation of the two-component vector associated with the left termination point with that of the right termination point, as the matrix product of a succession of such matrices. It follows that

$$\begin{pmatrix} \phi_R \\ \phi_R' \end{pmatrix} = M(\text{Whittaker}) \begin{pmatrix} \phi_L \\ \phi_L' \end{pmatrix},$$

where the subscripts R and L denote the right and left termination points. Matching the logarithmic derivative ϕ_L'/ϕ_L to that for the KP solution decaying to the left, and the logarithmic derivative ϕ_R'/ϕ_R to that for the KP solution decaying to the right leads to a solution for E as a function of λ. The minimum E obtained by variation of λ leads to the ground state of the superlattice with the impurity. The difference between this value and the lowest E for the KP solution is our ground state binding energy of the donor impurity.

As shown in Figure 4.4, the binding energy of dopant or impurity in units of Rydberg, $Ry^* = \hbar^2/2m^*(a_B^*)^2$ is plotted versus the barrier width normalized to the Bohr radius for various normalized well widths. For GaAs, the Bohr radius is increased from 0.053 to 9.843 nm by the factor $\varepsilon_r/m^* \sim 10$ nm and the Rydberg is reduced from 13.6 eV for a H atom to 5.6 meV. Note that for a barrier width of thickness $b/a_B^* > 1$, the solution approaches that for the isolated quantum well, where E_b increases with decreasing well width. Our results for the quantum well are within 1% of those obtained numerically (Mailhiot et al., 1982; Greene and Bajaj, 1983). Owing to the spread of the wave function to neighboring wells with thin barriers, the reverse happens and E_b decreases with the decrease in well width.

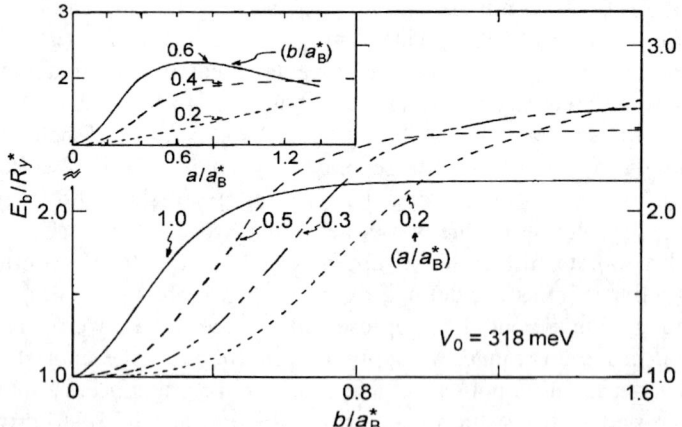

Figure 4.4 Binding energy E_b in units of Ry^* (5.6 meV) versus the barrier width b and various well widths a, normalized to a_B^*. Inset shows E_b versus well width for several barrier widths. Only at large b does E_b increase with a decrease of well width.
Source: After Ioriatti and Tsu (1986), with permission.

Our solution does converge to the H atom in three dimensions as the barrier width diminishes. In the inset of Figure 4.4, the binding energy is plotted versus the well width for several barrier widths. For $b \sim 2$ nm, E_b increases with the increase in the well width and at $b \sim 6$ nm, there is a peak for E_b at $a \sim 5-6$ nm. Beyond this point, E_b decreases with an increase of well width like the case for an isolated quantum well. I thought I could predict almost everything that happens to the superlattice and quantum well. When Ioriatti showed me the computed results, I thought that there was an error until repeated checks—even going as far as repeating the computer program—gave the same results and we realized the true meaning of the band states. When we use such textbook approximations as the particle in the box, we must realize that the validity of these simple models must be carefully examined. In particular, *ab initio* calculations cannot be trusted without repeated tests.

4.3   Summary

The dielectric constant of a superlattice has been calculated. The value is about 20% lower for a typical superlattice. The economy presented to the problem of shallow donor states through the use of the Whittaker functions is obvious. Even without treating the superlattice case using the transfer matrix and Krönig–Penney model, the question of mixing of closely spaced quantum well states is automatically taken care of by the variational approach. What is satisfying is that we discover this powerful method rather by chance in the sense that we did not envision the way it turned out. One wonders why we want to bother with the binding energy of shallow dopants in a superlattice. After all, quantum well doping is detrimental in most cases. To answer this, we must simply look at the difference between an isolated molecule and a polymer chain. We dope the chain to acquire transport; similarly, we dope the superlattice to acquire current transport. In fact, in quantum cascade lasers for a superlattice to serve simply as a wire with the capability of matching the output from the nth section to the input of the $(n + 1)$th section, it must be doped.

References

Bastard, G., 1981. Phys. Rev. B 24, 4714.
Greene, R.L., Bajaj, K.K., 1983. Solid State Commun. 45, 825.
Ilegems, M., Pearson, G., 1966. Phys. Rev. 149, 715.
Ioriatti, L., Tsu, R., 1986. Surface Sci. 174, 420.
Kahen, K.B., Leburton, J.P., Hess, K., 1985. Superlattices Microstruct. 1, 289.
Levine, J.D., 1965. Phys. Rev. 140, A586.
Mailhiot, C., Chang, Y.C., McGill, T.C., 1982. Phys. Rev. B 26, 4449.
Mendez, E.E., Chang, L.L., Landgren, G., Ludeka, R., Esaki, L., Pollak, F., 1981. Phys. Rev. Lett. 46, 1230.
Merzbacher, E., 1970. Quantum Mechanics, Wiley, New York.
Pollak, F., Cardona, M., 1966. J. Phys. Chem. Solids 27, 423.

Tsu, R., Ioriatti, L., 1985. Superlattices Microstruct. 1, 295.

Tsu, R., Koma, A., Esaki, L., 1975. Appl. Phys. Rev. Lett. 33, 827.

Whittaker, E.T., 1968. In: Abramowitz, M., Stegun, C.A. (Eds.), Handbook of Math Functions. Dover, New York, p. 505.

Ziman, J.M., 1988. Principles of the Theory of Solids, Cambridge University Press, Cambridge, p. 313.

5 Quantum Step and Activation Energy

5.1 Optical Properties of Quantum Steps

I spent a few summers at Fort Monmouth at the US Army EDTL Laboratory. In 1989, Paul Shen showed me some intriguing structures in the optical data involving AlGaAs at the surface of GaAs. He told me that these structures were considered by some to be surface crud. I could not believe that surface crud could be so regular and repeatable. We embarked on a joint study of quantization, not with a quantum well, but with a quantum step. We developed ways to calculate the density of states (DOS) of quantum steps allowing us to sum the optical transitions and correlate them with the photoreflectance measurements.

At energies below the barrier, the k-vector in the region of the barrier is purely imaginary, the discontinuity in potential due to the step is greater and the DOS is more discrete. The calculated peaks of the photoreflectance for a GaAlAs step on GaAs agree well with experiments on energy. For optical absorption, the usual DOS does not apply because momentum is only conserved in the transverse direction. New understanding relating to band-edge alignment may be studied with quantum steps. In particular, it should be possible to utilize the quasi-discrete nature of the DOS of a quantum step in a variety of photo-assisted processes.

5.1.1 Density of States of a Quantum Step

Following Shen et al. (1990), we start with the potential profile:

$$
\begin{aligned}
V &= -V_1, & z &< 0, \\
&= 0, & 0 &< z < L, \\
&= \infty, & z &> L.
\end{aligned}
\tag{5.1}
$$

Since the work function of GaAs/GaAlAs is much greater than the potential step V_1, for simplicity, the surface at $z = L$ is represented by an infinite barrier. Without intentional doping, the depletion width is at least several micrometers, and with a

Superlattice to Nanoelectronics. DOI: 10.1016/B978-0-08-096813-1.00005-9

step of thickness $\sim 5-50$ nm, depletion may be totally ignored. The boundary conditions at $z = 0$ are the continuity of ψ_i and ψ_i'/m_i. The reflection coefficient is simply a phase change, $R = \exp(i\partial\varepsilon\phi)$ with $\phi = 2 \tan^{-1}(\eta \tan k_2 L + \pi)$, in which $\eta = k_1 m_2/k_2 m_1$, where $k_i^2 = 2m_i(E_L + V_i)/\hbar^2$, with $i = 2$ for the step and E_L being the energy along the quantum step. The DOS in terms of E_L, $n_L(E_L)$, is given by (Wigner, 1955):

$$n_L(E_L) = \frac{1}{2\pi L} \frac{\mathrm{d}\phi}{\mathrm{d}E_L} = \frac{1}{2\pi L} \frac{F}{1 + F^2} \frac{1}{E_L} \left(\frac{2k_2 L}{\sin(2k_2 L)} - \frac{V_1}{E_L + V_1} \right), \tag{5.2}$$

where $F \equiv \eta \tan k_2 L$. With $V_1 = 0$, $n_L(E_L) = (\mathrm{d}k_2/\mathrm{d}E_L)/\pi$, the usual DOS of a free particle. In terms of the total energy $E = E_L + \hbar^2 k_t^2/2m_2$, after integration in the transverse coordinates, the DOS $n(E)$ is given by

$$n(E) = \frac{2}{4\pi^2} \int_0^\infty n_L(E_L)\mathrm{d}E_L \int_0^\infty \delta\left(E - E_L - \frac{\hbar^2 k_t^2}{2m_2} \right) 2\pi k_t \mathrm{d}k_t$$

$$= \frac{1}{2\pi^2 L} \left(\frac{2m_2}{\hbar^2} \right) \tan^{-1} F. \tag{5.3a}$$

Again for $V_1 = 0$, $m_1 = m_2$ and $n(E) = (1/2\pi^2)(2m/\hbar^2)^{3/2}E^{1/2}$, the usual three-dimensional DOS for free particles. Let us evaluate Eq. (5.3a) for $V_1 \neq 0$ and $L \to \infty$, applying to a simple barrier,

$$n(E) = \frac{1}{2\pi^2} \frac{2m}{\hbar^2} \lim_{L\to\infty} \frac{\tan^{-1}F}{L} = \frac{1}{2\pi^2} \left(\frac{2m}{\hbar^2} \right)^{3/2} E^{1/2} \lim_{L\to\infty} G(k_2 L), \tag{5.3b}$$

where

$$G(k_2 L) = \frac{\eta}{\cos^2 k_2 L + \eta^2 \sin^2 k_2 L},$$

which oscillates between η and $1/\eta$, and averages to 1 as $L \to \infty$. The calculated $n_L(E_L)$ versus E_L for various thicknesses of the quantum step with $V_1 = 0.15$ eV, $m_1 = 0.066 m_0$, and $m_2 = 0.076 m_0$, corresponding to the conduction band of a $Ga_{0.84}Al_{0.16}As$ quantum step on GaAs, is shown in Figure 5.1. Unlike in quantum well, these peaks oscillate with decreasing amplitudes. The thicker the quantum step, the farther apart in energy are the peaks. The DOS of the heavy hole $n_{Lv}(E_{Lv})$ can be calculated similarly.

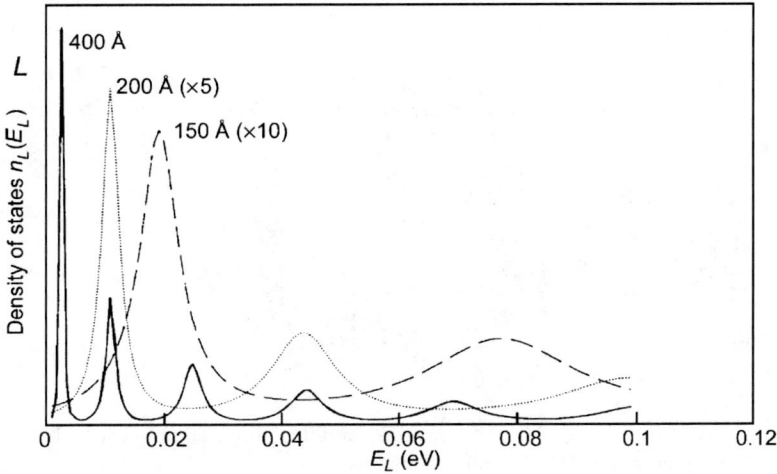

Figure 5.1 DOS $n_1(E_L)$ versus E_L for various thicknesses from Eq. (5.2). Solid line, 50 nm; dotted line, 20 nm; dashed line, 15 nm. Unlike the quantum well, these peaks oscillate with decreasing amplitudes.
Source: After Shen et al. (1990), with permission.

5.1.2   Matrix Element between the Valence and Conduction Bands

The matrix element between the valence band and the conduction band is

$$M_{vc} = M^B_{vc} M^{Env}_{vc} \delta(\mathbf{k}_{tc} - \mathbf{k}_{tv}),$$ (5.4a)

where the subscripts c and v denote conduction and valence band, respectively. The Bloch component $M^B_{vc} = \langle \psi^B_c | \mathbf{e} \cdot \mathbf{P} | \psi^B_v \rangle$ and the envelope component

$$
\begin{aligned}
M^{Env}_{vc} &= \int_{-\infty}^{L} \psi_c \psi^*_v \; dz \\
&= \frac{1}{2\pi} A_c A^*_v \int_0^L \sin[k_{2c}(z - L)]\sin[k_{2v}(z - L)]dz + \exp\left(i\frac{\phi_c - \phi_v}{2}\right)\delta(k_{1c} - k_{1v}),
\end{aligned}
$$ (5.4b)

where $|A_{c(v)}|^2 = 4\eta^2_{c(v)}/[\eta^2_{c(v)}\sin^2(k_{2c(v)}L) + \cos^2(k_{2c(v)}L)]$. The notation c(v) in $|A_{c(v)}|^2$ indicates either c or v in A_c or A_v. The second term in Eq. (5.4b) contributes to the optical constant originating from GaAs, which gives the square root singularity of the absorption coefficient at the energy of the GaAs direct gap. Since we are only interested in the vicinity of the AlGaAs gap, this second term is neglected in the

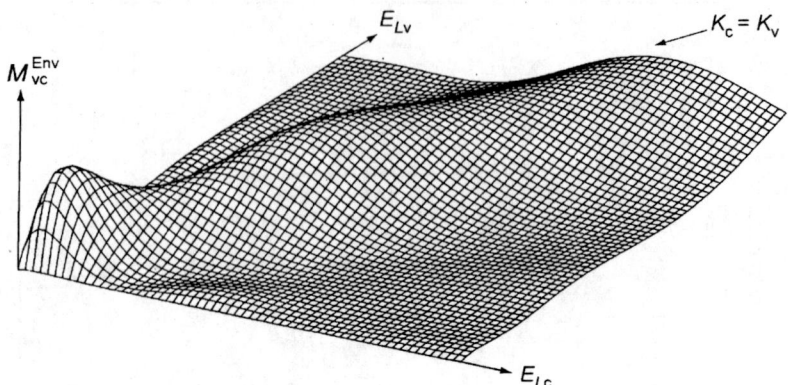

Figure 5.2 Matrix elements $|M_{vc}^{Env}|$ versus dE_{Lc} and dE_{Lv} for a 20 nm $Ga_{0.84}Al_{0.16}As$ Q step. *Source*: After Shen et al. (1990), with permission.

discussion that follows. Figure 5.2 shows M_{vc}^{Env} with $V_c = 0.15$ eV, $V_v = 0.082$ eV, $m_{1c} = 0.066m_0$, $m_{2c} = 0.076m_0$, $m_{1v} = 0.34m_0$, $m_{2v} = 0.36m_0$, and $L = 20$ nm. The conduction band offset used is 65%. Note that M_{vc}^{Env} peaks at $k_{2c} = k_{2v}$. For $V_c = V_v = 0$, and $L = \infty$, $|M_{vc}^{Env}|^2/L \to \partial (k_{2c} - k_{2v})$. For a finite L, $|M_{vc}^{Env}|$ spreads out so that it is not possible to express optical transitions in terms of a simple joint DOS.

The absorption coefficient of the quantum step is given by

$$\alpha = (4\pi^2 e^2 / ncm_0^2\omega)|M_{cv}^B|^2 J_{Env}, \qquad (5.5)$$

with

$$J_{Env} = \left(\frac{2}{8\pi^3}\right)^2 \frac{1}{L} \int dk_c\, dk_v\, dk_{tc}\, dk_{tv}\delta\left(E_g + E_{Lc} + E_{Lv} + \frac{\hbar^2 k_{tc}^2}{2m_c} + \frac{\hbar^2 k_{tv}^2}{2m_v} - \hbar\omega\right)$$
$$\times \delta(k_{tc} - k_{tv})|M_{vc}^{Env}|^2$$

$$(5.6)$$

Substituting $n_{Lc(v)}dE_L$ for $dk_{c(v)}/\pi$,

$$J_{Env} = \frac{1}{2\pi^2}\left(\frac{2\mu}{\hbar^2}\right)\frac{4}{L}\int n_{Lc}(E_{Lc})n_{Lv}(E_{Lv})|M_{vc}^{Env}|^2\theta(\hbar\omega - E_g - E_{Lc} - E_{Lv})dE_{Lc}\, dE_{Lv},$$

$$(5.7)$$

where μ is the reduced mass and θ is the Heaviside step function. For $L \to \infty$, $|M_{vc}^{Env}|^2/L \to \partial (k_c - k_v)$, and J_{Env} reduces to a joint DOS. J_{Env} plotted against $(\hbar\omega - E_g)$ is shown in Figure 5.3, with various thicknesses L of the quantum step.

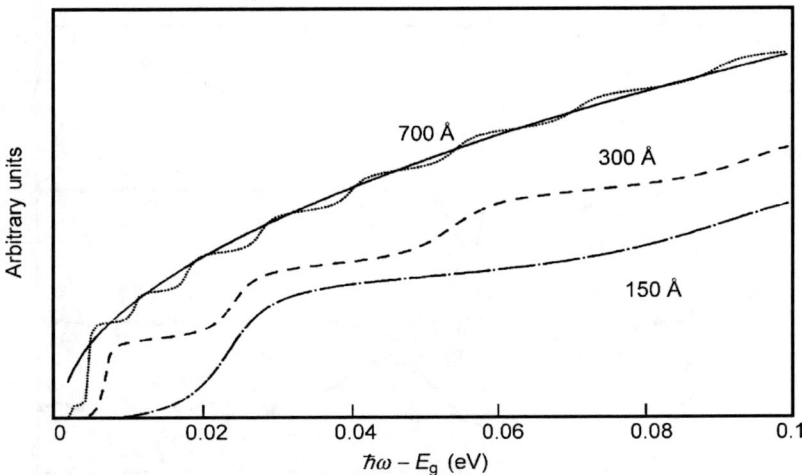

Figure 5.3 J_{Env} versus ($\hbar\omega - E_g$) for various thicknesses as marked. For large L, J_{Env} approaches the joint DOS of the bulk GaAlAs shown by the solid line as given by Eq. (5.3b). *Source*: After Shen et al. (1990), with permission.

As $L \rightarrow \infty$, $J_{\text{Env}} \propto E^{-1/2}$. The solid line in Figure 5.3 is the DOS of the bulk GaAlAs. Note that the absorption spectrum of the 70 nm GaAlAs slab resembles closely the bulk GaAlAs.

5.1.3 Electroreflectance from a Quantum Step

A quantum step of $Ga_{0.84}Al_{0.16}As$ on a GaAs buffer was etched in 0.02% NH_4OH, 0.02% H_2O in 1 part of DI water at an etching speed of 20 nm min^{-1} for various thicknesses for the photoreflectance measurements. Some details of the determination of the step thickness of GaAlAs on GaAs have been described by Shen et al. (1990 and references therein). An He−Ne laser was used in the photoreflectance spectra shown in Figure 5.4. A good account of the apparatus for photoreflectance is detailed in Shen et al. (1987) and the details of line-shape fitting are described by Aspnes (1980).

In Figure 5.4, the photoreflectance spectra, $10^5 \Delta R/R$ at room temperature is shown for three thicknesses: 5, 15, and 50 nm. The smooth curves are line-shape fits with marked arrows showing the values obtained for the energies. They give excellent agreement except for the thin 5-nm sample, which has a much broader linewidth. The small discrepancy in the position of the peak may be due to an error in the determination of the thickness. The calculated peak position and linewidth are obtained by differentiating J_{Env} in Figure 5.3. The intrinsic linewidth at room temperature from a GaAlAs bulk sample is used in the fitting process of these spectra. If we had increased all the thicknesses by 5 nm, then the fit for the energy positions for all samples would have been perfect.

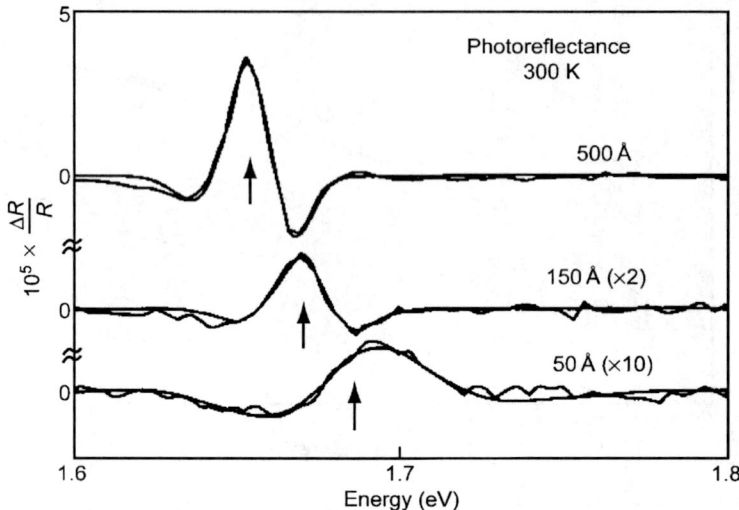

Figure 5.4 Measured photoreflectance spectra (dotted lines with more wriggles) of a quantum step of various thicknesses as marked. The solid lines are line-shape fittings. The values obtained for energy gap are shown by arrows.
Source: After Shen et al. (1990), with permission.

The photoreflectance spectra are sensitive to a quantum step thickness of at least 5 nm, perhaps down to 2 nm. Therefore, the technique should be viewed as one of the most sensitive optical characterizations of heterojunctions. If the measurement can be done at moderately low temperatures, for example at 77 K, excitonic effects should show up as in quantum wells.

I knew about the technique of finding the delay time and the derivatives of the phase with frequency when I was at BTL. While working on the tunneling time with Zypman (see Chapter 2), I showed him how we could obtain the DOS from the delay time. A few days later, he showed me the paper written by Wigner in 1955 on the same subject. It is surprising that such a powerful technique has not been described in any textbook such as Merzbacher (1970). At least, with regard to the quantum step, this technique offers distinct simplicity. We know that the quantum step offers another way of characterizing the heterojunction systems. Unlike quantum wells where many quantum devices have been explored and developed into routine applications like oscillators, lasers, and amplifiers, it is conceivable that the quantum step may be developed into useful devices. The primary difference between the two lies in the degree of trapping of electrons in a given region of space, which is far more for a quantum well. On the other hand, since the trapping time in a step is much less, it is conceivable that a series of quantum steps arranged in phase may offer a mechanism for application.

5.2   Determination of Activation Energy in Quantum Wells

The concept of activation energy E_a is extremely useful particularly in chemical reactions and crystal growth. The usual way to obtain E_a is using the Arrhenius plot, involving, for example, a current, size of a crystallites, and so on, versus inverse temperature. For a current $J \propto \exp(-E_a/k_B\theta)$, plotting J versus $1/k_B\theta$ gives E_a as the slope. Or in general,

$$E_a(V,\theta) = -\frac{\mathrm{d}ln\,J}{\mathrm{d}(1/k_B\theta)}. \tag{5.8}$$

Since the current is $J(V,\theta)$, with V and θ being the voltage applied and the temperature, the activation energy from the Arrhenius plot is also a function of V and θ. Normally, there are several regions where a straight line may be fitted for a particular value of the activation energy. Even if we use such an approach, we would need to answer: what is this fitted value that has units of energy and what is the relationship to other known parameters characterizing the system? This is what we derived (Ding and Tsu, 1997). The system we chose is a silicon-based double-barrier resonant tunneling (DBRT) structure, with two barriers on each side of a quantum well (Tsu, 1993; Tsu et al., 1997). After developing the theory and proving our results by measurement, we found that the effective barrier height is nothing other than the ground quantum state of the quantum well. The activation energy E_a in Eq. (5.8) generally does not have much meaning, except after extrapolation to zero for both V and θ. Let us start with the usual tunneling formula (Tsu and Esaki, 1973) for the current per energy, E_1, in the longitudinal direction,

$$i(E_1) = \frac{em^*k_B\theta}{2\pi^2\hbar^3}|T^*T|ln\left(\frac{1+\exp[(E_F-E_1)/k_B\theta]}{1+\exp[(E_F-E_1-eV)/k_B\theta]}\right), \tag{5.9}$$

where $|T^*T|$, E_1, E_F, V, and k_B are the transmission coefficient, the longitudinal energy, the Fermi level in the left contact, the applied voltage between the contacts to drive the current, and the Boltzmann constant, respectively. Next, let us define the average transmitted energy of the electrons flowing between the contacts,

$$\bar{E}_a = \frac{\int_0^\infty i(E_1)E_1\,\mathrm{d}E_1}{\int_0^\infty i(E_1)\mathrm{d}E_1}. \tag{5.10}$$

In order to arrive at a close form for $\bar{E}_a$, we replace the transmission coefficient in Eq. (5.9) by

$$|T^*T| \approx B(E_1) + L(E_1), \tag{5.11}$$

where the background $B(E_1)$, which is primarily due to tunneling through both barriers as if no well were present (see Chapter 2), may be approximated by

$B(E_1) = e^{-a/E_1^m}$, with a and integers m for the best fit, $L(E_1)$ the resonant part approximated by a Lorentzian lineshape with linewidth Δ, and the center energy $E_1(V)$,

$$L(E_1) = \frac{1}{1 + [E - E_1(V)]^2/\Delta^2}. \tag{5.12}$$

For symmetrical structures, $E_1(V) = E_1 - eV/2$. Using $|T^*T|$ in Eq. (5.11) and at a limited voltage range $eV \le 2(E_1 - E_F)$, we obtained

$$\bar{E}_a \approx \begin{cases} C_1 k_B T, & \text{for the low temperature region,} \\ E_1 - eV/2 + C_2 k_B T, & \text{for the high temperature region,} \end{cases} \tag{5.13}$$

in which C_1 and C_2 are determined by the background $B(E_1)$. Note that the use of $|T^*T|$ in Eq. (5.11) is for convenience only to avoid a lengthy numerical solution of $\bar{E}_a$, which is really not necessary.

Next, we numerically obtain E_a defined in Eq. (5.8) with J given by

$$J(V, \theta) = \int_0^\infty i(E_1)dE_1. \tag{5.14}$$

E_a is plotted versus the temperature θ in Figure 5.5 for the DBRT structure shown in the inset. The steps are as follows:

1. Calculate the energy E_1 using the $|T^*T|$ from the usual matrix procedure in Chapter 2 for the structure. The structure was fabricated using the process that will be described in

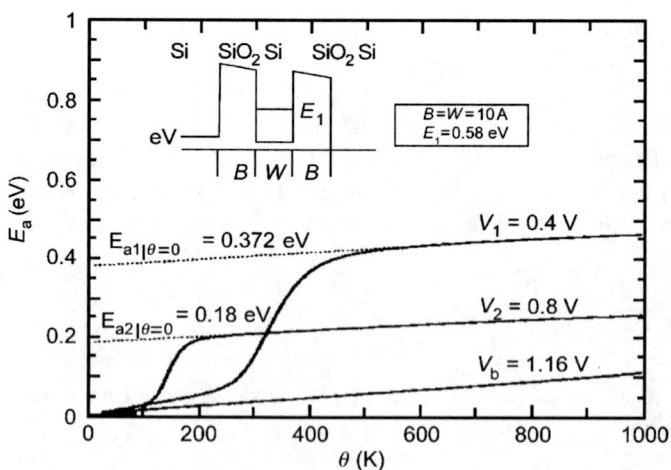

Figure 5.5 $E_a(V, \theta)$ versus $\theta(K)$ for the structure shown at the top with $E_1 = 0.58$ eV obtained by the usual transfer matrix formulation. In the extrapolation to $\theta(0)$, $E_a(V, 0)$ is used to obtain $E_a(V = 0, \theta = 0)$, the activation energy.
Source: From Ding and Tsu (1997), with permission.

Chapter 6 on the semiconductor/atomic superlattice, which consists of epitaxially grown Si/O multilayers with a monolayer of oxygen adsorbed between epitaxially grown silicon (Tsu et al., 1997). The precise thickness of the barrier is 1-unit cell of Si/1-monolayer of oxygen/1-unit cell of Si, serving as a barrier, followed by 2-unit cells of Si as the quantum well together with a second barrier forming the DBRT structure. Therefore, the effective thickness of the barrier is ~ 1.1 nm and the well is ~ 1.1 nm deep. The calculated $E_1 \sim 0.58$ eV for the energy state of the lowest resonant state.

2. From the position of E_1 in Eqs. (5.11) and (5.12), we compute $\bar{E}_a$ from Eqs. (5.9) and (5.10). We also compute directly $J(V, \theta)$ from Eqs. (5.11) and (5.14).
3. Next we compute $E_a(V, \theta)$ from Eq. (5.8), also numerically, and plot the results in Figure 5.5 at various applied voltages.
4. Extrapolating to $\vartheta = 0$, for $E_a(\theta = 0)$.
5. Next, we plot $E_a(\theta = 0)$ versus V shown in Figure 5.6 as circles. A straight line fit through the circles intercepts the ordinate at 0.58 eV, which is the final activation energy $E_a(\theta = 0, V = 0)$, equal to E_1, proving that the lowest quantum state E_1 is the activation energy.

To show why E_1 is equal to $E_a(\theta = 0, V = 0)$, let us take the case in Figure 5.5 for $V_2 \sim 0.6$ V, with $E_a(\theta = 0) = E_1 - eV_2/2 = 0.58 - 0.6/2$ eV $= 0.28$ eV ~ 0.275 eV. And $E_a(\theta = 0, V = 0)$ is precisely 0.58 eV. Another way of looking at these numbers is to use the results obtained by Barker and Gruodis (1967), where the maximum peak value of

$$J(V, T_1, T_2) \equiv [J(V, T_1) - J(V, T_2)]/J(V, T_2)$$

is located very nearly at $eV = E_b$, with E_b being the height of a square barrier (Duke, 1969). Then for our example, putting $eV = E_b$ in $E_a(\theta = 0) = E_b - eV_2/2 = E_1 - E_b/2$ again gives rise to $E_b \sim E_1$. Thus, what we know intuitively that the current must pass through the structure via the lowest quantum level is indeed correct. One may ask why we need to go through all this just for that. The implications are enormous.

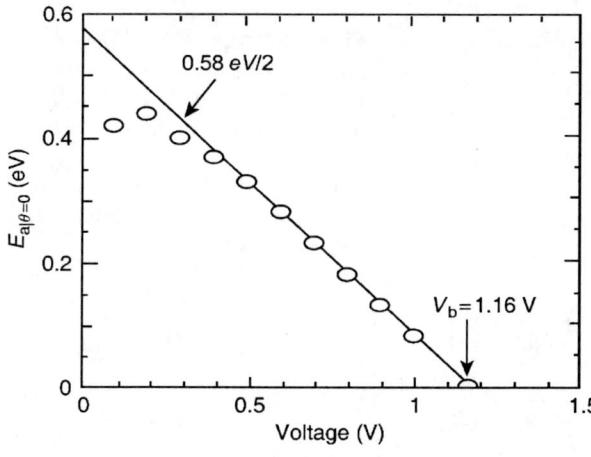

Figure 5.6 $E_a(\theta = 0, V)$ versus the applied voltage V. The best straight line through the circles intercepts the ordinate at $E_a(\theta = 0, V = 0)$, the activation energy. *Source*: From Ding and Tsu (1997), with permission.

Before, the only way to find the quantum state was to have a good peak current located at the resonant energy of the structure. Now we know that we can find the lowest level of the quantum well state even without conditions for good NDC to locate the peak, and therefore, the resonant state. We can simply measure the current versus the applied voltage at various temperatures as shown in Figure 5.7. Using Eq. (5.8), we obtained a plot of $E_a(V, \theta)$ shown in Figure 5.5. Values extrapolated to $E_a(\theta = 0, \propto)$ are plotted as in Figure 5.6 resulting in Figure 5.8. The best possible straight line through the circles gives intercepts at $0.51-0.57$ eV as the two limits of the procedure. We arrive at the energy state $E_1 \sim 0.51-0.57$ eV for the resonant state of the structure shown in Figure 5.8. This is the basis we used to estimate the position of the quantum state ~ 0.5 eV for the Si/O superlattice. There will be a full treatment of this subject later (Tsu et al., 1997).

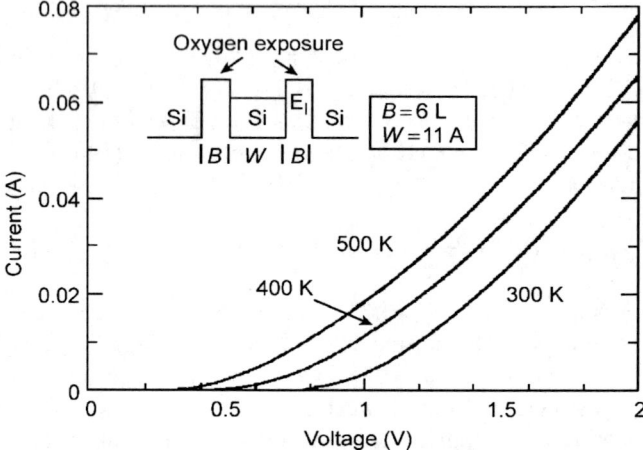

Figure 5.7 The I versus V measurement of a Si/O DBRT structure for $\theta = 300-500$ K. *Source*: From Ding and Tsu (1997), with permission.

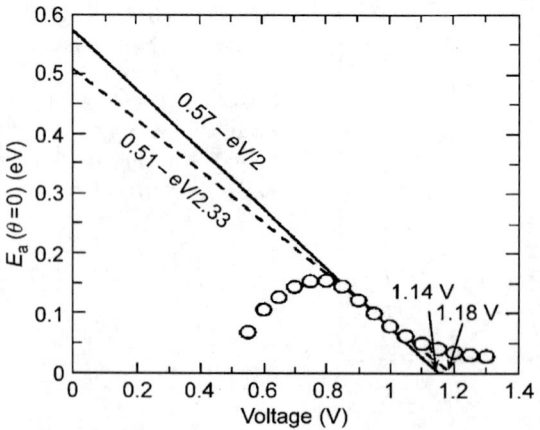

Figure 5.8 $E_a(\theta = 0, V)$ obtained numerically from Figure 5.7 versus V. The extrapolated straight line through the circles to $V = 0$ gives the activation energy, $E_a(\theta = 0, V = 0)$
Source: From Ding and Tsu (1997), with permission.

Figure 5.8 shows $E_a(\theta = 0, V)$ obtained numerically from Figure 5.7 plotted against the applied voltage V. Compared to Figure 5.6, it is clear that the structure is much worse in the sense that the better resonant structure, i.e., the one with a NDC, would have plots similar to Figure 5.6. The determined $E_1 \sim 0.51 - 0.57$ eV would have shown a NDC at $V \sim 1$ V. Failure to show the NDC indicates that the effective mobility or the mean free path is poor.

In conclusion, our method of obtaining the position of the quantum states based on measurement is quite similar to the determination of the activation energy. It allows determination even in the absence of a NDC. Mark Reed suggested writing a longer version of the Ding and Tsu (1997) paper because the procedure is really quite complex. I hope that in this more detailed version of the paper, it will be much easier to follow the steps. Last, it is probably obvious to the reader that the measurement and computation required are much more involved than the simple $I-V$ plot locating the peak current. However, in the absence of a current peak, this method is capable of providing the barrier height.

5.3   Summary

It is well known that a confined system has a localized DOS, but it is not generally recognized that at an energy above but near a barrier, the DOS is also localized, although not as completely. This is simply because wave functions do not change discontinuously. Therefore, a quantum step can also provide various parameters that characterize a quantum system. He considered me as somewhat adventurous while I considered him a someone quite cautious. I have worked with Pollak on several joint publications. For these reasons, I told Shen that I am very sure that the electroreflectance data support quantum confinement. In retrospect, even tunneling should exhibit localization at an energy just above the barriers.

Activation energy is usually determined by using the Arrhenius plot of some variable such as current, particle size, and so on, versus $1/k_B\theta$. We developed a method to find the barrier height of a RTD even without the appearance of a current maximum by plotting the average energy transport through the structure, extrapolated to zero applied voltage and $\theta = 0$, because we have shown that this gives the activation energy precisely and therefore, the barrier height may be determined. As shown in Section 5.2, the process is much more complicated than locating the current peak and extracting the barrier height from the calculation of this current peak. Therefore, this procedure for the determination of barrier height is for cases where resonant tunneling does not provide a current peak. With regard to accuracy, in Figure 5.8, we have extrapolated a small amount of available data all the way to $V = 0$ for the barrier height. To enlarge the range of data, it is necessary to use a larger range of temperature θ. Therefore, accuracy relies on how high a temperature the structure can withstand, and how reliable measurements can be made at lower temperatures. This method for the determination of an effective barrier height is not in any book, illustrating my earlier statement that consulting books is necessary, but certainly not sufficient for new ideas.

References

Aspnes, D.E., 1980. In: Moss, T.S. (Ed.), Handbook on Semiconductors, vol. 2. North-Holland, Amsterdam, p. 109.

Barker, R.C., Gruodis, A.J., 1967. Solid State Commun. 5.

Ding, J., Tsu, R., 1997. Appl. Phys. Lett. 71, 2124.

Duke, C., 1969. Tunneling in Solids, Academic Press, New York, p. 64.

Merzbacher, E., 1970. Quantum Mechanics, Wiley, New York.

Shen, H., Parayanthal, P., Liu, Y.F., Pollak, F.H., 1987. Rev. Sci. Instrum. 58, 1429.

Shen, H., Pollak, F.H., Tsu, R., 1990. Appl. Phys. Lett. 57, 13.

Tsu, R., 1993. Nature 364, 19.

Tsu, R., Esaki, L., 1973. Appl. Phys. Lett. 22, 562.

Tsu, R., Filios, A., Lofgren, J.C., Ding, J., Zhang, Q., Morias, J., et al., 1997. In: Cahay, M.M. (Ed.), Proceedings, ECS 97-11, Quantum Confinement IV: Nanoscale Materials, Devices and System, Montreal, May 4−7.

Wigner, E.P., 1955. Phys. Rev. 98, 145.

6 Semiconductor Atomic Superlattice (SAS)

Superlattices and quantum wells were introduced as man-made quantum structures to engineer the quantum states for electrical and optical applications (Esaki and Tsu, 1970; Tsu and Esaki, 1973). The idea relies heavily on the availability of good heterojunctions, lattice-matched systems (Cho, 1971), and later, strain-layer systems (Matthews and Blakeslee, 1976). To realize quantum states in a given geometry, the size must be smaller or comparable to the coherence length of electrons, in order to exhibit quantum interference. This requirement eliminates doping as an effective means of achieving confinement, except at low temperatures (Döhler et al., 1981), because doping comes from charge separation which results in dimensions generally far exceeding the coherence length of electrons at room temperatures. On the other hand, band-edge alignment of a heterojunction provides an abrupt barrier height. This short-range potential is the consequence of higher-order multipoles in the atomic potentials as discussed in Chapter 1. The requirement of lattice match to below 1% eliminates most materials except notably III–V semiconductors.

A new type of superlattice was proposed, the epilayer doping superlattice (EDS), consisting of, for example, a couple of layers of Si in AlP (Tsu, 1988). The idea is fundamentally different from atomic plane- or δ-doped superlattices (Zrenner et al., 1985), where only a small fraction of the plane is occupied by doping or substitution. Basically, if the entire layer is involved, unlike doping, disorder is eliminated. Another type of superlattice designed to incorporate extremely localized interactions that is most promising for silicon was introduced by Tsu (1993), consisting of an effective barrier to silicon, formed by a suboxide with a couple of monolayers of oxygen atoms. To overcome the problem of structural robustness associated with porous silicon, p-Si (a brief account is treated in this chapter), it was proposed that nanoparticles of silicon with a size in the range of several nanometers sandwiched between thin oxide layers to form a superlattice may solve the problem of mechanical robustness while retaining the features of quantum confinement as in the case of porous silicon (Tsu et al., 1995), where the name "interface adsorbed gas-superlattice" (IAG-superlattice) was introduced. This name originates from the idea that oxygen is introduced *via* surface adsorption in order to prevent the formation of a very thick oxide. However, there is no attempt to circumvent the polycrystalline nature of the structure. At this stage, it was recognized that perhaps

Superlattice to Nanoelectronics. DOI: 10.1016/B978-0-08-096813-1.00006-0

it is possible to adsorb one monolayer of atomic species, such as oxygen onto silicon, beyond which epitaxial growth of silicon may be continued. Such a system used as a barrier for silicon is treated in detail theoretically and realized experimentally (Tsu et al., 1996, 1998a,b). Let us describe in more detail the structure of this kind of superlattice, which is quite different from any superlattices including the type-III, in fact the proper description may be "multiple quantum well barrier" or "superlattice barrier." Originally the idea involves thin Si with a monolayer of oxygen together forming Si/O/Si/O/Si as an effective superlattice barrier for a silicon-based superlattice—a superlattice with Si quantum wells having a barrier consisting of Si/O/Si... *It is important to recognize that such a Si/O/Si barrier is not expected to have NDC or Bloch oscillation.* On the other hand, we shall see in what follows that the Si/O/Si structure itself is very interesting in its own right because the combination has resulted in photoluminescence (PL) and electroluminescence (EL) in energy ~2.2 eV. Again a new type of superlattice emerged, particularly, it shares an almost identical function with the graphene/Si superlattice discussed at the end of this chapter (Zhang and Tsu, 2010). We have discussed that the type-III superlattice was discovered for quite a different reason, trying to push the point of inflection, the change of the sign of the effective mass, to a point in *k*-space nearer the zone center. And this new superlattice is very different in concept: the monolayer of oxygen does not play the role of a barrier, rather, as in any molecule, introducing a *local molecular complex*. Localized interaction in a man-made quantum system is not really new, for example, resonant tunneling involving localized defects was reported by Dellow et al. (1992). A detailed account dealing with what is new in superlattice is fully addressed in Chapter 10.

6.1   Silicon-Based Quantum Wells

The electronic industry is overwhelmingly dominated by the silicon integrated circuit (IC). For the III−V (columns III and V of the periodic table)-based lattice-matched systems, and even with strain-layer systems (Matthews and Blakeslee, 1976), quantum devices remain as research topics and speciality applications. In 1993, I proposed that the concept of a strain-layer superlattice be pushed to the limit by a Si−SiO_2 system, perhaps with the thickness of the SiO_2 no more than one unit cell thick. Figure 6.1 shows the proposed quantum well with a barrier consisting of one unit cell thick of Si in between the two SiO_2 barriers.

Basically, the concept of a strain-layer superlattice is that with a sufficiently thin epitaxial layer, the strain energy in each layer is below the energy needed for the growth of point defects or dislocations. Since dislocations have an activation energy for nucleation and a lower activation energy for growth, it is possible to exceed the energy requirement without generating defects. The metal−oxide−silicon (MOS) device owes its success to the low defect density at the Si/amorphous SiO_2 interface with a barrier height of 3.2 eV. It is not possible to grow Si epitaxially on the amorphous SiO_2, but it is possible to continue epitaxy if the SiO_2, or perhaps

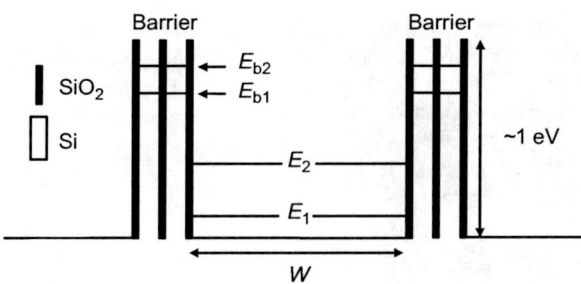

Figure 6.1 Scheme of a Si quantum well with thickness w between two barriers formed by a Si/SiO$_2$/Si superlattice. *Source*: After Tsu (1993), with permission.

more appropriately the SiO$_x$ layer, is only one or two monolayers. This is the thinking that led to my proposed scheme. To my mind, defects should have a hard time nucleating in a very small system with a much-reduced degree of freedom. And this is good for us trying to build a nanoelectronic quantum device, although a major problem is the difficulty in defining voltage, input−output, and so on, subjects we shall treat in more detail later.

6.2   Si-Interface Adsorbed Gas (IAG) Superlattice

During the process of converting our III−V molecular beam epitaxy (MBE) system to silicon MBE, Jonder Morais came from Brazil to pursue a Ph.D. with me. He had worked as a machinist and knew my theoretical work on amorphous superlattice and quantum wells from when I was in Brazil. He challenged me to consider nanocrystalline silicon with SiO$_2$, or SiO$_x$, by letting him build a simple system for the deposition of silicon with the introduction of oxygen *via* gas adsorption. This work originated the term "interface adsorbed gas superlattice" (IAG-SL). Figure 6.2 shows the transmission electron microscopy (TEM) of nine periods of IAG-SL that has an intended period ∼18 nm. We only know that the SiO$_2$ layer is thin. The process is as follows: (1) we set the thickness of the Si deposition, for example, after reaching 18 nm; (2) we interrupt the deposition by introducing O$_2$ or O$_2$ + H$_2$ into the deposition chamber, followed by Si deposition, all at a substrate temperature of ∼30°C, so that the deposited Si is mostly amorphous before annealing at ∼850°C. An estimate of the thickness showed a 10% variation between the center portion defined by a circle 2 cm in diameter and the edge at 4 cm diameter. It is obvious that the thickness control is poor in this simple system.

Figure 6.3 shows the Raman spectra, excited by a 459.7-nm Ar laser, of two periods of Si−IAG-SL (O$_2$ + H$_2$). The curve c−a is marked by Γ_a on the low-energy side of the half-width and on Γ_b the high-energy side of the half-width, with the center at 521.5 cm^{-1}, which is the peak expected by Raman for crystalline silicon. The particle size ℓ is determined from the half-widths to be 10 nm (Tsu et al., 1995). For $\ell \gtrsim 10$ nm, Si should not luminescence efficiently because the corresponding wave vector q is only 10% of q at the Δ-point of the Brillouin zone (BZ), so that the optical transition is still dominated by a phonon-assisted transition, resulting in a very weak transition.

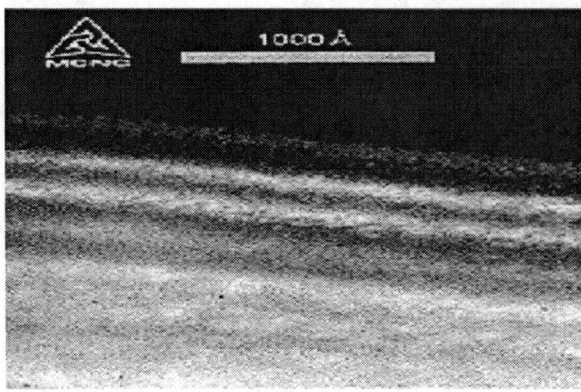

Figure 6.2 Cross-section TEM of a typical nine-period IGA-SL with a period $d \sim 18$ nm. *Source*: From Tsu and Zhang (2002), with permission.

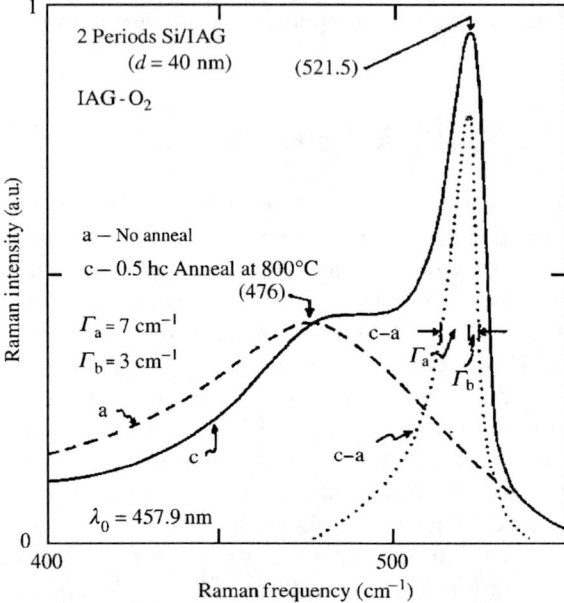

Figure 6.3 Raman spectra using a 459.7-nm Ar laser, of two periods of Si-IAG/SL ($O_2 + H_2$) with $d = 40$ nm the thickness of each Si layer at a substrate temperature T_s of 20°C. Note that without annealing, spectrum c is basically that of a-Si. *Source*: After Tsu et al. (1995), with permission.

Figure 6.4 shows the PL of two samples, a pronounced peak at ~ 1.7 eV for the nine-period sample ($d \sim 10$ nm) and 2.34 eV for the five-period sample, both after $(O_2 + H_2)$ annealing at 10^{-7} Torr at 850°C for 30 min. The PL intensity is increased more than 10 times after annealing. The PL is more than a factor of 3 weaker when annealing in O_2 without H_2. In a later paper, it was found that the best annealing temperature is 420°C (Tsu et al., 1997). The PL peak at 2.34 eV becomes far more dominant with a rather narrow linewidth <0.3 eV. The maximum period that was tried for the superlattice structure was only nine periods. The reason to keep the total thickness low is dictated by the need to passivate the interface defects further, annealing in the presence of gas mixtures. In particular, annealing in H_2 or $H_2 + O_2$ gives better results, as shown in Figure 6.5. It was found in

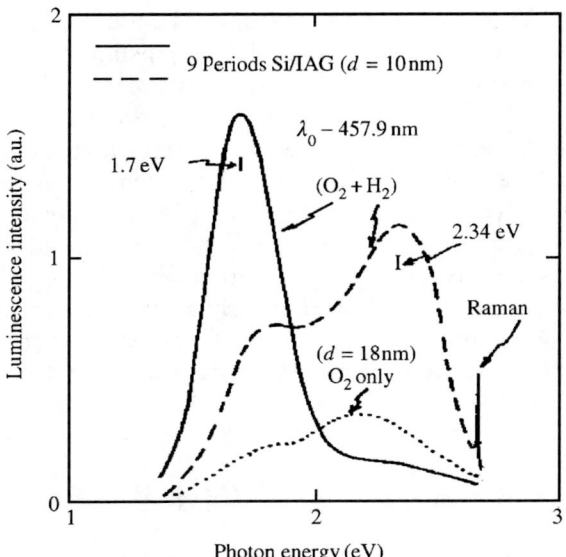

Figure 6.4 PL versus photon energy excited by the 457.9-nm Ar laser line. Annealing, for example $(O_2 + H_2)$, is usually at $800-850°C$ for half an hour. The Raman line is put there for calibration purposes, to estimate efficiency and the photon energy of PL.
Source: After Tsu et al. (1995), with permission.

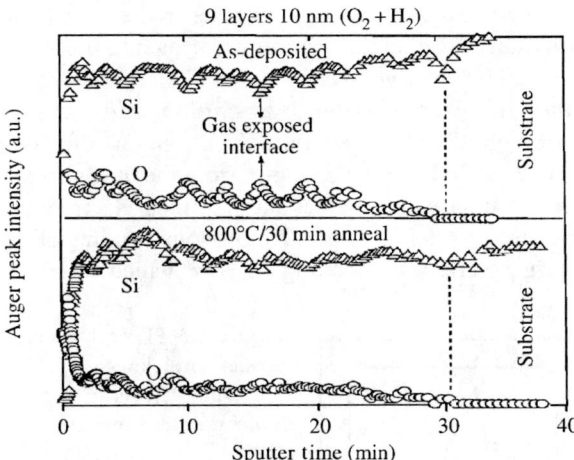

Figure 6.5 Surface Auger of the sample in Figure 6.4, with and without annealing. Note that the dips in the Si count coincide with the peaks in the O count.
Source: Taken from the unpublished Ph.D. thesis of J. Morais et al.

surface Auger (Morais, 1995), that the oxygen peaks coincide with the silicon dips, indicating that the structure indeed consists of silicon separated by regions of high oxygen content. Moreover, the 2.34 eV peak is attributed to a localized $Si-O$ complex at the interface (Morais, 1995). This brings us to an important point. In devices dictated by bulk, surface or interface regions are considered undesirable. As the particle size shrinks to the nanometer scale, surface or interface regions become significant or even dominate over the "bulk"; thus we need to reorient our views, so that surface or interface regions become the focus of our considerations.

Most of our samples that luminesce well have a grain size of ~ 3 nm. The mechanism controlling the grain size is quite involved. We shall only touch on the salient points here. Basically, unlike the amorphous—crystalline phase transition in bulk, in very thin structures, the phase transition is controlled by proximity effects rather than simple temperature effects (Tsu et al., 1997). These considerations prevented us from using extremely thin silicon layers as recently demonstrated in the work of Lu et al. (1995). In other words, if we were to crystallize their structure, we would have to use heat much beyond the usual crystallization temperature of bulk a-Si to $T_c \sim 700°C$ (Gonzalez-Hernandez et al., 1984). We shall postpone discussion of what happens when we use a much-reduced thickness for the silicon layers until the treatment of the epitaxially grown Si—O superlattices.

Finally, in Figure 6.5 surface Auger of a typical sample shows that the dips in the Si count coincide with the peaks in the O count. Since the annealing time is longer than the deposition time, the diffusive movement toward the substrate shown is a real effect.

6.3 Amorphous Silicon/Silicon Oxide Superlattice

Lu et al. (1995) and Baribeau et al. (1995) reported strong visible PL from the amorphous Si/SiO_2 superlattice. The PL peak shown in Figure 6.6 can be shifted from 1.7 to 2.3 eV when the thickness of amorphous silicon layer in the superlattice is 1—3 nm. The visible light emission was explained in terms of quantum confinement of electrons in the two-dimensional silicon layers.

It is appropriate to discuss the difference between this a-Si/SiO_2 and the Si—O superlattice that was discussed in more detail in Section 6.2. As long as the amorphous silicon thickness is below that of the coherence length, from the quantum phenomenon point of view, it is not so important whether the confined electrons can interfere in an amorphous layer or in a crystalline layer. Amorphous bonding allows more flexibility resulting in a larger variety of bonding defects, which ultimately

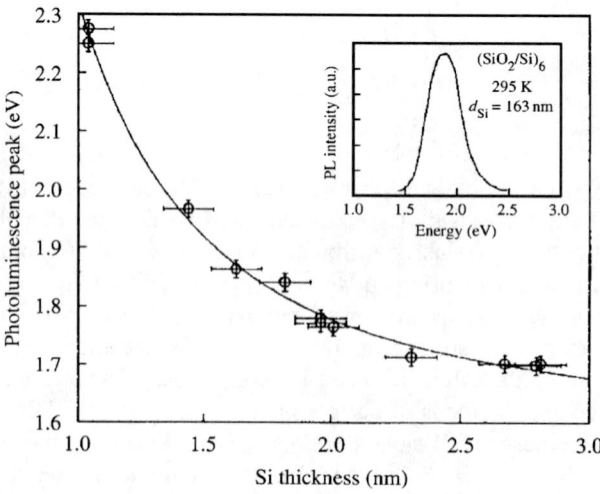

Figure 6.6 PL peak energy versus Si thickness of a-Si/SiO_2 superlattice.
Source: After Lu et al. (1995), with permission.

manifest in increased recombination and scattering centers, compared to the crystalline case. However, hydrogenated amorphous silicon photovoltaic materials have been steadily improved; any premature judgment is somewhat risky at this stage.

6.4   Silicon—Oxygen (Si—O) Superlattice

We introduced the term "semiconductor atomic superlattice" (SAS) to describe a superlattice system consisting of epitaxial layers of a semiconductor such as silicon, for example, sandwiched between adsorbed atomic species such as oxygen, carbon, and even CO. Several years ago, while searching for a barrier system for silicon, where there is no lattice-matched heterojunction except in the Si_xGe_{1-x} system (Bean, 1985; Meyerson, 1986; Feldman et al., 1987), it was proposed that perhaps the best and simplest way to build a barrier into silicon is to utilize the concept of strain-layer superlattice with sufficiently thin silicon layers (Tsu, 1993). Subsequently, it was realized that the best way to limit the thickness of the oxide that introduces disorder is to limit the supply of oxygen by surface adsorption (Tsu et al., 1996, 1998a,b). This is because after monolayer coverage of oxygen on a clean silicon surface, further oxygen adsorption is not possible without substantial heating to drive in the oxygen *via* diffusion. This method is, therefore, in the realm of self-organized crystal growth (Fukuda et al., 1997).

SAS is the outgrowth of the originally proposed barrier for silicon. Basically, the concept of a strain-layer superlattice is that, with a sufficiently thin epitaxial layer, the strain energy in each layer is below the energy needed for the growth of point defects or dislocations. The effective barrier height of a Si monolayer in an oxygen—Si system, as we will show, may be higher than 1 eV, so that it is possible to design an effective barrier height much greater than k_BT at room temperature. The thickness at which SiO_2 can be tolerated for continuous epitaxy is the key for this kind of superlattice. The debate is about whether the monolayer of SiO_2 on Si is ordered or disordered. Epitaxial growth of silicon may be continued after interruption by oxide growth as shown by Meakin et al. (1988). The explanation involves the natural seeding provided even by the reconstructed Si(100) at a low-pressure chemical vapor deposition (CVD) at a low flow rate. Based on grazing angle X-ray diffraction (Rabedeau et al., 1991), we have found evidence for a low coverage 2×1 epitaxial structure at the SiO_2/Si interface for dry oxides grown on ordered Si surfaces at room temperature. However, the 2×1 structure does not survive thermal annealing.

Subsequently, the Si—O superlattice was formed and current—voltage (I—V) measurements showed a barrier height of >0.5 eV (Ding and Tsu, 1997), which is sufficiently high for a variety of electronic and optoelectronic applications. Basically, the method involves the adsorption of oxygen onto a clean Si surface followed by Si deposition 1—2 nm thick, using *in situ* reflection high-energy electron diffraction (RHEED) to monitor the 2×1 surface reconstruction as a measure of the restoration of a clean silicon surface before the next growth step. More detailed work about defects and degree of epitaxy has been carried out with oxygen exposure (Tsu et al., 1998a,b).

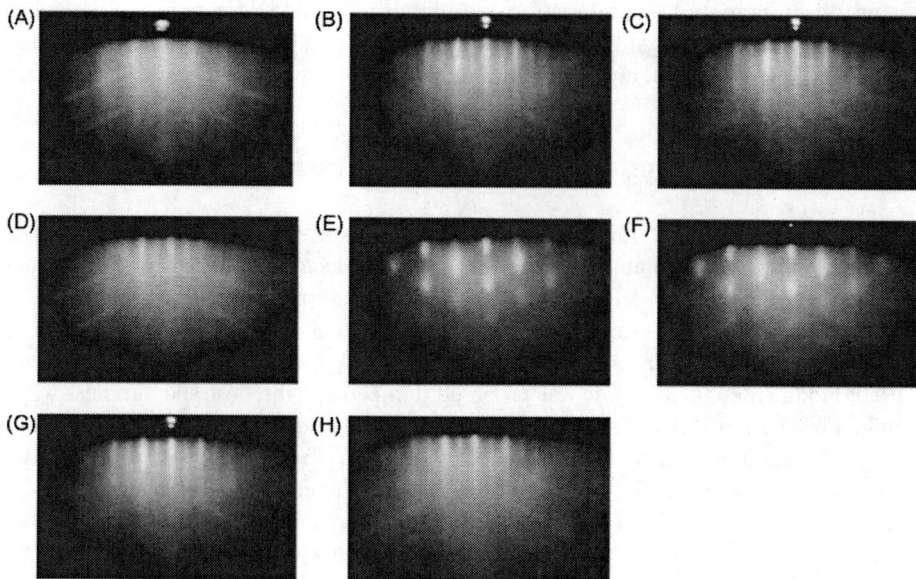

Figure 6.7 *In situ* RHEED for epitaxial Si MBE beyond oxygen adsorption: (A) initial RHEED after HF dipping (cleaning); (B) 1×2 reconstruction appears after 60 min at 850°C; (C) the pattern is further improved after 20 nm of Si buffer is deposited at 550–650°C at 0.03 nm s^{-1}; (D) first 10 L of oxygen exposure, at 30°C still shows 1×2 reconstruction; (E) 1.1 nm of Si is deposited at 550°C at 0.03 nm s^{-1} showing degradation of two-dimensional pattern and loss of the 1×2 reconstruction; (F) a second exposure to 10 L oxygen is followed by the repeat of (E); (G) 1×2 reconstruction reappears after 4 nm of Si is deposited at 550–650°C at 0.03 nm s^{-1}; and finally, (H) the pattern is almost completely recovered after a repeat of (G).

Figure 6.7 shows the typical *in situ* RHEED used to monitor epitaxy. After HF treatment Figure 6.7(A), 2×1 dimerization, the half-order, is clearly shown in Figure 6.7(B) from surface reconstruction of Si(100). With a 2-nm Si buffer, the two-dimensional RHEED pattern is and should be almost perfect in Figure 6.7(C). Exposure to 10 L (1 L is 10^{-8} Torr of oxygen for 100 s) of oxygen at 30°C does not eliminate the 2×1 structure, shown in Figure 6.7(D), until 1.1 nm of Si is deposited, shown in Figure 6.7(E), where the transformation from a two-dimensional to a three-dimensional pattern is quite evident. A second exposure at 10 L does not substantially alter the RHEED shown in Figure 6.7(F), however, the 2×1 reappears after 4 nm of Si is deposited, shown in Figure 6.7(G), and approaches the original pattern, shown in Figure 6.7(H). This series of *in situ* RHEED demonstrate that epitaxial growth is continued beyond the adsorbed oxygen layer. It may be argued that RHEED favors structures so that few clusters can in principle dominate the patterns, although we know that this is not the case, because we can scan the e-beam over the sample and no big changes were observed. Besides, RHEED was used to monitor the progress of the growth, and TEM and other measurements were used to corroborate the *in situ* RHEED.

A sample subjected to four times repeated sequence of 10-L oxygen exposure followed by 1.1 nm of Si deposition at 550°C has been fabricated for high-resolution cross-section TEM (HRX-TEM). In this sample, after the 2×1 reconstruction was brought back, a layer of 12 nm was deposited serving to restore the epitaxial structure as well as capping for protection. A Si buffer of 12–13 nm (above the two bottom arrows) was first deposited at 550°C on a 0.01 Ω cm (100) n-type silicon wafer with RCA cleaning and hydrogen stabilization followed by 30-min annealing at 850°C. The active part, which is the region of interest, is formed by two 10-L oxygen exposures separated by 1.1-nm Si deposition at 550°C. The active layer is capped by 8 nm of Si, mainly used for the study of epitaxial recovery and for protection.

Figure 6.8 shows the HRX-TEM of this sample. Note that the top of the lower vertical arrow points to a region showing cluster formation. On the other hand, with lower resolution TEM, we found that the so-called cluster is actually quite uniform. This is further assurance that a few clusters containing oxygen are not available so that the other portions will allow the continuation of epitaxial growth. This point is further clarified when we show the plane view TEM.

Obviously, the sample used for the HRX-TEM shown in Figure 6.8 was much improved in comparison to the earlier sample shown in Figure 6.9, where stacking

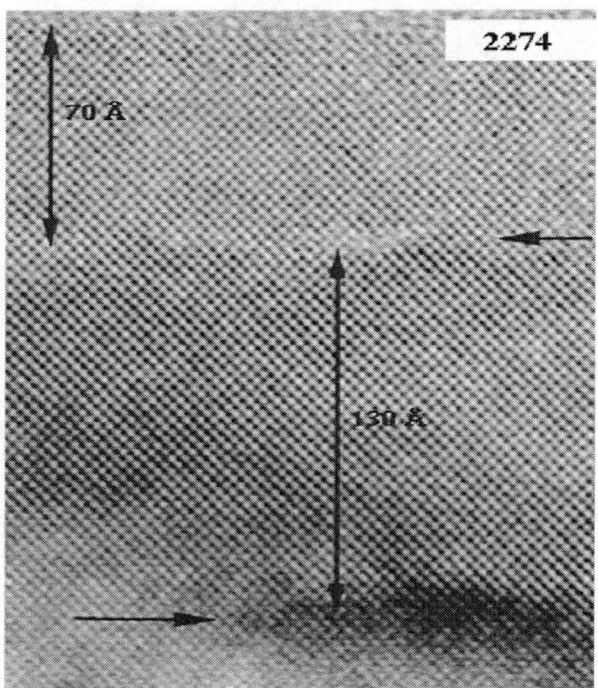

Figure 6.8 HRX-TEM of the first successful epitaxial growth after two 10-L oxygen adsorptions. Note that the top of the lower vertical arrow points to a whitish region showing possible cluster formation.
Source: From Tsu et al. (1998a,b), with permission.

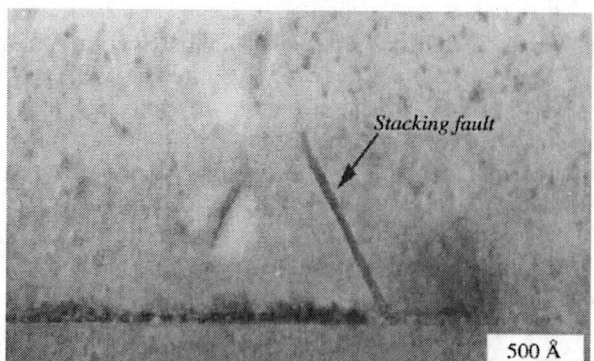

Figure 6.9 Early Si–O superlattice showing stacking fault defects without 850°C annealing under UHV for 30 min or more. Note that in lower resolution, the superlattice layer is continuous.

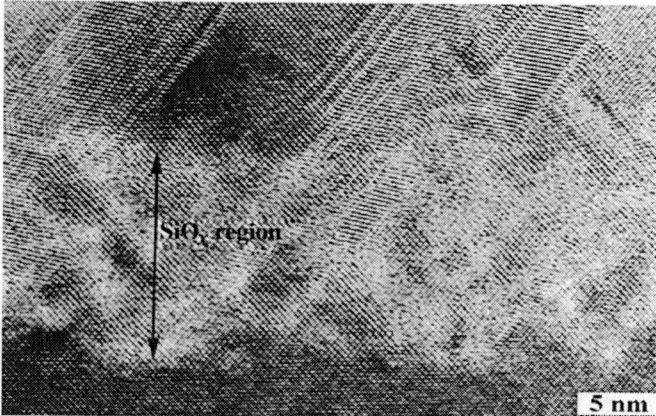

Figure 6.10 Low-defect epitaxial growth is not possible with oxygen remaining in the deposition chamber. Stacking faults dominate and persist beyond 100 nm of Si growth.

fault defects dominate. Most of the TEM work was by Dovidenko. At the time she was pursuing her Ph.D. under Professor Hren, who thought that the origin of these observed defects may very well be externally introduced rather than caused by the adsorbed oxygen. More careful cleaning and etching with thermal annealing under high vacuum showed a drastic improvement. However, a repeated superlattice structure was not observed for several years, a subject that will be treated more fully later.

Figure 6.10 shows several attempts to introduce oxygen during growth instead of using adsorption as a separate step. Only after Si deposition of several micrometers thickness does the stacking fault subside. Therefore, oxygen introduced *via* gas phase adsorption seems to be the only viable technique for incorporating monolayers of oxygen.

Let us now describe the process in more detail. Typically, we expose the clean silicon surface to oxygen at 10 L, and epitaxial silicon is grown at a deposition temperature of 550–600°C at a rate 0.4 Å s^{-1}. (The substrate temperature has been increased to 700°C in a more recent work.) Deposition and exposure are done in a growth chamber and an analysis chamber, respectively, with a low base pressure of 10^{-10} Torr. Our structure typically consists of 12 nm of Sb-doped silicon buffer on 0.01 Ω cm (100) n-type silicon wafer, followed by oxygen exposure and 1.1-nm undoped Si deposition. The use of 1.1 nm is to approximate two silicon unit cell thicknesses. A second oxygen exposure followed typically by 5–8-nm silicon completes the barrier structure. For $I-V$ measurements, the last Si deposition is again Sb doped followed by aluminum contact. The diode size varies between 10×10 and 40×40 μm. For barrier height determination, the barrier thickness usually consists of several repeats up to a maximum of four, with various exposures to oxygen, initially between 1 and 6 L. An initial attempt was made to fit the $I-V$ (Tsu et al., 1996), but was found to be grossly inadequate. The use of a generalization of the Arrhenius procedure that is fully discussed in Chapter 5 (Ding and Tsu, 1997) results in a more consistent value for the barrier height of the Si–O superlattice barrier. Figure 6.9 shows the specifics of the determination of the effective barrier height varying almost linearly with oxygen exposures of 1, 2, and 6 L (Figure 6.11).

We have exposed up to 100 L. The effective barrier height saturates at 20 L. In most of our later work, 50 L is used because we discovered that the superlattice region becomes more uniform laterally. The asymmetry $I-V$ is due to a Schottky barrier at the Si–metal contact, as well as to the possibility of oxygen diffusion toward the front contact. John Sullivan of Sandia National Laboratory pointed out to us that localized defects such as dangling bonds can serve to pin the Fermi level resulting in a Schottky-like depletion region that serves as a barrier close to the Si–O interface. Furthermore, this Fermi level pinning may in fact serve to smooth out the possible clustering effects. I felt as if I were working as a one-man crusade, when obviously the active participation of many researchers was required. Nevertheless, as we shall see, positive, though small, incremental progress has been made in the realization of the SAS.

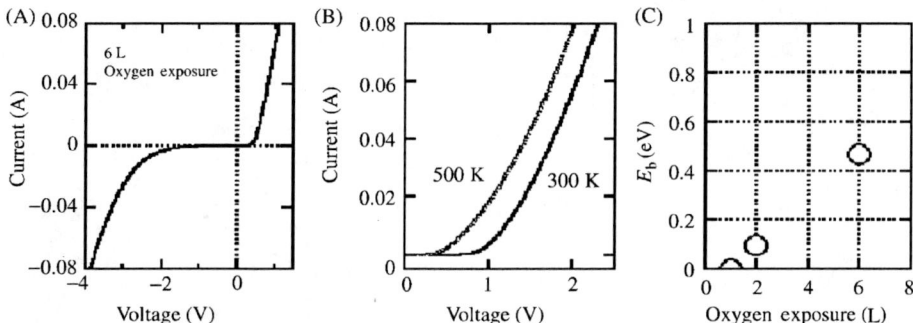

Figure 6.11 (A) Current versus voltage of a superlattice barrier for 6-L oxygen exposure, (B) temperature-dependent $I-V$, and (C) effective barrier height E_b versus oxygen exposure between 1 and 6 L.
Source: After Tsu et al. (1998a,b), with permission.

6.5 Estimate of the Band-Edge Alignment Using Atomic States

Even before we built a ball-and-stick model structure to estimate the strains in SAS, and subsequently density functional theory (DFT), computations as well, we embarked on models that allowed us to estimate the band-edge alignment between Si and Si−O superlattice. Before we go into more detail, I would like to share my experience in working with solids. In the early 1980s, Morrel Cohen visited me at Energy Conversion Devices. I was showing some of my work on the band structure of a-Si to him while mentioning the importance of the role of atomic silicon. He saw the plots of the ε_1 and ε_2 of Si and a-Si pinned to my board. He said that I should put the plot of atomic-Si side by side with these two plots. I pointed out the importance of the two silicon peaks at 3.5 and 4.6 eV due to the Van-Hove singularities. He told me that these are details and that I should look at the main peak at ∼4 eV, which is largely from the atomic transition. At that point he mentioned how most people fail to recognize the similarities rather than the vast differences between the atomic states and the band structures of solids. This greatly affected my realization of the importance of atomic states and transitions. Figure 6.12 shows the energy states of atomic silicon and oxygen together with the energy bands of SiO_2.

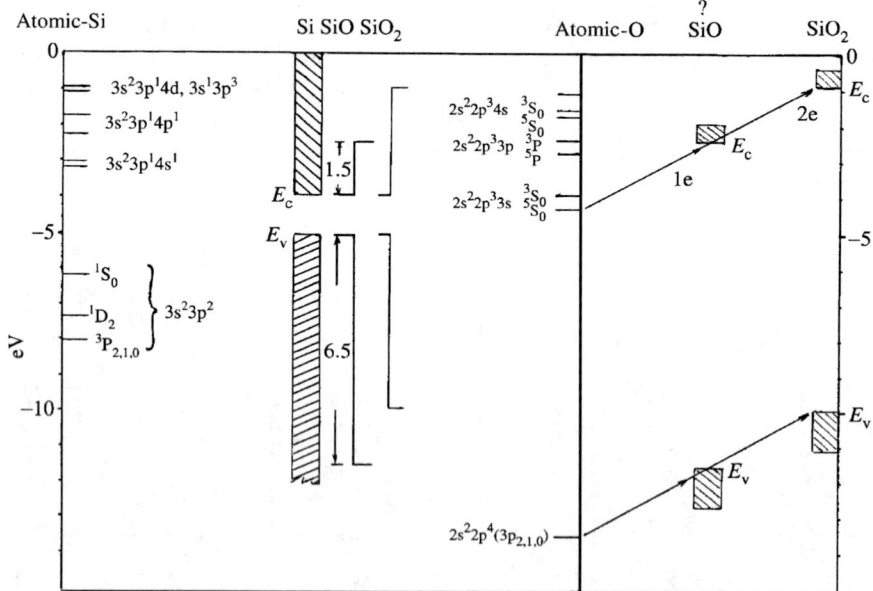

Figure 6.12 Taken from the following: Si arc spectrum: Radziemski and Andrew (1963); ionization energies $\varepsilon_i(Si) = -8.149$ eV, $\varepsilon_i(O) = -13.615$ eV, and $\varepsilon_i(SiO_2) = -10$ eV: Vier and Mayer (1944); O spectrum: Herzberg (1944); SiO_2: Li and Ching (1985). SiO is estimated to have band-edge offsets from silicon at $\Delta E_c = 1.5$ eV and $\Delta E_v = 6.5$ eV.

Figure 6.12 shows the ground state of atomic Si: $^3P_{2,1,0}, ^1D_2, ^1S_0$ denoted by the term $3S^23P^2$, and the excited states denoted by the terms $3S^23P^14S^1$, and so on, with an average energy separation of the ground state from the lower excited states of ~ 4.1 eV, which is close to the ε_2 of silicon as discussed. The ground state of atomic O is $3P_{2,1,0}$ denoted by $2S^22P^4$ separated from the excited state $2S^22P^33S-^5S_0$ by 9 eV from the ground state and 0.4 eV from the next higher $-^3S_0$ state. The bands for SiO_2 are shown at the extreme right of the figure with an ionization energy of -10 eV which is above E_v and $\chi = -0.95$ eV, which is below E_c. Two arrows are drawn from $^5S_0$ to E_c and from $3P_{2,1,0}$ to E_v. Two electrons are transferred from Si to SiO_2, but only one electron per Si atom is transferred to SiO; therefore, we take the midpoint as the energies for SiO, leading to the band-edge offset from silicon $\Delta E_c = 1.5$ eV and $\Delta E_v = 6.5$ eV, shown in the middle of Figure 6.12. These values suggest that the Si—O superlattice barrier is derivable from the use of 1.5 eV for the band-edge offset from silicon. Next, we shall use the HOMO—LUMO to estimate these offsets.

6.6 Estimate of the Band-Edge Alignment with HOMO—LUMO

Instead of forming coupled states with the ground states and the lower excited states of Si and O atoms, for the lowest approximation, we take the highest occupied Si state, $3P_{2,1,0}$ at -7.95 eV as E_2, coupled to the lowest unoccupied O state, $^5S_0$ at -4.3 eV as E_1 with coupling $\alpha_0 = 5.11$ eV to fit the band-edge offset $\Delta E_c = 3.2$ eV between Si and SiO_2.

In other words, the secular determinant is adjusted to the coupling constant α_0, that is

$$\det \begin{vmatrix} E_1 - E & \alpha \\ \alpha & E_2 - E \end{vmatrix} = 0, \tag{6.1}$$

then

$$E_\pm = E_{av} \pm (\Delta^2 + \alpha^2)^{1/2}, \quad E_{av} \equiv (E_1 + E_2)/2 \quad \text{and} \quad \Delta \equiv (E_1 - E_2)/2. \tag{6.2}$$

In Table 6.1, ΔE_c and ΔE_v are calculated for α/α_0 as a parameter. For two electrons transferred from Si to O_2, $\alpha = \alpha_0 = 5.11$ eV. However, for SiO, there is only one electron per Si atom for each O atom. Therefore, we take $\alpha/\alpha_0 = 0.5$ given in

Table 6.1 Calculated ΔE_c and ΔE_v Values for α/α_0 Using $\alpha_0 = 5.11$ eV and $\alpha/\alpha_0 = 0.5$ for SiO

α/α_0 (eV)	1.0	0.75	**0.5**	0.25
LUMO (eV)	-0.70	-1.88	**-2.98**	-3.90
HOMO (eV)	-11.55	-10.37	**-9.27**	-8.35
ΔE_c (eV)	3.2	2.32	**0.92**	0.01
ΔE_v (eV)	6.55	5.37	**4.27**	3.34

bold in Table 6.1. Note that $\Delta E_c = 0.92$ eV, which is lower than the previous estimate of 1.5 eV. However, the energy gap $11.55 - 0.7 = 10.8$ eV is 20% larger than the correct value of 9 eV for SiO_2. Therefore, the corresponding $\Delta E_v = 4.27$ for SiO is probably also 20% too high. A better estimate should include more than 2×2 in the secular determinant, for example, it should include $3S^2 3P^1 4S^1$, the excited state of Si, and $^3P_{2,1,0}$, the ground state of O. From the activation energy measurements, an effective barrier height for a sample with >20 L oxygen adsorption is found to be in the range of 0.5 eV, corresponding to an offset of close to 1 eV. We have been using 1 eV as the band-edge offset for the Si—O superlattice. We found that it was not possible to alternate monolayers of Si and O. The minimum thickness of the Si layer for each O adsorption is probably a unit cell, or at least every two silicon layers for one oxygen layer, otherwise the system will become a random alloy. I want to emphasize here once more that the whole idea is to avoid a method like atomic layer epitaxy (ALE), because whenever the situation is such that silicon cannot dominate oxygen and oxygen cannot dominate silicon, an amorphous SiO_2 structure will result. The whole idea of our scheme relies on the principle that Si forms a structure and dictates at what position the O atom can sit. Thus, the logic of our scheme will require at least two Si layers for every one O layer. This is the reason we do not think the thickness can be reduced significantly from 1.1 nm. Perhaps the minimum is ~ 0.55 nm, the unit cell of silicon.

Figure 6.13 shows the calculated HOMO—LUMO (Tsu, 2000a) giving an estimate of the band-edge offset between Si and Si—O superlattice.

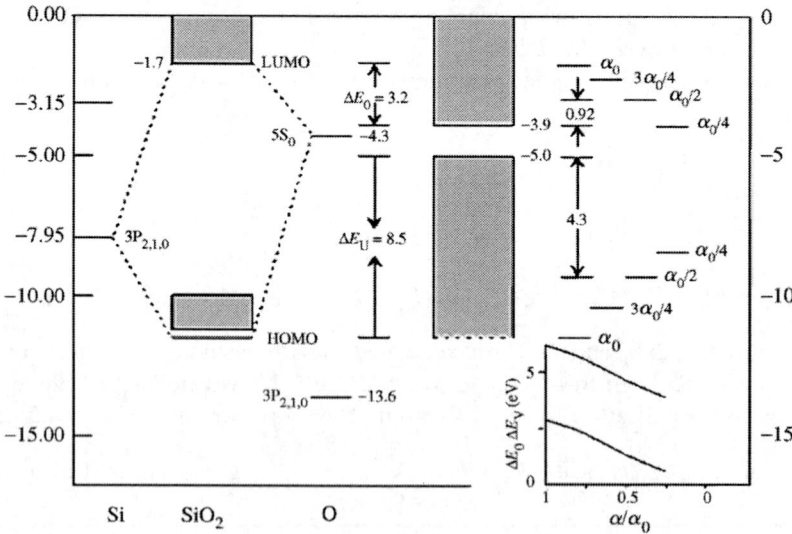

Figure 6.13 HOMO—LUMO used to estimate the band-edge offset between Si and Si—O superlattice. $\alpha_0 = 5.11$ eV is taken to fit SiO_2. For SiO, it is assumed that $\alpha/\alpha_0 = 0.5$ giving $\Delta E_c \sim 1$ eV and $\Delta E_v = 4.3$ eV. We are comfortable with ΔE_c, but ΔE_v is probably closer to 3.4 eV. The shaded regions denote the conduction and the valence bands.
Source: Tsu (2000a), with permission.

6.7 Estimation of Strain from a Ball-and-Stick Model

I have stated on several occasions that whenever an expert says that something cannot be done, one should not take it as gospel, but if he gives a specific outline on how to do it, one should take it seriously. I believed in this statement, and once more I was reinforces by the work of Distler and Zvyagin (1966) and Henning (1970) and it encouraged me to pursue the idea of fabricating SAS epitaxial structures. Now, we need to construct a working model using hand-built ball-and-stick models. We start with the simple idea that the bond lengths Si−Si and Si−O are fixed and the coordination number of Si should be preserved. In addition, beyond the next layer of Si atoms, the usual tetrahedral structure should prevail. The first models with one O-layer and two O-layers are shown in Figures 6.14 and 6.15, respectively.

Figure 6.14 Ball-and-stick model for Si−O SL structure with one monolayer of oxygen.

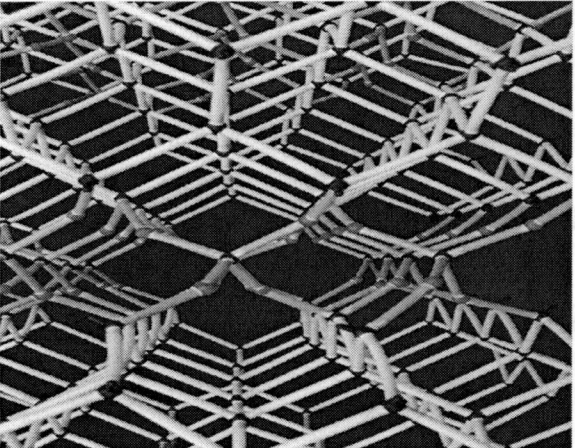

Figure 6.15 Ball-and-stick model for Si−O SL structure with two monolayers of oxygen.

It is generally difficult for anyone to visualize a three-dimensional relationship. The best way is to use the ball-and-stick model kit, although the directions of the bonds are preset in these chemistry sets so that the angles cannot be further adjusted, resulting in a less-than-ideal hand-built model. Nevertheless, the visual perception that these models provide is invaluable. Next, to get the coordinates, we simply put strings into the model as shown in Figure 6.16. With the locations of these atoms assigned, we can calculate the strains and later give these coordinates to DFT calculators for more rigorous computations.

Figure 6.17 shows all the coordinates of a planar view of the Si−O superlattice with one monolayer of O. Two planes, marked by 4 and 3, are shown on the right. Basically, the bottom part of the Si structure is shifted up. The position of P' is at the midpoint between (134) and (113), and P'' is the midpoint between (314) and (333) and P is the midpoint between P' and P''. In addition, the two vectors giving the direction along which the oxygen atoms can sit are marked by J_1 and J_2 as

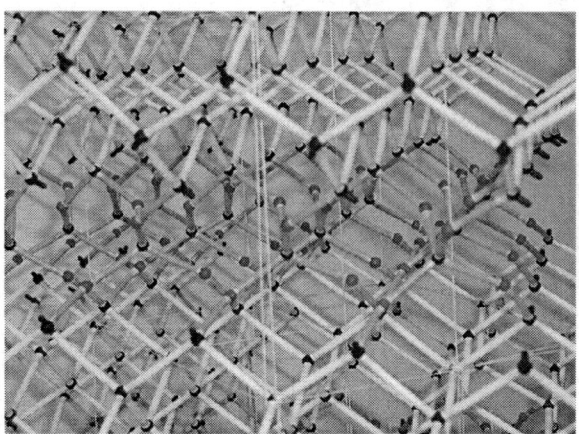

Figure 6.16 Ball-and-stick model for Si−O SL structure with strings added for the coordinates of atoms involved in strain estimate and subsequent DFT computations.

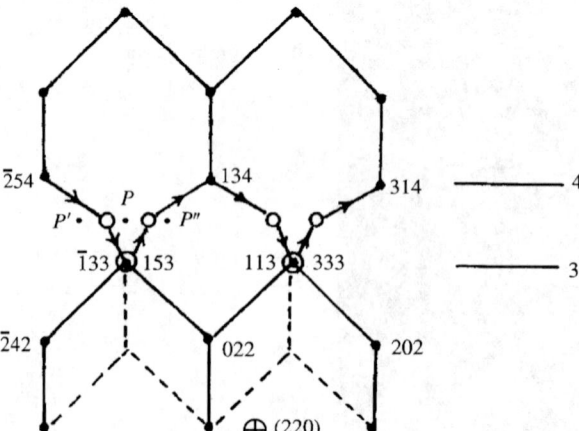

Figure 6.17 A planar view of the Si−O superlattice with one layer of O, showing the coordinates. The arrows indicate the direction in going from one atom to the other; for example, from Si(134) to Si(113) with arrows pointing down is distinguished from Si(333) to Si(314) with arrows pointing up. Dots and circles are used for Si and O atoms, respectively. Two planes, marked by 4 and 3, are shown on the right. For points P, P', and P'', see text.

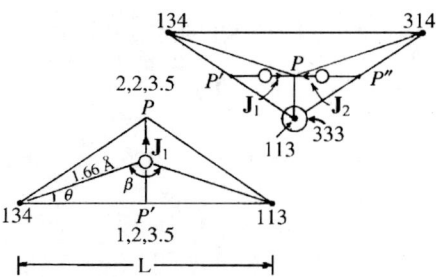

Figure 6.18 P' is the midpoint between (134) and (113). P'' is the midpoint between (314) and (333). The midpoint between P' and P'' is P. The vectors $\mathbf{J}_1$ and $\mathbf{J}_2$ are shown with the O-atom located in the direction of these vectors. The coordinates, bond angles θ and β, together with the bond length of 1.66 Å for Si–O are shown in the lower part.

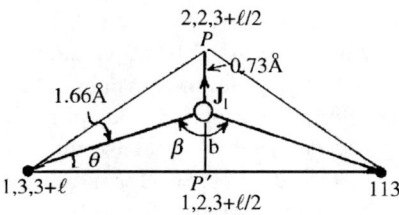

Figure 6.19 P is now $(2, 2, 3 + 0.5\ell)$, with ℓ adjusted for the minimum strain energy.

shown at the top of Figure 6.18 and the bond angles θ and β, together with the bond length of 1.66 Å for Si–O are shown in the lower part of Figure 6.18. These figures are to facilitate the calculation of the strain and minimization of the strain energy that will be presented next.

If the lattice constant for Si is $a = 5.43$ Å, then $L/2 = \sqrt{5}a/4 = 1.52$ Å (Figure 6.18), giving a bond angle $\theta = \cos^{-1}(L/2)/1.66 = 23.9°$, so that the distance P'–O is 0.67 Å. From these, we have the bond-bending strain $S_\beta = 6.4\%$ using the bond angle for Si–O of 141°, and a bond length between the two oxygen atoms of 1.37 Å, resulting in the stretching strain $S_\alpha = 6.2\%$ using our bond length for O–O of 1.46 Å. Next, we allow the separation between the planes P-3 and P-4 to be 2ℓ and minimizing the strain energy with respect to ℓ, we shall show that much lower strains result.

Instead of using the bottom figure of Figure 6.18 in the minimization of the strain energy by letting the separation be adjustable, Figure 6.18 is reconfigured as Figure 6.19, with ℓ as a variable to be adjusted for the minimization of the total strain energy.

Specifically, our procedure of minimizing the strain energy is to fix the bond length of Si–O at 1.66 Å, but letting the angles θ and β depend on ℓ, resulting in varying the distance O–O as well. The strain energy ε_s is then

$$\varepsilon_s = k_\alpha S_\alpha^2 + k_\beta S_\beta^2. \tag{6.3}$$

From the variation of $d\varepsilon_s = 0$ with respect to ℓ, we obtain

$$k_\alpha S_\alpha \frac{\partial S_\alpha}{\partial \ell} = -k_\beta S_\beta \frac{\partial S_\beta}{\partial \ell}, \tag{6.4}$$

in which the two Keating constants (Keating, 1966) for stretching and bending are k_α and k_β, respectively. For silicon, $k_\alpha = 27k_\beta$, writing $\ell \equiv 1 + \Delta$ and neglecting terms $\sim \Delta^2$, the condition for minimum strain energy is given by

$$k_\beta \times 10^{-2}[120(1 - 38.5\Delta) + 2.3(1 - 2.8\Delta)] = 0, \text{ giving } \Delta = 0.026. \quad (6.5)$$

We need to be careful; although Figure 6.19 is drawn in a plane, it is really a three-dimensional picture. For example, the distance between the points $P' - (1, 3, 3 + \ell) = \sqrt{1 + (\ell/2)^2}(a/4)$, therefore, for $\ell \equiv 1 + \Delta$, the length $P' - O \equiv b$, so that $b^2 + (a/4)^2 \times [1 + (\ell/2)^2] = 1.66^2$. These values are not too different from those of the density functional calculation (DFC), which will be discussed later. To summarize, at least our model shows a value for strain, not negligible, but perhaps tolerable. There are several models possible making similar assumptions; keeping the bond length fixed and allowing the bond angle to be adjusted while conserving the coordination number. To accommodate the presence of oxygen, for minimum strain, the distance between two planes given by Eq. (6.5) increased by only 2.6%, resulting in the strain $S_\alpha = 0.05\%$ and $S_\beta = 5.9\%$. The HRX-TEM is not able to show the 2.6% increase, however, using a sample with 20−22 layers of Si for each monolayer of oxygen, the strain pattern does show a superlattice structure in Figure 6.20.

Figure 6.20 shows HRX-TEM of a sample with a four-period Si−O superlattice. This is the first time a periodic superlattice structure has been seen in TEM.

Following this success, a nine-period Si−O superlattice having the same thicknesses was examined by HRX-TEM in more detail (see Figure 6.21) indicating that the strain pattern caused by the introduction of oxygen spreads significantly beyond the monolayer (Gurdal, unpublished). At this stage, we realized that the strain pattern should spread several atomic distances in a similar manner to the theoretically calculated GaP on GaAs (Tsu et al., 1989), which will be treated next.

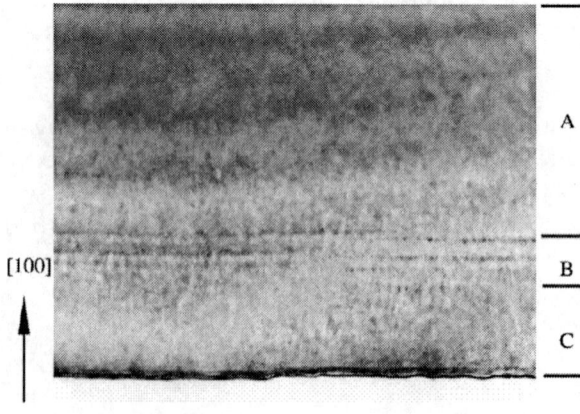

[100]

A

B

C

Figure 6.20 A four-period Si−O superlattice with (A) part of a 235-nm Si buffer, (B) four periods of Si−O superlattice with each period consisting of one monolayer of adsorbed oxygen sandwiched between five unit cells of Si on each side, giving a total thickness per period of ~11.5 nm, and (C) an epitaxial Si cap of ~24 nm. Note the presence of long-range coherent layering.
Source: Tsu et al. (2000b), with permission.

Figure 6.21 Eight periods of Si–O superlattice with each period consisting of nine monolayers of Si with one monolayer of adsorbed oxygen with a total thickness per period of ~1.25 nm, making a total thickness in the superlattice region of ~10 nm, on top of 5 nm of undoped buffer. Note that the strain propagates to a couple of monolayers on each side of the adsorbed oxygen, shown as dark lines. *Source*: From Gurdal (unpublished).

6.7.1   Charge Transfer on Strain-Layer Epitaxy

Replacing a Si atom by an O atom with a transfer of charge from the Si to O is similar to the replacement of As by a P atom in GaAs, calculated by Tsu et al. (1989). When an atom of As is replaced by a P atom in GaAs, owing to the difference in ionicity, local puckering is affected by charge transfer. The total energy stored $E_{ex} = E(\text{strain}) + E(\text{interface}) \equiv E_s + E_{int}$. For several atomic layers of GaP on GaAs, the dominant stored energy is the electrical dipole–dipole energy at the interface. The creation of misfit dislocations may release the strain energy (Matthews and Blakeslee, 1976), although it cannot affect the interface dipole term. Only interface diffusion can lower E_{int}. It has been further argued (Dodson, 1988) that the dislocation sources are inhomogeneous in nature resulting from local stress concentrations near the interface. Strain energy is responsible for the creation of misfit dislocations, while the dipole energy drives diffusion. The excess energy of L, the number of strained epilayers of GaP on a GaAs substrate, is calculated from a minimization of the total internal energy $U = U(\text{repulsive}) + U(\text{electrostatic})$. In Figure 6.22, a GaP epilayer is confined to the GaAs substrate in the plane of the interface, the (100) plane, leaving the vertical dimensions $\delta_l = \delta_1, \ldots, \delta_L$, to be determined *via* a minimization of U. Instead of using the Madelung sum of the electrostatic term, we have chosen a unit tetrahedron to calculate the electrostatic energy, $U(\text{electrostatic}) = U(\text{tetrahedron}) + U(\text{dipole–dipole}) + U(\text{higher order})$, where the higher-order multipoles are neglected.

This procedure is similar to finding a faster convergence to the Madelung sum (Evjen, 1932). Table 6.2 gives the charges in various III–V of interest. In calculating the effective charge e^*, the starting point is the ionicity f. Using a simple heteronuclear diatomic model, $f = (1 - b^2)/(1 + b^2)$, in which b is the normalization for the coupled modes A and B, with wave functions, $\psi_{\pm} = (1 + b^2)^{-1/2}[\psi_A \pm \psi_B]$.

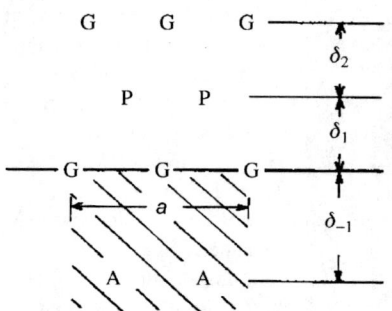

Figure 6.22 Two epilayers of GaP are shown on a (100) GaAs substrate. Note that the distance in the (100) plane is fixed by the lattice constant of GaAs, while the vertical distances are adjusted *via* a minimization of the total internal energy. *Source*: Tsu et al. (1989), with permission.

Table 6.2 Constants for Various III−V Compounds for the Repulsive Energy U_R

	a (Å)	$Z\lambda$ (10^3 eV)	ρ (Å)	e^*
AlAs	5.66	8.29	0.33	1.91
GaAs	5.64	6.26	0.345	1.97
GaP	5.45	6.09	0.34	2.02

Applying the 8-N rule for GaAs, the effective charge $e^*(\text{Ga}) = 3/(1 + b^2) = 1.97e$ and $e^*(\text{As}) = -1.97e$ for $f = 0.31$, from Phillips (1973). Further, the repulsive energy is given by Kittel (1973), by

$$U_R = NZ\lambda \exp(-R/\rho), \tag{6.6}$$

in which R is the nearest neighbor distance. The parameters $Z\lambda$ and ρ are obtained from the relationship between the bulk modulus and the energy $U = U_R - \alpha q^2/R$, with α being the Madelung constant. These values for various III−V compounds are listed in Table 6.2.

Further simplification uses a fixed value δ for the δ_l averaged at thickness L, while taking the lattice constant of GaAs, a, for the in-plane separation of Ga−P. In other words, δ is the variational parameter for the minimization of the energy. Then the repulsive energy

$$U_R = Z\lambda \exp[-(1 + 8(\delta/a))^2]^{1/2}(a/2.828\rho)(L + 1)/2. \tag{6.7}$$

The energy of the tetrahedron, U_{tet} is given by

$$
\begin{aligned}
U_{\text{tet}} = {} & \frac{(L-1)(2.02e)^2}{2a}[(1 + 16(\delta/a)^2)^{-1/2} - 5.657(1 + 8(\delta/a)^2)^{-1/2}] \\
& + \frac{(1.97)2.02e^2}{4a}[(5/16) + 0.5(\delta/a) + (\delta/a)^2)]^{-1/2} \\
& - \frac{(1.97 + 2.02)2.02e}{1.414a}(1 + 8(\delta/a)^2)^{-1/2}.
\end{aligned}
\tag{6.8}
$$

The last two terms in Eq. (6.8) represent the electrostatic energy of a unit tetrahedron located just below the interface. The dipole–dipole term is

$$U_{d-d} = \frac{(1.97e)(2.20e)a\delta}{4(0.25a + \delta)^3}.$$

(6.9)

Next, we minimize the sum of the three terms (Eqs. (6.7)–(6.9)) with respect to δ. To find the excess energy, we must subtract $(a/4)$ (GaP) for the vertical distance and a (GaP) in the plane. In other words, we are looking for the unstrained GaP without the interface dipole term. Therefore, the excess energy E_{ex} includes the dipole and the strain terms because we have subtracted the inherent energy of the crystal.

Figure 6.23 shows a comparison of the classical stored strain energy for a film that has two equal strains in the (100) plane and a stress-free direction normal to the interface. The stored strain energy for this case is

$$E_s = 0.5(C_{11} + 2C_{12})(1 - C_{12}/C_{11})S^2,$$

in which S for GaP on GaAs is 0.036. Note that the excess energy is normalized to Å^2. For a given (100) plane, there are two Ga atoms in an area of $5.65^2 \, \text{Å}^2$.

Figure 6.23 shows the calculated excess energy plotted against the number of epilayers for GaP on GaAs. Considering that no adjustable parameters were used in the calculation, the 20% higher asymptotic value compared to the classical value is really quite reasonable. Note that the excess surface energy, primarily due to the dipole energy from charge transfer, extends to a couple of epilayers. From Figures 6.20 and 6.21, it is quite clear why we failed to observe the superlattice pattern in HRX-TEM for all of our samples before we significantly increased the thickness of each period. Spreading the strain on both sides of the oxygen layer prevented the TEM pattern from distinguishing between the regions with or without the oxygen.

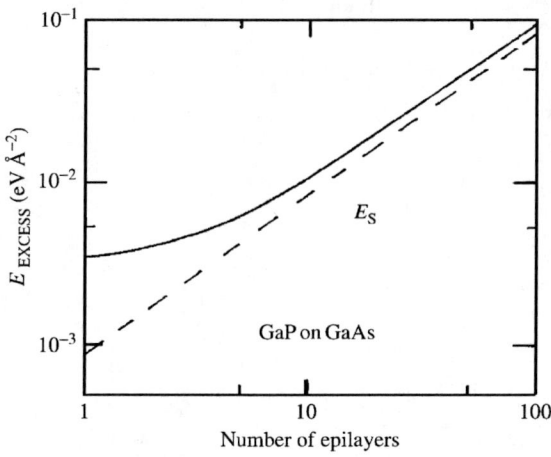

Figure 6.23 Calculated excess energy in eV Å^{-2} versus the number of epilayers. The dashed line represents the classical strain energy. Most of the surface energy is dominated by the electrically repulsive dipole–dipole energy at the interface. Note that the strain energy at the surface extends to a couple of epilayers.
Source: Tsu et al. (1989), with permission.

As Figure 6.15 shows the ball-and-stick model of the Si—O superlattice structure with two monolayers of oxygen; the corresponding detail for the coordinates of Si and O atoms is shown in Figure 6.24. Basically, there are now two monolayers of O atoms separated by a monolayer of Si. Obviously, nothing else is possible. Four planes are marked, with O atoms in between P-1 and P-2 planes, and P-2 and P-3 planes.

To enable visualization, Figure 6.25 gives a three-dimensional arrangement of Si and O atoms shown as dots and circles. Three planes, P-1, P-2, and P-3 in Figure 6.25 are also marked. The coordinate axes are placed at (000) of the Si unit cell, which is shown as a dotted cube.

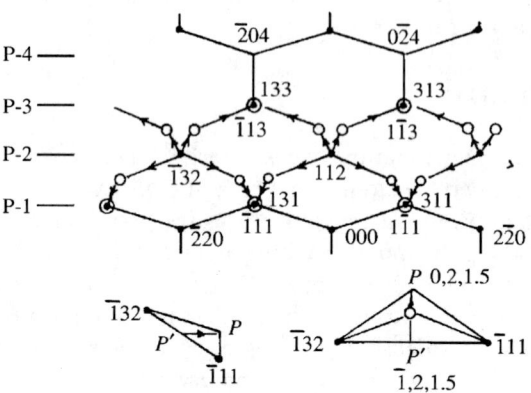

Figure 6.24 Coordinates for the two layers of O atoms separated by a monolayer of Si atoms. The explanation of the arrows is similar to Figure 6.17. Four planes are marked, with O atoms between P-1 and P-3. The bottom part of the figure shows the detail of the locations of the O atoms, along the vector P'—P, with respect to Si atoms.

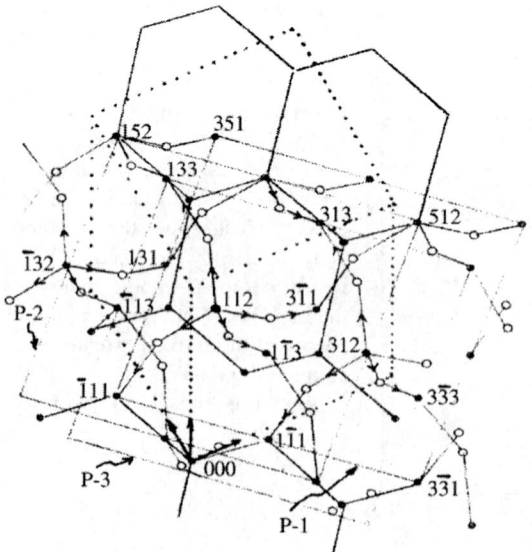

Figure 6.25 Three-dimensional arrangement of Si and O atoms for the case of two monolayers of O as in Figure 6.23.

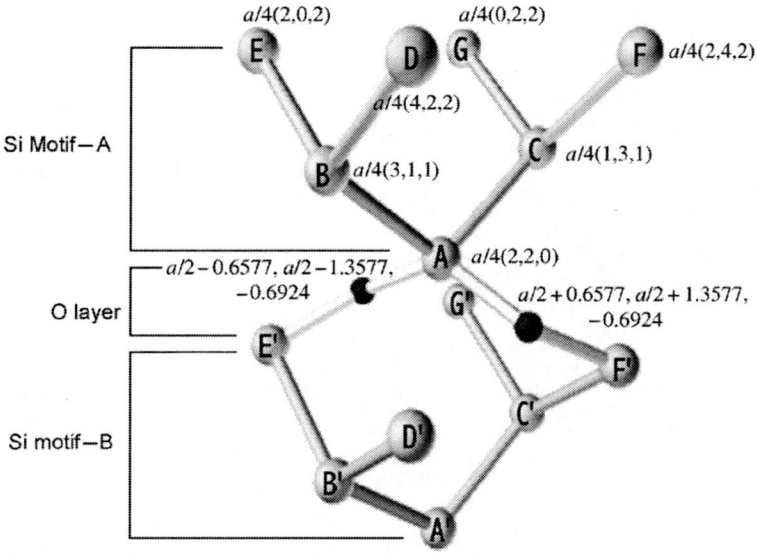

Figure 6.26 Coordinates used by Tsu and Lofgren (2001) to calculate strain without minimizing the strain energy giving $S_\beta \sim 6.4\%$.
Source: After Tsu and Lofgren (2001), with permission.

Several versions of the coordinates were used. For example, the center Si atom is placed at A (2, 2, 0) $a/4$ in Figure 6.26. Obviously, one can simply use a translational vector $T = (1, \bar{1}, \bar{4})$ to transform, for example, the Si atom at (1, 3, 4) in Figure 6.17 to the Si atom at A (2, 2, 0), used by Tsu and Lofgren (2001), in Figure 6.26.

Before Babic proceeded with his DFT computation (Figure 6.27), he used the computer-generated model shown in Figure 6.28(A), the original, Si motif where the darkened Si atoms are designated for replacement by two O atoms as shown in Figure 6.28(B), along the (100) plane.

Babic found that the best ℓ is 1.3, which is significantly larger than the 1.026 using the minimization of the total strain given before (Table 6.3). The shear value S of 1% is below the 6.4 and 0.05% obtained before. But his tensile S value of 10% is huge, coming from the dipole energy resulting from the transfer of charge e^*, which was not accounted for in the previous minimization of the strain energy. Both strains decay very rapidly when moving perpendicularly away from the interface. The minimum thickness of the epitaxial silicon layer between the two oxygen interfaces is given in the fourth column as ~ 1 nm, which is near the thickness we used, 1.1 nm or two unit cells of the Si lattice, for most of our Si–O superlattice structure. The charge transfer induces a repulsive electric dipole at the interface with the energy shown in the fifth column. The charge transfer e^*, from the Si atom to the oxygen, produces a slight ionicity. $\Delta(eV)$ is the most important column, giving the increase in the binding energy per atom, and reaching a maximum

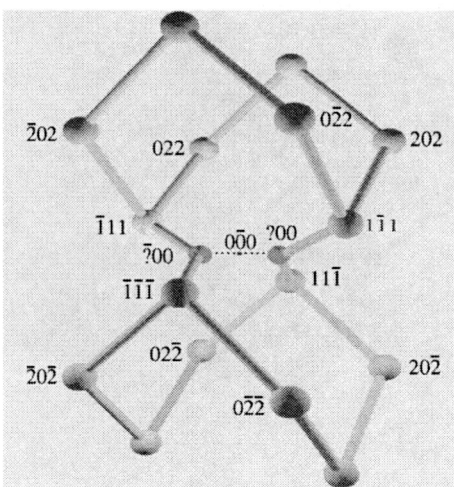

Figure 6.27 The coordinates used by Babic for the DFT computation of the best possible strain, shear $S \sim 1\%$, but tensile $S \sim 10\%$, and an increase in the binding energy per atom of 1.13 eV.
Source: From Babic (unpublished).

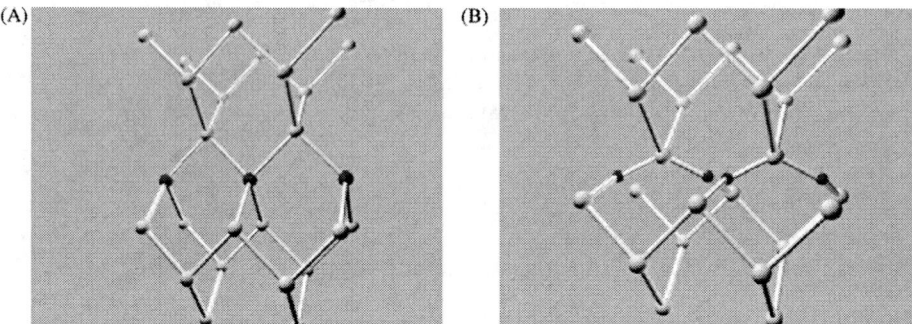

Figure 6.28 (A) Computer-generated model of crystalline silicon lattice where the darkened silicon atoms are designated for replacement by two oxygen atoms as shown in (B) along the (100) plane.

Table 6.3 Babic's DFT Computation, with an Adjustable Parameter for the Layer Thickness Given by $(a/4)\ell$ in the First Column, Charge Transfer e^* from the Si Atom to the O Atom, the Increased Binding Energy per Atom Δ and the Barrier Height, BH

ℓ	Shear S (%)	Tensile S	Minimum thickness (nm)	Dipole E (eV)	e^*	Δ (eV)	BH (eV)
0.8	8.5	45	0.6–0.7	0.11	0.095	0.08	0.95
1.0	6	25	0.6–0.7	0.33	0.12	0	1.22
1.2	3	19	0.7–0.8	0.45	0.15	0.8	2.30
1.3	**1**	**10**	**0.8–0.9**	**0.29**	**0.16**	**1.13**	**0.5–1.74**
1.4	3	10	0.9–1.0	0.35	0.18	1.03	4.06

From Babic (unpublished).

at $\ell = 1.3$, which coincides with the minimum shear strain of 1%. This provides the stability of the SAS system. Consequently, the silicon epitaxy beyond the oxygen interface is stable. The last column gives the barrier height as the sum of the lowest excitation energy and the dipole energy. The range of the barrier height of $0.5-1.74$ eV depends on the dielectric screening used. If a dielectric constant of 12 is used, we obtain BH $= 0.5$ eV, and if we simply use the average value of SiO_2 and Si, we obtain 1.74 eV. In other words, the lower the dielectric constant, the higher the barrier height. We know that the dielectric constant is lowered in a short-period superlatttice (Chapter 4) and an increase in ionicity results in a decrease in screening. Therefore, we come to the best estimate for the BH at ~ 1 eV, consistent with my estimate from the activation measurements. We used this value all along, even before the DFT was done, demonstrating the importance of coordinating theory and measurement.

We learn from the strain estimate and the DFC calculation that the tensile stress is large from the DFC results. Although it decays rapidly when moving perpendicularly away from the interface, the important question is whether defects are generated by this large repulsive dipole–dipole energy, creating a tensile strain perpendicular to the layers. The only way to affect this energy is *via* diffusion to alter the charge transfer. As long as we assume that diffusion does not occur, this energy is dormant. My intuition is that it can augment generation, but not create defects or direct participating in the nucleation of defects. According to the Matthews and Blakeslee (1976) criterion, the stored strain energy is simply not enough to reach the threshold. As long as we only construct a one-oxygen monolayer or even a two-oxygen monolayer superlattice, we are dealing with the strain-layer superlattice (Osbourn, 1982), or at least we think so.

6.7.2   Defects in the Si–O Superlattice

The HRX-TEM of a Si–O sample: Si (buffer)/(O–Si (1.1 nm)–O–Si (1.1 nm)–O)/ Si on Si(100) is shown in Figure 6.29(A). The "whitish" part of the figure may indicate where the oxygen "cluster-like" region is located. Epitaxy is evident beyond and around this "whitish" region. We have also superceded the SAS by Si (111). Generally, it is slightly more defective than Si(100). Although we have a relatively continuous layer of oxygen, as pointed out before (Tsu et al., 1999a,b), even discontinued clusters can serve as a barrier because electrons, like de Broglie waves, cannot pass through a region of space smaller than the wavelength. This is a good place to emphasize that there are two mechanisms, step in energy and/or step in geometrical shape, both of which give rise to an effective barrier to electrons. In Figure 6.29(B), plane view TEM shows rather low defect densities, somewhat below 10^9 cm^{-2}. Both of these TEM pictures were obtained by Dovidenko, who assured me that this cluster-like region is not a cluster in the usual sense of a region, definable by its own lattice structure. We are not too sure why HRX-TEM shows "cluster-like" regions, but the plane view is rather uniform throughout.

We know that without the oxygen adsorption process our epitaxially grown silicon has much lower defects, and strain is really not the cause of defects. We also

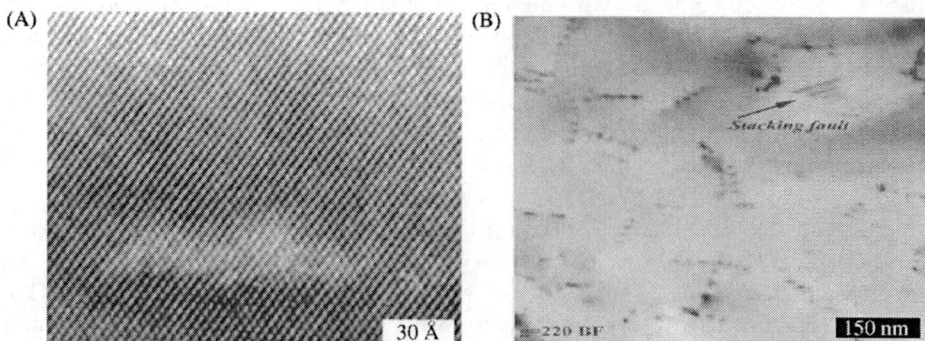

Figure 6.29 (A) HRX-TEM of Si(100). Moving from the bottom of the figure toward the top shows Si (buffer)/($2 \times$ O−Si (1.1 nm)−O)/Si-capping. (B) Plane view TEM of the same sample. The defect densities is below 10^9 cm^{-2}.
Source: After Tsu et al. (1999a,b), with permission.

know that heavily Sb-doped silicon, even without the oxygen adsorption, has much more defects. Therefore, I am forced to accept the idea that the source of the defects is already in the silicon wafer, or is introduced externally.

We have decreased the number of defects by almost two orders of magnitude during the couple of years since we launched this work. We are optimistic that further reduction should be possible. Note that the defect density at the Si/SiO$_2$ interface for most MOS gates is generally higher. At this stage, we think the defects may still be extrinsic rather than from the strain. Certainly, more research may provide an answer. As pointed out previously (Tsu et al., 1999a,b), whenever the oxygen leak valve is left on during the silicon deposition, horrendous numbers of defects are generated, although the three-dimensional diffraction pattern still persists. The defects cannot be covered up even after 200 nm of Si deposition on top of the SiO$_x$ layer, as shown in Figure 6.10. It is generally known that it is not possible to grow good Si epitaxially on an oxide layer, but at least we have shown that it is possible using the Si−O superlattice.

6.8 Electroluminescence and Photoluminescence

Figure 6.30(A) shows a schematic of a nine-period EL device with a Si−O superlattice as the active layer. EL covers the whole contact with bright EL around the edges. The voltage applied across the Schottky diode is between -20 V reverse bias and as low as -6.8 V reverse bias. This point will be discussed in more detail when the life-test result is shown. Annealing in H$_2$ + N$_2$ (1:10) at 420°C for 10 min leads to a greater EL intensity.

The spectra of both EL and PL were obtained with a U-1000 Yvon-Jobin monochromator and detected by a R-943 Hamamatsu PMT. The 457.9-nm line of the Ar laser was used for the PL spectra. Figure 6.31 shows typical EL and PL. Although

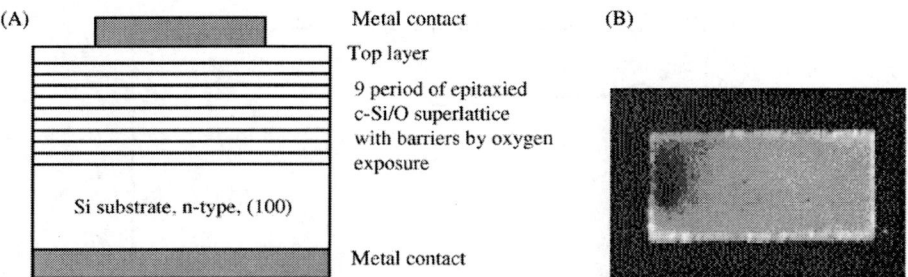

Figure 6.30 (A) This is a schematic of the EL device with a nine-period Si—O superlattice as the active layer. EL from the top through a partially transparent Au electrode with dimensions of 0.5 × 1.2 mm is shown in (B). The dark spot is caused by the wire contact. *Source*: Tsu (2000a,b), with permission.

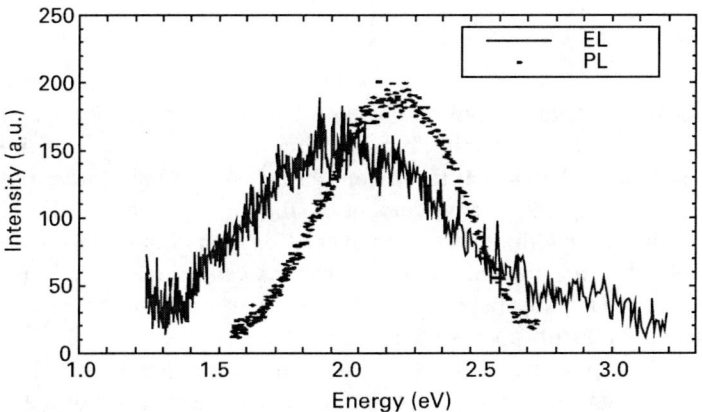

Figure 6.31 EL and PL from a typical Schottky diode with nine-period Si—O superlattice. The 457.9-nm line of the Ar laser was used for the PL spectra. The great intensity of the PL is due to greater collection of the emitted light. The PL cutoff is due to the Ar laser. *Source*: After Tsu et al. (1999a,b), with permission.

the main peak is located at 2 eV, the emitted light appears to be greenish because the EL spectrum extends to a photon energy beyond 3.5 eV. From the activation energy determination (Figure 6.11), the ground state in the Si—O superlattice is $\sim$0.5 eV, and an estimate for the valence band is $\sim$0.4 eV, giving a total, with the 1.2 eV band gap of silicon added, of $\sim$2 eV for the optical transition. Therefore, the peak measured at $\sim$2 eV may be due to the quantum state. The photons in EL above 3.5 eV may possibly be due to higher quantum states, or may be attributed to a Si—O localized complex as in porous silicon. Although the yellowish-green EL emission is visible with the naked eye through a low-power microscope, it is not very bright. Even at 1.1 nm, the indirect nature of the silicon energy band is still operative so that the optical process is still dominated by phonon-assisted

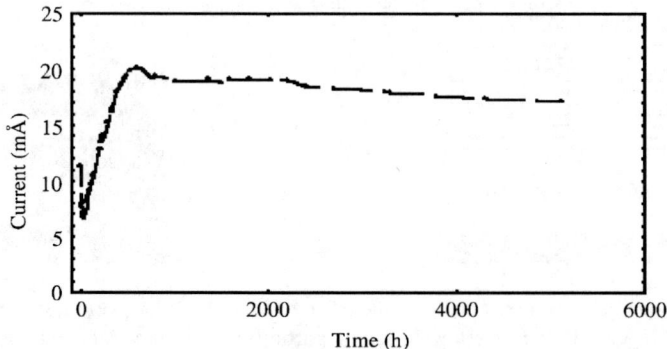

Figure 6.32 Current versus time under reverse bias. For 30 days, -10.4 V was applied. The applied voltage was then cut to -6.8 V to maintain the same current approximately, which resulted in an apparent peak in current at ~ 750 h of operation, with further reductions at the 80th, 120th, and 150th days to -6 V.
Source: After Tsu (2000a), with permission.

transitions. After all, at 1.1 nm, the value of q is still a factor of two below the Δ-minimum of the Si conduction band. The cutoff due to the laser line is evident in the PL spectrum. The spectrum is quite broad. It is possible that the main peak originates from the quantum confinement of silicon and the broad line signifies that the effective barrier width of the monolayer of oxygen is quite thin. Of course, it is also possible that nonuniformity in the barrier causes the broad spectrum. The strong shoulder extending beyond 3.5 eV may even be larger because our spectrometer efficiency falls off above 3.5 eV.

Figure 6.32 shows one of our life tests with continuous monitoring of current. Since the light output is proportional to the current, the stability of the EL device is obvious. In fact, we have also performed under constant current. After the first 30 days, the applied bias was dropped to -6.8 V to keep the current approximately constant. Therefore, the apparent current peak is an artifact of reducing the applied voltage at 750 h. The longest operating time is over 1 year when my laboratory was required to relocate. The applied bias includes the voltage drop over the substrate. The drop from -10.4 to -6.8 V is probably due to current-induced annealing effects, commonly known as electrical forming. After the initial 30 days of operation, there is an increase of 50% observed in the light output (Tsu et al., 1999b). In other words, the light output is linear with current. As noted, the photon emission is certainly not as efficient as direct-gap semiconductors, but it is more efficient than silicon. And since it appears to be extremely robust, it may find applications in Si-based devices.

The PL of a four-period Si–O superlattice with an increase of the Si thickness from 1.1 to 2.2 nm has been fabricated in order to ascertain whether a lowering of the quantum state is possible. The spectrum consists of a broad peak at 1.85 eV, a shoulder at 2.15 eV, and a sharp peak at 2.2 eV, shown in Figure 6.33. Note that this sharp peak at 2.2 eV is close to the value of 0.92 eV above the Si conduction

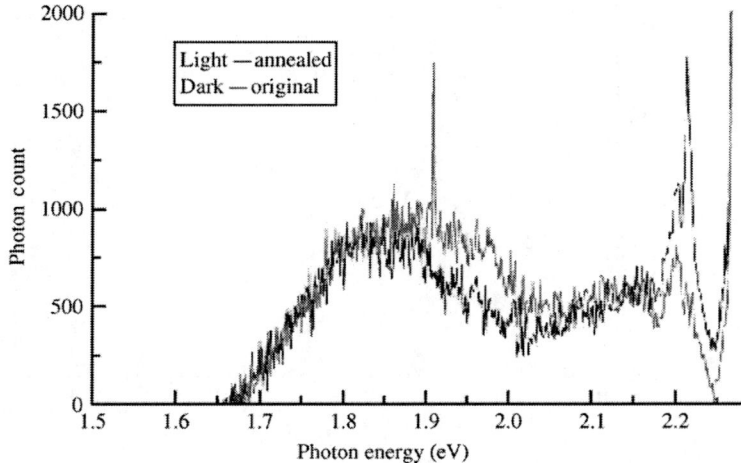

Figure 6.33 PL of a four-period Si(2.2 nm)—O superlattice complex excited by the second harmonic Nd—YAG laser. The increase in the Si thickness is to lower the quantum state. The spectrum was measured by A. Sevian.
Source: After Tsu (2000a), with permission.

band minimum for the LUMO calculated with $\alpha/\alpha_0 = 0.5$ in Table 6.1. If we consider the 2.2 eV structure as the down-shifted EL shoulder structure at 3.1 eV in Figure 6.31 due to increased thickness of the Si layers, there is a weak quantum confinement effect. The sharp peak decreases after annealing at 850°C, indicating that oxygen may have diffused into a different arrangement, supporting the interpretation of the Si—O complex as the origin of the emitted light. On the other hand, the 850°C anneal could have altered the Si—O superlattice structure in the first place. In essence, we think the origin of all these optical emissions consists of both man-made quantum effects and interface-localized complexes. Again, more research needs to be carried out to acquire more definitive answers.

In my laboratory, we do not have a setup for absolute calibration of efficiency. We have put sufficient effort into developing porous silicon, PSi, targeting improvement in the surface morphology of PSi using a gentle etching technique (Filios et al., 1996). This effort not only gave consistent results, but also obtained the best PL from PSi for a given thickness. Thus, for emission efficiency, we simply compare the emission from Si—O superlattice with our best PSi sample. Figure 6.34 shows the comparison of a typical nine-period Si—O superlattice with an active region of ~10 nm and a PSi sample with an active region of ~100 nm. The emission efficiency of our Si—O superlattice is at least as good as the PSi sample, giving the best reported efficiency of 0.1—0.2% (Tsybeskov et al., 1996). Considering how many people have been excited by the prospect of the application of PSi in optoelectronics that has dubious reliability, the availability of stable, robust Si—O superlattice is much more appealing. EL is another story. EL from PSi is simply not stable at all. The EL efficiency is estimated to be 2×10^{-4}%, by means of threshold voltage and minimum operating current, and experimentally

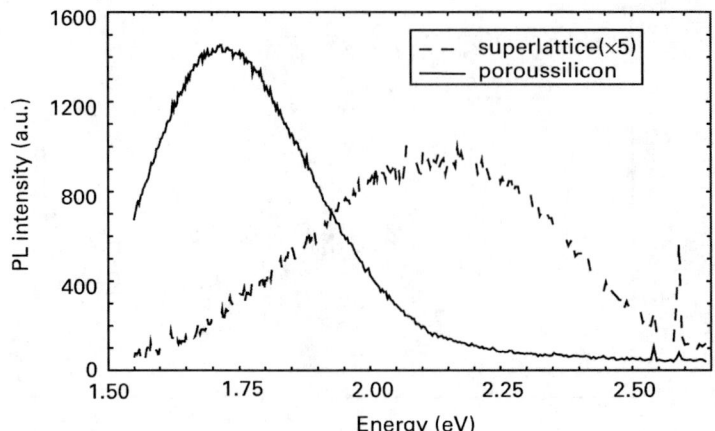

Figure 6.34 PL from both porous silicon and Si−O superlattice taken under an identical setup. The PSi sample is freshly made from p-Si(100), $1-2\,\Omega$ cm with a thickness of ~ 100 nm. Thus, the PL efficiency is at least similar or better than PSi.
Source: After Tsu et al. (1998b), with permission.

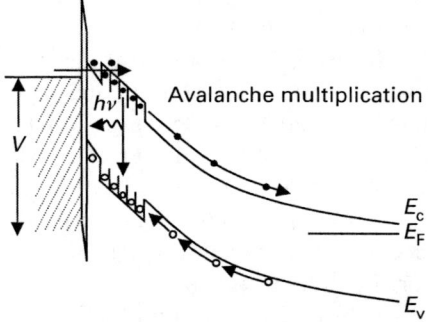

Figure 6.35 Band profile shows a Schottky barrier and the active region of the Si−O superlattice EL device under reverse bias. Electrons undergo avalanche multiplication providing holes for recombination with light emission.
Source: After Tsu et al. (1999a,b), with permission.

using the same procedure as for PSi. It is evident that EL efficiency is much lower than that of PL. Next, we shall give several reasons why this is so.

Note that the EL device is operated under reverse bias, which is rather different from the pn-junction light-emitting diode (LED), which is operated under forward bias. The holes for recombination are generated as a result of an avalanche of injected hot electrons, as shown in Figure 6.35.

The need for an avalanche, shown in Figure 6.35, obviously lowers the efficiency. We launched a project to provide double injection by placing the active region within the depletion region of a pn-junction. Serious difficulties were encountered in trying to fabricate the Si−O superlattice structure on fairly and heavily doped silicon. We even failed to grow a good epitaxial silicon buffer on a heavily Sb-doped buffer. We called off this attempt due to lack of experience, work force, and other resources.

In conclusion, we know that more repeats of the Si−O period result in higher emission, indicating that even if the emission is due to the hybridization of the oxygen−silicon arrangement, the complexes contribution to emission is not from some unwanted oxygen on the surface of the last interface. The most important feature for new applications lies in the fact that the EL device is stable and robust. In fact, the current-induced annealing effect shown in Figure 6.32 may be linked to the evening out of the large dipole−dipole energy at the interface. As pointed out by Tsu et al. (1989), the only way to affect this interface term is *via* diffusion, or more precisely, rearrangement of the positions of the oxygen atoms relative to the silicon atoms. Thus, with the application of a voltage, current-induced diffusion may be the cause of electrical annealing. The best part is the fact that photoemission has gone up over the months of operation. Whether the low efficiency can be significantly improved is another matter. And possible application depends on finding a niche in the world of solid-state devices.

6.9 Transport through a Si−O Superlattice

The Si−O superlattice was introduced as an epitaxial barrier for silicon quantum devices. A barrier height of ∼0.5 eV with a thin layer of silicon sandwiched between two adsorbed oxygen layers has been achieved. Since the current−voltage shows a fair degree of insulation, naturally it may serve as a replacement for the present amorphous silicon dioxide gate for metal oxide semiconductor field effect transistor (MOSFET), moving one step closer to the ultimate goal of a three-dimensional integrated circuit (3D-IC). Furthermore, the epitaxially formed insulating layer can also serve as a silicon on insulator (SOI) instead of the present MOSFET using the depletion of the pn-junction to isolate neighboring devices.

Although current−voltage measurements have been used from the very start of the project, detailed I−V measurement was undertaken by Seo et al. (2001). For electrical measurements, the buffer layer is doped with antimony to ∼10^{17} cm^{-3}, forming a relatively thin layer of 20 nm. After the buffer layer is completed, RHEED shows the typical 2×1 surface reconstruction. For a nine-period Si−O superlattice, the sample is moved back and forth between the growth chamber for silicon deposition and the analysis chamber for oxygen exposure, usually at 10^{-7} Torr. Therefore, the fabrication process is time consuming and quite tedious particularly because the procedures have not been computerized. Finally, an antimony-doped 4−5-nm silicon layer is deposited on top of the nine periods to cap the superlattice and facilitate the deposition of the electrical contact. Without this relatively light doping, the I−V will be totally dominated by the Schottky barriers at the contacts.

After the MBE deposition is completed, the sample is cut into smaller pieces for further processing, such as annealing and passivation. Specifically, our samples were annealed in $H_2 + N_2$ (1:10) at 420°C for 10, 20, and 30 min, respectively, mainly to remove dangling bond defects. For comparison, we have also annealed our sample in $O_2 + N_2$ (2:1) at 800°C for 10 min. Finally, a 200-nm-thick aluminum electrode is evaporated at 10^{-7} Torr followed by patterning to provide

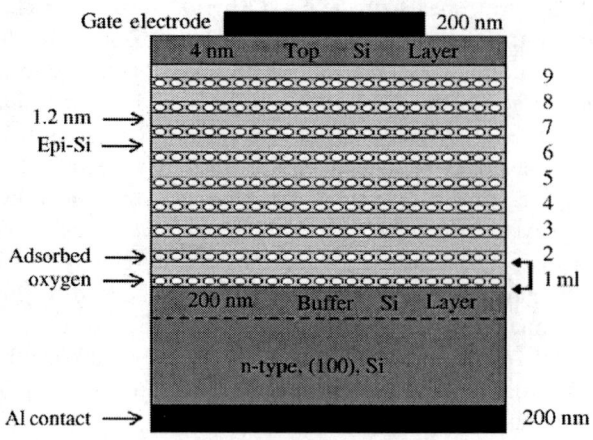

Figure 6.36 Schematic cross-sectional structure of a nine-period Si−O superlattice. *Source*: Seo et al. (2001), with permission.

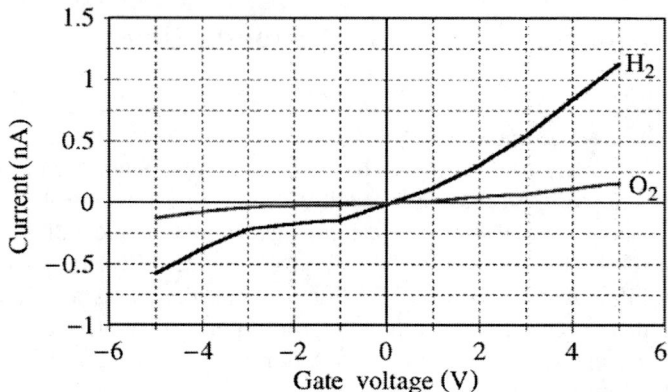

Figure 6.37 Current versus bias voltage for two samples, H_2 for hydrogen annealed, (420°C, 30 min) and O_2 for oxygen annealed (800°C, 10 min). *Source*: After Seo et al. (2001), with permission.

contacts and to define the individual devices. The back contact is also aluminum. Nicollian taught us that a simple way to remove the Schottky barrier on the back surface of the wafer is simply to rub the back surface of the wafer with fine sandpaper before metallization. Figure 6.36 gives a schematic view of the device used for the $I-V$ and $G-V$ measurements, having contact area $\sim 10^{-3}$ cm^2.

The DC $I-V$ measurements were performed with a Keithley 236 $I-V$ system. Figure 6.37 shows the comparison of the $I-V$ characteristics for hydrogen- and oxygen-annealed samples. It is worth pointing out the difference between this work and other work reported previously (Tsu et al., 1997). The previous single-barrier results in a rather unsymmetrical $I-V$ because the Schottky barrier was formed by n-doping while relatively heavier doping is used for this investigation.

Figure 6.38 shows results from similar two cases, H_2 and O_2 annealed at a bias voltage between $+30$ and -30 V. The device is stable in this range. Figure 6.39 shows the conductance in a detailed comparison of the H_2 and O_2 annealed cases.

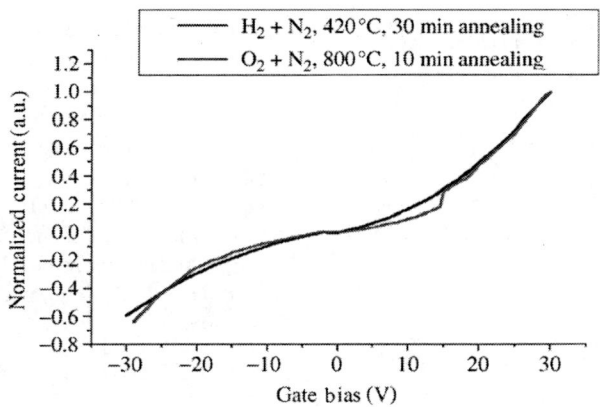

Figure 6.38 $I-V$ for the same type of samples with a larger range of bias voltage.
Source: Seo et al. (2001), with permission.

There is a negative conductance, which is probably due to hot carrier injection into the silicon capping region and the silicon buffer region undergoing avalanche multiplication, as discussed by Tsu (1998). Resonant tunneling *via* some inadvertent isolated defects may be responsible for the initiation of the process. The detail of the $G-V$ is really far more complex. A simple explanation based on resonant tunneling without a model including trapping mechanisms cannot offer a satisfactory understanding at this point. Perhaps, the domain formation, discussed in Chapter 1 of this text (Esaki and Chang, 1974) is involved.

We have shown that the Si−O superlattice can serve as an isolation for Si devices. Up to a bias of 30 V, the field inside the multilayer structure reaches $\sim 3 \times 10^7$ V cm^{-1}. There is no sign of breakdown, which is rather reassuring. However, the low-voltage isolation is not sufficient for implementation as a substitute for SOI. Drastically increasing the number of periods, for example, 50 periods, may offer the needed requirements. However, the appearance of jumps and negative resistance indicate the presence of electrically active defects and traps, which may be reduced by further investigation involving annealing, passivation, and optimization of parameters, such as the thickness of the thin silicon as well as the oxygen exposure, in addition to all the usual parameters for MBE or CVD such as temperature, base pressure, and deposition rates. Present interest in thin oxide gates for high-frequency MOSFET is running into similar problems, known as the soft breakdown and hard breakdown (Nigam et al., 1999). The nature of this investigation is believed to be tied to these studies. In summary, we have shown that the Si−O superlattice can serve as a barrier as well as an isolation for Si devices. Our experimental results show substantial promise.

6.10 A Si−O Superlattice and Other Si/Ge, Si/Co, Si/C Monolayer Superlattice

In searching for a barrier for silicon, an epitaxially formed Si−O strain-layer superlattice was launched. The necessary condition is that the oxygen layer is a monolayer. It

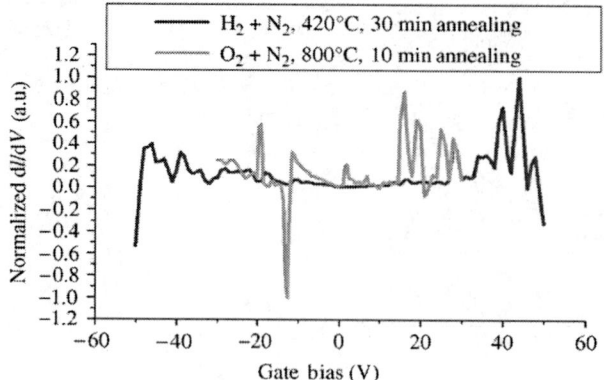

Figure 6.39 Conductance for the two types of samples: (A) for hydrogen annealed: dark and (B) for oxygen annealed: light. There is a negative resistance for the oxygen annealed. Instability sets in beyond 40 V for the hydrogen-annealed sample. *Source*: Seo et al. (2001), with permission.

is incorporated by adsorption. The interfacial strain is at most a few percent, which is really not too high for strain-layer superlattice. However, the interfacial dipole−dipole energy due to charge transfer is high enough to cause alarm. Nevertheless, we think the latter could be a problem with excitation such as a current. Oddly enough, a long-term operation passing a substantial current with a fairly high electric field for at least several months seems to have annealed out significant numbers of the defects. The band-edge alignment in the conduction band of the Si/Si−O superlattice is ∼1 eV, sufficiently high for most quantum device applications. The PL and EL are stable and robust, with efficiencies almost as good as the best porous silicon. The high field breakdown is excellent such that the Si−O superlattice may be a candidate oxide for use as an insulating gate for MOSFET, and even as an insulating, epitaxially grown SOI.

Other projects attempted for the SAS include Si−CO superlattice, which has a PL quite similar to Si−O superlattice, but with a narrower emission linewidth and Si−C superlattice, to which we have devoted more effort. The Si−C superlattice that was tried structurally seems to be on track, though the RHEED pattern is not as good as Si−O superlattice, and no visible light emission has been observed thus far. An initial estimate gives a band-edge alignment in the conduction band of ∼2 eV, which is consistent with why there is no visible light emission. The electronegativity of C, $\chi(C)$ is 2.55, significantly smaller than $\chi(O) = 3.44$, leading to a smaller effective charge transfer from the Si atom, giving rise to a smaller dipole energy and a smaller tensile strain. The bond length of Si−C, ∼1.9 Å, is also closer to Si than Si−O, ∼1.6 Å. Thus, the in-plane strain is also likely to be less than Si−O superlattice. All these figures seem to indicate that the Si−C superlattice should be better than the Si−O superlattice. However, Si and C are isoelectronic, meaning that C is structurally the same as silicon. And this may be the problem, because the likelihood of carbon atoms forming clusters is greater. *Fortunately, the ways in which we introduce carbon via gas adsorption may be sufficient to prevent the formation of clusters.* In recent years graphene has risen in popularity. Zhang and I using DFT (Zhang and Tsu, 2010) computed graphene/Si superlattice with very encouraging results: the strain is very small and the k-point degeneracy is preserved. These results are included in Chapter 10. Here again gives a good example of what was started ten years ago which led to something could

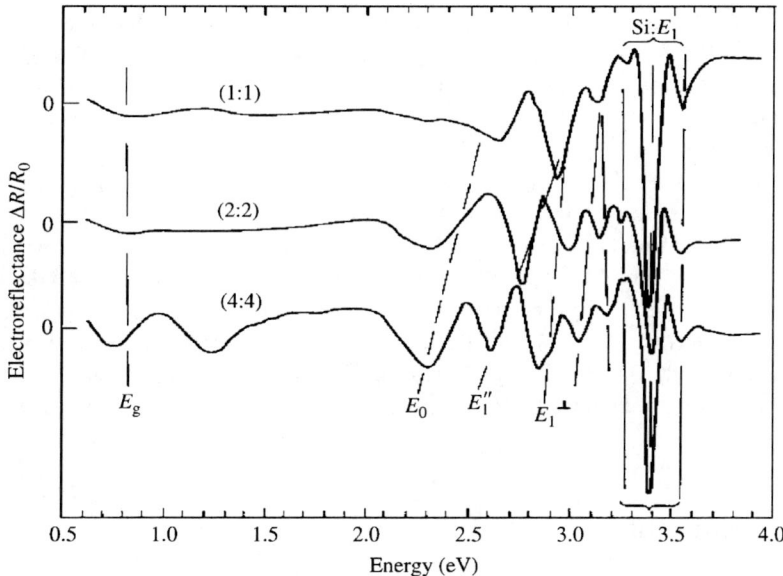

Figure 6.40 Electroreflectance spectra of (1:1)−(6:6) superlattice structures. The notation (1:1) means Ge(1)−Si(1). Notations (2:2) and (4:4) show superlattice characteristics; however, (1:1) is very similar to the random alloy of $Ge_{0.5}Si_{0.5}$.
Source: After figure 11 of Pearsall et al. (1998), with permission.

not be considered as worth doing may be coming back strong ten years later as the technical world at present is dealing with graphene related research at a feverish pace. The main problem then was the lack of light emission in Si−C superlattice considered as unworthy of effort with objectives involving light emission in Si-based materials. Neither could I justify, nor could I develop any enthusiasm in continuing the Si−C work.

To my mind, the work on the Ge−Si monolayer strain-layer superlattice (Pearsall et al., 1998) should be singled out to make a comparison with the Si−O superlattice. Before we do, I shall re-emphasize that the key point is to prevent the formation of a random alloy by letting one of the two, silicon in this case, dominate over the other. Oxygen consists of only one monolayer so that oxygen atoms have to sit on sites compatible with the surface structure, or more precisely, reconstructed surface structure.

Figure 6.40 shows electroreflectance spectra for three structures: Ge(1)−Si(1), Ge(2)−Si(2), and the Ge(4)−Si(4), taken from figure 11 of Pearsall et al. (1998). They emphasized that the superlattice transition at ~ 2.3 eV for (2:2) and (4:4) does not appear in the (1:1) structure and that the electronic band structure resembles the random $Ge_{0.5}Si_{0.5}$ alloy. Also, the spectra of the superlattice between 2.5 and 3 eV are more complex than those of (1:1). They also observed that for Ge(6)−Si(6) and beyond, a TEM of lower resolution clearly shows strain relaxation because the Matthews and Blakeslee (1976) criterion for pseudomorphic defect-free growth has been exceeded. Therefore, their results are consistent with my reasons why only one monolayer or at most two monolayers of oxygen may be considered in the SAS.

There is something very attractive about this work, in that Ge(1)—Si(1) is stable and defect-free growth is possible. Therefore, to my mind, it should serve as a good barrier for Si quantum devices—superlattices as well as quantum wells. This enormously important possible application, to my surprise, was not mentioned by Pearsall et al. (1998).

As we have explained in detail, the Si—O superlattice retains the epitaxial growth. The in-plane strain is larger than most III—V superlattice, but is well within the strain-layer superlattice. The tensile strain is rather large, creating conditions for diffusion which limit the operation temperature, otherwise oxygen could diffuse. However, we must recognize that if the silicon epitaxial layers beyond the Si—O layers were perfect, there would be nowhere for the oxygen to diffuse into. Thus oxygen diffusion outside the intended regions might not be possible. Diffusion experiments should be a priority in further development of the Si—O superlattice.

6.11 Summary

The concept of the SAS materializes in stages. Initially it was thought that if we made oxide thin enough, it may form a strain-layer superlattice. This did not work and we took the route of using polycrystalline silicon on polycrystalline silicon dioxide. Both PL and EL were observed. However, I knew that it would never "fly," since even in the polycrystalline case, oxides were formed with adsorbed oxygen to limit the oxide thickness. It occurred to us that epitaxial growth is very aggressive, however, what counts is how low the defect density is. It took us a while to learn how to grow epitaxial silicon on saturated adsorbed oxygen. We learned from John Hren that our results may be much improved with a better cleaning process for the substrate. Following his advice, the stacking faults disappeared and the defect density was drastically lowered. Our nine-period Si—O SAS was life tested with the monitoring of PL and EL for more than 1 year with stable operation. We think the visible light comes from a combination of quantum confinement and interface oxygen complexes. We then embarked on a theoretical study of hand-built models, calculating the strain and energy states and comparing with DFC, by Babic and Edwards (2000, unpublished). First of all, DFC calculations basically agree with simple calculations that the strain is surprisingly low. Moreover, $I-V$ measurements suggest the possibility of applying Si—O superlattice to a MOSFET having insulating SOI replaced by this epitaxially formed insulating layer. And this insulating layer can also serve as the gate oxide in such way that the whole MOSFET is epitaxial, allowing possible 3D-ICs.

This chapter gives a good example of how an idea was developed and metamorphosed into an all epitaxially formed structure. The process is fundamentally different from ALE, because in SAS, for Si—O superlattice, silicon dictates what sites oxygen can occupy consistent with the surface reconstruction, whereas in ALE, no one constituent dictates, resulting in a random alloy. Thus, SAS offers an extension to include cases where the strain is really too large to be accommodated.

In closing this section, I want to mention that I, like many others, am moving away from superlattices and quantum wells for one reason—federally funded research has moved away from these topics, although from the technology viewpoint, interest is just beginning owing to the success of, for example, terahertz resonant tunneling diodes (Sollner et al., 1983; Brown et al., 1989), the VCSELs (Huffaker et al., 1996), and the quantum cascade laser (Faist et al., 1994). Research into quantum dots (QDs), with bench-top proof of principle, may not materialize as useful devices. The reason is simple: nobody has the vaguest idea how to put diodes input/output into nano-scaled QD devices. Nevertheless, this is the subject I shall treat next.

References

Baribeau, J.-M., Lockwood, D.J., Lu, Z.H., 1995. Mater. Res. Soc. Symp. Proc. 382, Pittsburgh.

Bean, J.C., 1985. Science 230, 127.

Brown, E.R., Sollner, T.C.L.G., Parker, C.D., Goodhue, W.D., Chen, C.L., 1989. Appl. Phys. Lett. 55, 1777.

Cho, A.Y., 1971. Appl. Phys. Lett. 19, 467.

Dellow, M.W., Beton, P.H., Langerak, C.J., Foster, T.J., Main, P.C., Eaves, L., et al., 1992. Phys. Rev. Lett. 68, 1754.

Distler, G.I., Zvyagin, B.B., 1966. Nature 212, 807.

Jinli, D., Tsu, R., 1997. Appl. Phys. Lett. 71, 2124.

Dodson, B.W., 1988. Appl. Phys. Lett. 53, 394.

Döhler, G.H., Kunzel, H., Olego, D., Ploog, K., Ruden, P., Stoltz, H., et al., 1981. Phys. Rev. Lett. 47, 864.

Esaki, L., Chang, L.L., 1974. Phys. Rev. Lett. 33, 495.

Esaki, L., Tsu, R., 1970. IBM Res. Dev. 14, 61.

Evjen, H.M., 1932. Phys. Rev. 39, 675.

Faist, J., Capasso, F., Sivco, D.L., Sirtori, C., Hutchison, A.L., Cho, A.Y., 1994. Science 264, 553.

Feldman, L.C., Bevk, J., Davidson, B.A., Grossmann, H.J., Mannaerts, J.P., 1987. Phys. Rev. Lett. 59, 664.

Filios, A.A., Hefner, S.S., Tsu, R., 1996. J. Vac. Sci. Technol. B14, 3431.

Fukuda, M., Nakagawa, K., Miyazaki, S., Hirose, M., 1997. Appl. Phys. Lett. 70, 2291.

Gonzalez-Hernandez, J., Martin, D., Chao, S.S., Tsu, R., 1984. Appl. Phys. Lett. 45, 101.

Henning, C.A.O., 1970. Nature 227, 1129.

Herzberg, G., 1944. In: Spinks, J.W.T. (Ed.), Atomic Spectra and Atomic Structure. Dover Publications, New York.

Huffaker, D.L., 1996. IEEE Photon. Technol. Lett. 8, 974.

Keating, P.N., 1966. Phys. Rev. 145, 637.

Kittel, C., 1973. Introduction to Solid State Physics, third ed. Wiley, New York.

Li, Y.P., Ching, W.Y., 1985. Phys. Rev. B 31, 3172.

Lu, Z.H., Lockwood, D.J., Barlbeau, J.M., 1995. Nature 378, 258.

Matthews, J.W., Blakeslee, A.E., 1976. J. Cryst. Growth 32, 265.

Meakin, D., Stobbs, M., Stoemenis, J., Economou, N.A., 1988. Appl. Phys. Lett. 52, 1053.

Meyerson, B.S., 1986. Appl. Phys. Lett. 48, 797.

Morais, J., 1995. Doctoral Thesis, UNICAMP, Department of Physics, Campinas, Brazil.

Nigam, T., Degraeve, R., Groeseneken, G., Heyns, M.M., Maes, H.E., 1999. Structure and electronic properties of ultra-thin dielectric films on silicon and related structures. Material Research Society Symposium Proceedings T, 29 November–3 December, Boston.

Osbourn, G.C., 1982. J. Appl. Phys. 53, 1586.

Pearsall, T.P., Bevk, J., Bean, J.C., Bonar, J., Mannaerts, J.P., Ourmazd, A., 1998. Phys. Rev. B 39, 3741.

Phillips, J.C., 1973. Bonds and Bands in Semiconductors, Academic Press, New York.

Rabedeau, T.A., Tidswell, I.M., Pershan, P.S., Berk, J., Freer, B.S., 1991. Appl. Phys. Lett. 59, 706.

Radziemski, L.J., Andrew, K.L., 1963. J. Opt. Soc. Am. 55, 474.

Seo, Y.-J., Clay, J.J., Tsu, R., 2001. Appl. Phys. Lett. 79, 788.

Sollner, T.C.L.G., Goodhue, W.D., Tannenwald, P.E., Parker, C.D., Peck, D.D., 1983. Appl. Phys. Lett. 43, 588.

Tsu, R., 1988. Mater. Res. Soc. Symp. Proc. 102, 219.

Tsu, R., 1993. Nature 364, 19.

Tsu, R., 1998. Int. J. High Speed Electron. Syst. 9, 145.

Tsu, R., 2000a. Mater. Res. Soc. Proc. 592, 351–361.

Tsu, R., 2000b. Phys. Stat. Sol. (a) 180, 333.

Tsu, R., Esaki, L., 1973. Appl. Phys. Lett. 22, 562.

Tsu, R., Lofgren, J.J., 2001. J. Cryst. Growth 227/228, 21–26.

Tsu, R., Zhang, Q., 2002. In: Koch, C. (Ed.), Nanostructured Electronics and Optoelectronics Materials. Noyes Publications, Westwood, NJ.

Tsu, R., Zypman, F., Greene, R.F., 1989. Mater. Res. Soc. Symp. Proc. 148, 373.

Tsu, R., Morais, J., Bowhill, A., 1995. Mater. Res. Soc. Symp. Proc. 358, 825.

Tsu, R., Filios, A., Lofgren, C., Cahill, D., VanNostrand, J., Wang, C.G., 1996. Solid-State Electron. 40, 221.

Tsu, R., Filios, A., Lofgren, C., Ding Tsu, R., Zhang, Q., Morais, J., et al., 1997. In: Cahay, M. (Ed.), Proceedings of ECS. 97-11 Quantum Confinement: Nanoscale Materials, Devices and System, Montreal, May 4–7.

Tsu, R., Filios, A., Lofgren, C., Dovidenko, K., Wang, C.G., 1998a. Electrochem. Solid State Lett. 1 (2), 80–82.

Tsu, R., Zhang, Q., Filios, A., 1998b. SPIE 3290-31, 246.

Tsu, R., Dovidenko, K., Lofgren, J.C., 1999a. ECS Proc. 99-22, 294–301.

Tsu, R., Filios, A., Zhang, Q., 1999b. In: Vincenzini, P., Righini, G. (Eds.), Advances in Science and Technology, Innovative Light Emitting Materials, vol. 27. Techna Srl, Florence, Italy, p. 55.

Tsybeskov, L., Duttagupta, S.P., Hirschman, K.D., Fauchet, P.M., 1996. Appl. Phys. Lett. 68, 2058.

Vier, D.T., Mayer, J.E., 1944. J. Chem. Phys. 12, 28.

Zrenner, A., Reisinger, H., Koch, F., Ploog, K., 1985. In: Chadi, J.D., Harrison, W. (Eds.), 17th International Conference on Physics of Semiconductors. Springer, Berlin, p. 325.

Zhang, Y., Tsu, R., 2010. Nano Res. Let. 5(5), 805–808.

7 Si Quantum Dots

7.1   Energy States of Silicon Quantum Dots

Resonant tunneling *via* silicon quantum dots has been observed by Ye et al. (1991). The system consists of nanocrystalline silicon particles with dimensions of $3-10$ nm embedded in an amorphous SiO_2 matrix. In order to understand the measured conductance as a function of the applied gate voltage, we need at least the positions of the energy states of a three-dimensionally confined quantum dot (3D-QD). The energy state of an idealized quantum box is simple enough and is covered in most elementary books. However, the energy surface is not spherical in general, the barrier height is not infinite, and the geometry is not a cube. In this section the calculation of the energy state for a silicon sphere with different effective masses is presented in more detail than published in Tsu (1990) to illustrate some of the issues we need to consider when moving on a step from the usual particle in a box.

We shall first calculate the energy state of a silicon cube oriented along several high-symmetry axes, assuming that the barrier height is infinite. In fact this assumption is not bad at all because the band-edge alignment between Si and SiO_2 is 3.2 eV and we are considering dimensions in the range of $3-10$ nm, so that the energy involved is much less than the barrier height. Since the transverse mass m_t and the longitudinal mass m_l are appropriate for a valley in the $\langle 100 \rangle$ direction, the usual $m_t = 0.19 m_e$ and $m_l = 0.916 m_e$ may be used for a cube along the $\langle 100 \rangle$ direction, or

$$E_{m,n,p} = \frac{\hbar^2 \pi^2}{2d^2} \left(\frac{m^2}{m_t} + \frac{n^2}{m_t} + \frac{p^2}{m_l} \right). \tag{7.1}$$

However, along the $\langle 110 \rangle$ and $\langle 111 \rangle$ directions, it is not readily obvious which effective mass should apply. Stern and Howard (1967) considered the appropriate masses for a sheet oriented with respect to $\langle 110 \rangle$ and $\langle 111 \rangle$ in consideration of the inversion system. Since our cube has the same symmetry, the masses given by Ando et al. (1982) are used. Table 7.1 gives values along the three principal orientations.

For the convenience of the reader, the low-lying energies are given in Table 7.2 for a cube with sides $d = 75$ Å. The calculated energy states for a 7.5-nm cube are shown in Figure 7.1. The notation $\langle 111 \rangle_4$ indicates our results for a cube oriented

Superlattice to Nanoelectronics. DOI: 10.1016/B978-0-08-096813-1.00007-2

Table 7.1 Conduction Band Effective Masses in the Orientations $\langle 110 \rangle$ and $\langle 111 \rangle$

	$\langle 100 \rangle$	$\langle 110 \rangle$	$\langle 111 \rangle$
m_t/m_e	0.19	0.19	0.19
m_l/m_e	0.92	0.31	0.26

Values are taken from Ando et al. (1982).

Table 7.2 Energy below 0.3 eV of a Cube with Sides $d = 75$ Å for the Three Orientations from Eq. (7.1) Using the Masses Given in Table 7.1

m	n	p	$E \langle 100 \rangle$ (eV)	$E \langle 110 \rangle$ (eV)	$E \langle 111 \rangle$ (eV)
1	1	1	0.078	0.092	0.079
1	1	2	0.100	0.156	0.157
1	1	3	0.137	0.264	0.300
1	2	1	0.184	0.198	—
2	1	1			
1	2	2	0.206	0.263	0.235
2	1	2			
1	2	3	0.243	0.37	—

along $\langle 111 \rangle$ with a degeneracy of 4. In other words, the other two degeneracies should have the same energy as $\langle 100 \rangle_6$.

For a spherical silicon particle, the wave equation with principal masses of m_t and m_l, satisfying the boundary condition of a sphere, does not have simple solutions. We shall use a variational approach, which represents a generalization of the work by Efros and Efros (1982). We take a product wave function to calculate the expectation value of the energy. By setting the first variation to zero, we generate a new differential equation involving only one of the three variables. The approach taken here is entirely similar to the calculation of the ground state energy for a shallow impurity state in a superlattice by Ioriatti and Tsu (1986), presented in Chapter 5. Let the Hamiltonian

$$H = H_0 + H_1$$

in which

$$H_0 \equiv -\hbar^2 \left[\left(\frac{1}{\rho} \frac{\partial}{\partial \rho} \left(\rho \frac{\partial}{\partial \rho} \right) + \frac{\partial^2}{\rho^2 \partial \phi^2} \right) \right] \bigg/ 2m_t, \tag{7.2}$$

and

$$H_1 \equiv -\hbar^2 (\partial^2/\partial z^2)/2m_l$$

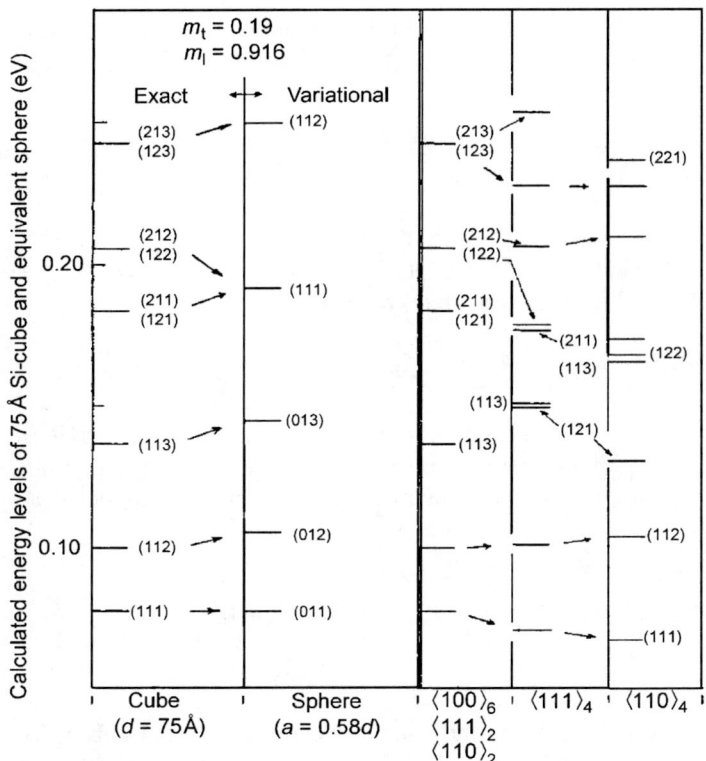

Figure 7.1 Calculated energy state for a $d = 7.5$ nm Si cube. The right side shows the other directions. The notation $\langle 100 \rangle_6$ indicates the orientation in $\langle 100 \rangle$ with a degeneracy factor of 6. The variational solution is for a sphere with $a = 0.58d$, chosen to align with the ground state. *Source*: After Tsu (1990), with permission.

in which $m_t = 0.19m_e$ and $m_l = 0.916m_e$ for the case of $\langle 100 \rangle$. Let $\Psi = \psi(\rho, \phi)Z(z) = NJ_m(k_t\rho)e^{im\phi}Z$ in which, N is to be determined by normalization, $k_t^2 = 2mE_t/\hbar^2$, and J_m is the Bessel function, Math Functions (Abramowitz and Stegem, 1964) and

$$H_0\psi(\rho, \phi) = E_t\psi(\rho, \phi). \tag{7.3a}$$

From $\int_0^{2\pi} d\varphi \int_0^{\rho_0} \rho \ \partial\rho|\psi(\rho, \varphi)|^2 = 1$, setting $\rho = \rho_0 t$, where $\rho_0 = \sqrt{a^2 - z^2}$, we obtain

$$\psi(\rho, \phi) = (\pi)^{-1/2}J_m'(\alpha_{mn})^{-1}|(a^2 - z^2)^{-1/2}|J_m(k_t\rho)e^{im\phi}, \tag{7.3b}$$

with α_{mn} being the zero values of $J_m(\alpha_{mn})$ because of the simple boundary condition that the wave function goes to zero at the surface of the sphere. From

$E = \langle \Psi | H | \Psi \rangle$ and $\delta E = 0$, we obtain an equation in $Z(z)$. Specifically, $\langle \Psi | H_0 + H_1 - E | \Psi \rangle$ becomes

$$E_t \int_{-a/2}^{a/2} Z^* Z \, dz - \frac{\hbar^2}{2m_1} \int_{-a/2}^{a/2} Z^* \sqrt{a^2 - z^2} \frac{d^2}{dz^2} \frac{Z}{\sqrt{a^2 - z^2}} dz = E \int_{-a/2}^{a/2} Z^* Z \, dz.$$

$$(7.4a)$$

The boundary condition $J_m(k_t\rho) = 0$, at $\rho = \rho_0 = \sqrt{a^2 - z^2}$, results in $k_t\sqrt{a^2 - z^2} = \alpha_{mn}$, the zeros of Bessel function. Therefore, we obtain

$$E_t = \hbar^2 \alpha_{mn}^2 / 2m_t(a^2 - z^2).$$

$$(7.4b)$$

Note that the energy E_t depends on z and the zeros of the Bessel function α_{mn}. Setting the variation $\delta E = 0$ with respect to Z^*, we obtain a new differential equation in E. Assuming the wave function vanishes on the surface of a sphere with radius a, we found that

$$Z'' + (2z/(a^2 - z^2))Z' + \left[2m_1 E/\hbar^2 - m_1\alpha_{mn}^2/m_t(a^2 - z^2) + \frac{(a^2 + 2z^2)}{(a^2 - 2z^2)} \right] Z = 0.$$

$$(7.5)$$

The eigenvalue E is given by the boundary condition $Z(\pm a) = 0$. If the wave function for transition probability is required, Eq. (7.5) must be solved. However, if our objective is to find the eigenvalue E, we may use the trial wave functions,

$$Z = \begin{cases} Z_e = a^{-1/2} \cos\dfrac{p\pi z}{2a} & p = 1, 3, 5, \dots \\[2mm] Z_0 = a^{-1/2} \sin\dfrac{p\pi z}{a} & p = 1, 2, 3, \dots \end{cases}$$

$$(7.6)$$

satisfying the boundary condition at $z = \pm a$. The eigenvalue E is now given by

$$E = \int_{-a}^{a} dz \left\{ E_t Z^* Z - \frac{\hbar^2}{2m_1} Z^* \left[Z'' + \frac{2z}{(a^2 - z^2)} Z' + \frac{a^2 + 2z^2}{(a^2 - z^2)^2} Z \right] \right\}.$$

$$(7.7)$$

We shall simplify the notation by writing the quantum number (m, n, l), with l such that

$$l = \begin{cases} 1, & \text{for } Z_e(p = 1) \\ 2, & \text{for } Z_0(p = 1) \\ 3, & \text{for } Z_e(p = 3) \\ 4, & \text{for } Z_0(p = 2) \\ \text{etc.} \end{cases}$$

The first few zeros of J_{mn} are listed

m	n	$\alpha(m, n)$
0	1	2.404
1	1	3.832
0	2	5.520
1	2	7.016
2	2	8.417

Note that the dimension of a cube d and the radius a of the sphere are not related. However, if we could relate them, for example, using equal volume, then $d/a = 1.612$ and the variational solution gives a higher value for the lowest energy. We must recognize that the quantum numbers for the sphere have nothing to do with those of the cube. However, if we can also set the lowest energy level for the variational solution equal to the lowest energy level of a cube, then $d/a = 1.71$. For comparison, we use the latter by setting the lowest level of the variational solution equal to that of the cube. For Z_e,

$$\left(\frac{E}{\hbar^2/2m_t a^2}\right) = \alpha_{mn}^2 F_{ep} + \frac{m_t}{m_1}\left[\left(\frac{p\pi}{2}\right)^2 + \left(\frac{p\pi}{2}\right)G_{ep} - H_{ep}\right] \tag{7.8}$$

where

$$F_{ep} = \int_{-1}^{1} \frac{\cos^2(p\pi y/2)}{1 - y^2}\, dy, \quad G_{ep} = \int_{-1}^{1} \frac{y\sin p\pi y}{1 - y^2}\, dy, \quad \text{and}$$

$$H_{ep} = \int_{-1}^{1} \frac{(1 + 2y^2)\cos^2(p\pi y/2)}{(1 - y^2)^2}\, dy.$$

And for Z_0,

$$\left(\frac{E}{\hbar^2/2m_t a^2}\right) = \alpha_{mn}^2 F_{op} + \frac{m_t}{m_1}\left[(p\pi)^2 + (p\pi)G_{op} - H_{op}\right], \tag{7.9}$$

where

$$F_{op} = \int_{-1}^{1} \frac{\sin^2 p\pi y}{1 - y^2}\, dy, \quad G_{op} = \int_{-1}^{1} \frac{y\sin(2p\pi y)}{1 - y^2}\, dy, \quad \text{and}$$

$$H_{op} = \int_{-1}^{1} \frac{(1 + 2y^2)\sin^2 p\pi y}{(1 - y^2)^2}\, dy. \tag{7.10}$$

Below we shall list the first few energies:

m	n	$\alpha^2(mn)$	$\alpha^2(mn)F_{e,1}$	$\alpha^2(mn)F_{e,2}$	$\alpha^2(mn)F_{e,3}$
0	1	5.78	7.04	10.11	8.96
1	1	14.68	17.88	25.69	22.61
0	2	27.05	32.95	47.34	41.66
1	2	49.22	59.95	86.1	75.80

p	F_{ep}	G_{ep}	H_{ep}	F_{op}	G_{op}	H_{op}
1	1.22	1.41	2.72	1.55	-1.600	9.2
3	1.75	1.18	9.86			

Table 7.3 Tabulation of Energies of a Sphere with the Effective Masses Given in $\langle 100 \rangle$, for the Lowest Energy Aligned with That of a Cube, That Is $d/a = 1.71$

m	n	P	$E\langle 100 \rangle$ (eV)	$E\langle 110 \rangle$ (eV)	$E\langle 111 \rangle$ (eV)	E(sphere) (eV)
0	1	1	0.078	0.086	0.07	0.079
0	1	2	0.106	0.13	0.11	
0	1	3	0.145	0.187	0.154	0.161
1	1	1	0.192	0.20	0.216	
1	1	2	0.25	0.273	0.22	0.266

In Table 7.3 we list the energies for the sphere using the variational solution for the sphere. For comparison, we set the ground state energy of the sphere using the masses for $\langle 100 \rangle$ equal to those of the cube oriented in $\langle 100 \rangle$ at 0.78 eV. Note that the highest levels for the sphere with the variational solution are higher than those for the cube $\langle 100 \rangle$.

The last column lists three energies of a sphere using the solution of the sphere assuming $m_t = m_l = 0.26m_e$, the roots of the spherical Bessel functions $j_m(\alpha_{mn})$, and E_{mn} in eV. For $m_t/m_l \neq 1$, the variational solution is better, although a cylinder with a height equal to the radius, having simple solution, might actually be better at representing the QD.

n	$j_0(\alpha_{0n})$	$j_1(\alpha_{1n})$	$j_2(\alpha_{2n})$	E_{0n}	E_{1n}	E_{2n}
1	π	4.495	5.69	0.078	0.161	0.266
2	2π	7.725	9.09	0.316	0.475	
3	3π					

The computed energies for a sphere of radius a with all levels empty are shown in Figure 7.1. We set the ratio $a/d = 0.58$ in order to line up the ground state for the cube, E_{111}, to coincide with the ground state E_{011} for the sphere using the masses for $\langle 100 \rangle$. It is interesting to note that for $a/d = 0.62$, the volume of a sphere with radius a is equal to that of a cube with sides d. When I first showed the result to Stern, he thought the small difference of $<7\%$ from the equal volume case was indeed very interesting. The quantum number for the sphere (011) of E_{011} stands for $m = 0$ and $n = 1$, with $p = 1$, where α_{01} is the lowest zero of the Bessel function and $p = 1$ denotes the lowest-order mode that satisfies the boundary condition at $z = \pm a$.

The Coulomb energy from electron occupation that is consistent with Pauli's exclusion principle owing to its small size is quite significant, particularly with a voltage applied (Ye et al., 1991). We shall show that our experimental results indicate that the effective mass anisotropy is overwhelmed by the induced image charge which apparently has nearly spherical symmetry. This fact was brought home previously from entirely different experimental evidence—the volume percolation of microcrystalline silicon grains in an amorphous silicon matrix is 0.18, which is close to the theoretical value of 0.16 for spheres (Tsu et al., 1982). Therefore, the microcrystalline structure obtained by annealing deposited amorphous silicon has an essentially spherical shape (Tsu et al., 1986). Note that the spherical solution using the variational approach is in fact more general than the cubic model, because for orientations of a cube other than the three high-symmetry axes, the cubic model involves the full effective mass tensor. In short, the simplicity of the cubic solution is deceptive; the tabulated results are for high-symmetry directions. Since most of the experimentally prepared QDs are closer to spherical shape, the size-dependent energy of a silicon sphere is more useful because orientation is not an issue. High-resolution TEM pictures also show an almost perfect spherical shape. That being the case, why did we go to the trouble of using the longitudinal and transverse effective masses? The answer is the fact that models can only be justified by comparison with measurements. We shall show later in conjunction with conductance steps that the energy states in silicon QDs are indeed closer to being spherically symmetric, at least for particle sizes under 5 nm. A nonspherical energy surface results in states with degeneracies.

The calculated energy levels of a silicon QD apply only to a neutral Si particle, similar to the so-called empty lattice band structures in solids, with an idealized boundary condition—that the wave function is zero at the surface and Coulomb terms are included. As soon as the ground state designated by (011) is occupied by two electrons with spin-up and spin-down, even using the idealized boundary condition, the extra charges induce image charges depending on the difference between the dielectric constants in the silicon and the matrix. This subject will be treated in Chapter 8 on quantum capacitance where the problem becomes extremely complex. Nonetheless, I want to point out some salient features that are useful for gaining some physical insight: the extra energy from the Coulomb interactions is quite significant as the particle size is reduced to a few nanometers. The smearing actually can bring the QD closer to a spherically symmetric "man-made atom." In other words, the three widely separated states in Figure 7.1, (012), (013), and (111) for the nanoscaled sphere due to the orientation-dependent masses, may merge toward the p-state of a spherically symmetric system. Using the measured conductance data, we found that the first two excited states are merged, but the third may be distinct. If optical emission is produced by electron injection, a rather complicated spectrum will also result depending on the occupation of the levels. Therefore what applies to nanoscaled QDs requires additional rules, which we normally accept for solid-state and atomic physics; it is, as the popular expression states, "a whole new ball game."

7.2   Resonant Tunneling in Silicon Quantum Dots

Resonant tunneling *via* nanocrystalline silicon, nc-Si, embedded in an amorphous silicon dioxide, a-SiO$_2$, matrix has been exploited (Tsu et al., 1990), using a thin layer of deposited a-Si at low temperature, followed by crystallization after annealing. These nanoscale silicon QDs are then embedded in an oxide matrix after subsequent annealing in an oxygen-rich environment. Resonant tunneling *via* nc-Si with Si$_x$C$_{1-x}$:H barriers has been reported by Fortunato et al. (1989). Takagi et al. (1990) have reported quantized size effects in PL with the size of Si particles in the range 4.5 nm using microwave plasma deposition with SiH$_4$, a sort of inverse of the work of Lu et al. (1995) discussed in Chapter 6.

In the present scheme, nc-Si is produced by crystallization from the amorphous phase and barriers are formed by thermal oxidation of silicon. It has been pointed out that the defect density of the c-Si/a-SiO$_2$ system under metal oxide semiconductor (MOS) technology is really low enough to apply to quantum well structures (Tsu et al., 1989). Figure 7.2 shows the grain size of crystallized silicon after annealing at 800°C plotted against the deposition temperature T_s (Tsu, 1985).

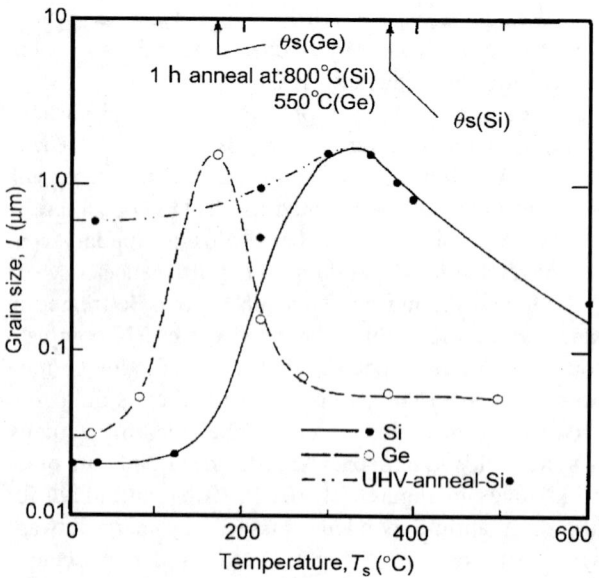

Figure 7.2 Grain size versus substrate temperature after crystallization. Note that under an UHV anneal, the huge drop in grain size at low substrate temperature does not occur. *Source*: After Tsu et al. (1986), with permission.

There is a peak at $T_s \sim 330°C$, although it is not commonly known to most researchers that this temperature is actually the crystallization temperature of a-Si annealing in a vacuum. Another fact unknown to most researchers involves the effect of the interface. Bulk amorphous Ge crystallizes at 180°C. However, a 5-nm-thin layer of a-Ge sandwiched into a thick layer of a-Si does not crystallize until the temperature is raised to above 550°C when a-Si begins to crystallize. Similarly, a thin a-Si layer sandwiched into a-SiO$_2$ does not crystallize until the annealing temperature is raised above 900°C. How thin is thin? When the thickness involved is below 10 nm, the interface dominates (Allred et al., 1988). Why is it important to make these points? I knew about these results when I was working at Energy Conversion Devices (ECD), although none of these results was published because the interface created an inhibition to phase changes which may apply to the chalcogenide switches used at ECD.

I actually forgot this important point so that initially Ye failed to crystallize a-Si layers ~ 8 nm thick, until she doubled the thickness, a development discussed in detail by Tsu (1998). Another important point is the fact that silicon tends to form a spherical shape after crystallization. The nc-Si formed after crystallization from a-Si are spherical (Tsu et al., 1982). Therefore, the particles are never pancake shaped, whereas Ge tends to crystallize into a columnar or cylindrical shape (Hernandez et al., 1984). Since the oxidation process consumes silicon, typically if the target is set for 10-nm nc-Si particles, the thickness of the a-Si layer should be increased to ~18 nm. A schematic cross section of the structure is shown in Figure 7.3. The substrate is a silicon wafer with $n = 3.5 \times 10^{16}$ cm^{-3}. Starting with a thermally grown field oxide 100 nm thick at 1050°C in dry oxygen, an active device varying in size from 40 µm × 40 µm down to 10 µm × 10 µm is formed by

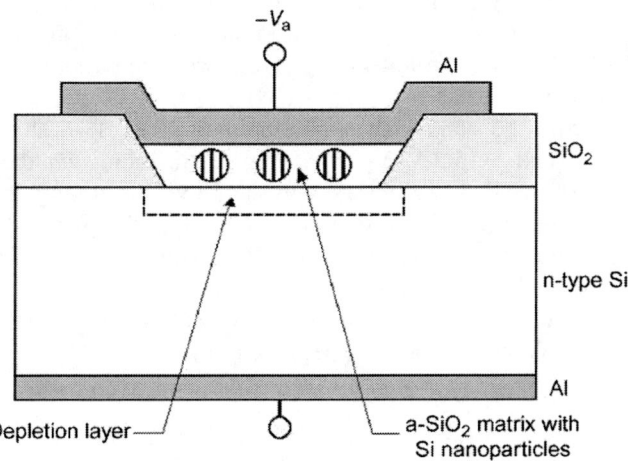

Figure 7.3 Sketch of a resonant tunneling diode structure. Nanocrystallites are shown as shaded circles embedded in a-SiO$_2$ matrix.
Source: After Ye et al. (1991), with permission.

etching a window photolithographically. A thin undoped a-Si layer ∼15−18 nm is deposited by an e-beam, followed by crystallization and oxidation at 800−900°C in 3:1 dry $N_2 + O_2$ at atmosphere pressure. Why we chose this rather low temperature is fully discussed by Tsu (1998). Aluminum gate and back contacts are vacuum deposited. Typically our samples consist of ∼5−10 nm crystallites surrounded by 2.5−3-nm oxide serving as barriers.

There is an important point dealing with tunneling measurements that needs to be emphasized. In normal resonant tunneling *via* quantum wells, the contacts are n+ -doped leading to the negative differential conductance (NDC) whenever the applied voltage is such that the quantum state of the well moves below the source of electrons, from the contact into the forbidden gap. However, with a metal contact, the Fermi sphere is very large compared to all these quantum states involved. When the applied voltage is such that the state involved moves below the conduction band edge, there is tunneling from the metal contact as shown in Figure 7.4. This leads to a conductance peak, but no current peak. Therefore, an NDC should never appear. In Figure 7.4(A), V_1 represents the voltage across the QD. Owing to deep depletion (for an explanation of deep depletion, see Nicolian and Brews 1982) in the silicon substrate at the right contact, there is an additional voltage V_s, so that the alignment of the quantum state with the Fermi level requires a voltage $V_a - V_s$, shown in Figure 7.4(B), together with a sketch showing the conductance jump by a peak. More detailed discussion is delayed until Chapter 11, where the wave impedance of an electron is introduced. In Figure 7.4(C), owing to the charge Q stored in the QD, the applied voltage for the onset of the conductance jump is further moved to a higher voltage and includes the term Q/C, the potential from charging the capacitance of the quantum dot including the two thin barriers.

The device in Figure 7.3 is characterized electrically by measuring the dc current and a small signal capacitance and equivalent conductance at 1 MHz as functions of the gate bias V_G, using a lock-in amplifier equipped with a current preamplifier (Model 410 C-V plotter). A variable voltage ramp, usually at a rate of $10\ mV\ s^{-1}$, provides the bias.

Figure 7.5 shows the conductance measured at 77 K. Two sharp conductance peaks at ∼−11.8 and −15.3 V are attributed to tunneling *via* the first two QD states. The steps shown on the right are attributed to the coupling of QDs at higher energy forming a two-dimensional-like layer. Note that the first and next state in a Si sphere of $d \sim 8$ nm are ∼0.05 and 0.11 eV, but the measured first and second sharp peaks are located at ∼ −11.8 and −15.3 eV.

How can we reconcile the data with the calculated energy positions? The difference between the voltage V_{n+1} and V_n per electron is

$$e(V_{n+1} - V_n) \equiv e\Delta V = E_{n+1} - E_n + e^2/2C, \qquad (7.11)$$

where C is the capacitance. For QDs with $d \sim 8$ nm, $E_{n+1} - E_n \sim 0.06$ eV, but $e^2/C \sim 0.32$ eV, so that $\Delta V \sim 0.44$ V for two electrons. But the difference is 3.8 or 1.9 V assuming symmetrical barriers, a factor of ∼5 larger than $\Delta V \sim 0.44$ V.

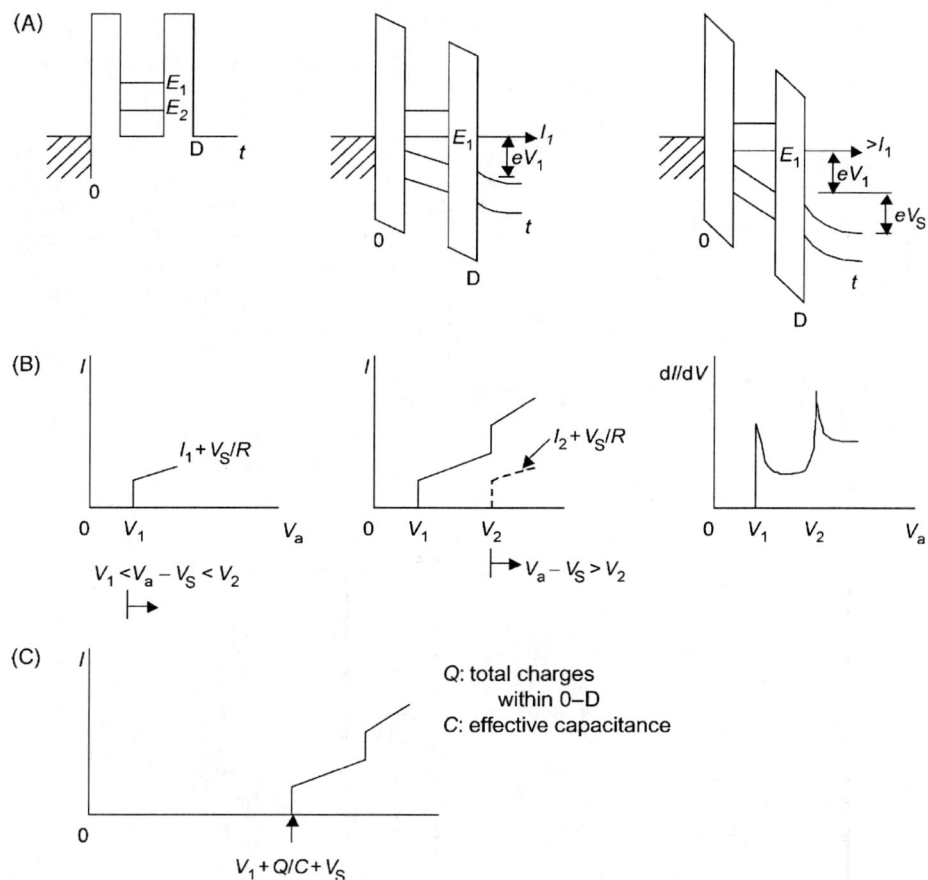

Figure 7.4 Schematic diagram of a Si-QD embedded in a SiO$_2$ matrix. Because of the large Fermi sphere of the metal contact, unlike conventional resonant tunneling with n+ contacts, the conductance peak but not the current peak appears: (A) the relationship between V_1, the voltage across the structure and V_S, the voltage due to deep depletion in the contact, the silicon substrate; (B) the current and conductance; and (C) the charge Q so that the onset of conductance jump appears at an applied voltage $V_a = V_1 + Q/C + V_S$.
Source: After Tsu (1993), with permission.

Thus, we know that only 20% of the applied voltage appears across the QD device and the rest appears across the substrate, as first pointed out in Ye et al. (1991).

Subsequently, more device configurations were obtained so that Nicollian decided to measure the parameters for the equivalent circuit, with results allowing us to fit together the data and estimate energy states more consistently (Nicollian and Tsu, 1993). The most convincing set of step-by-step measurements going from a-Si to nc-Si is shown in Figure 7.6. Of particular interest are the sharp conductance structures which show up when the temperature is reduced from 300 to 77 K.

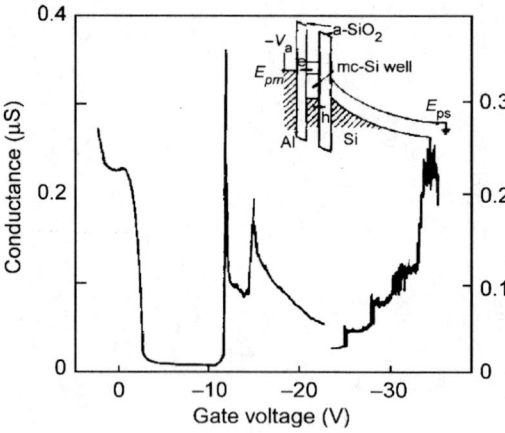

Figure 7.5 The conductance near zero bias is caused by electron capture and emission by traps at the a-SiO$_2$/nc-Si interface (Nicollian and Brews, 1982). The discontinuity between the left side and right side is due to two different samples adjacent to each other on the same wafer. The inset shows the deep depletion discussed in the text. *Source*: After Ye et al. (1991), with permission.

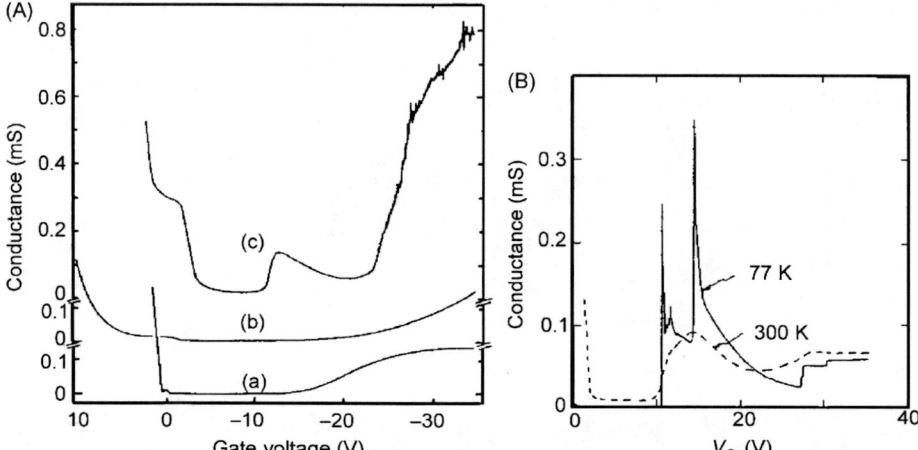

Figure 7.6 (A) Conductance versus gate bias at 300 K measured at 1 MHz. (a) Al/a-SiO$_2$/c-Si; (b) Al/a-SiO$_2$/a-Si/a-SiO$_2$/c-Si; (c) after annealing at 850°C of (b), Al/a-SiO$_2$/nc-SiO$_2$/a-SiO$_2$/c-Si. (B) Repeat of case (c) at 300 K compared to 77 K.
Source: After Ye et al. (1991), with permission.

Before I leave this section, it is reassuring that the linewidth of the conductance peak, obtained by Li, shown in Figure 7.7, is basically the same as $k_B T$, which indicates that all is well. At this stage Nicollian and I decided to apply for a patent for silicon-based functional quantum devices. Ye finished her postdoctoral assignment and the work was taken over by Li, who needed an M.S. thesis. Armed with a better understanding, I made a bold decision to fabricate a wide range of device configurations. In fact, Nicollian did warn me about opening a "can of worms" as he put it. Well, as we shall see in the next section, oscillations, hysteresis, and switching appeared in 5–10% of our samples, and even telegraph-like, apparently random structures.

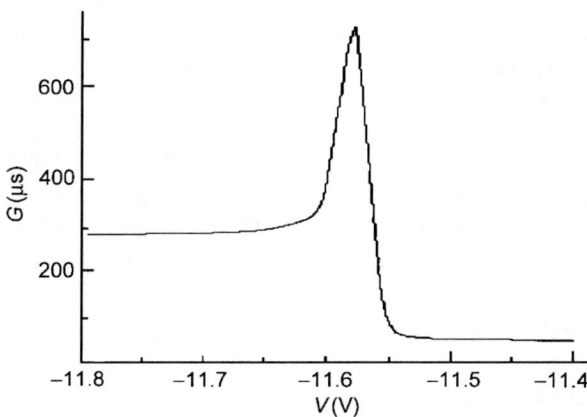

Figure 7.7 Conductance G versus V in volts showing the linewidth is $\sim k_B T$ at room temperature.
Source: Taken from Li's thesis (1993), published in Tsu (2003), with permission.

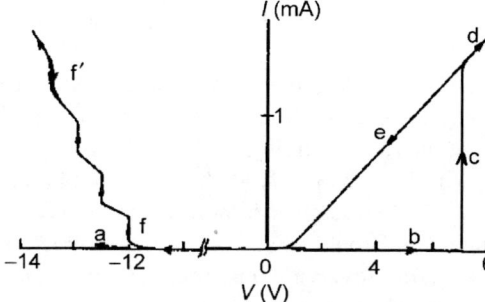

Figure 7.8 Electrical forming used to eliminate the unwanted tissue region of an annealed crystallized Si-QD. Paths (a) and (b) indicate only minute conduction until path (c) is reached. Subsequent paths (d)−(f) indicate structures after forming with (f′) often showing hysteresis.
Source: After Tsu et al. (1994), with permission.

Meanwhile, I gave a pre-annealed sample to S.Y. Chow, who thought all the results pointed to "breakdowns" (Chou, 1992). In a later article (Tsu, 1998, 2000), I pointed out that Chow (1992) was mainly working with the silicon that was still in amorphous phase, because he did not know that a nanometer thick a-Si, sandwiched in a-SiO_2 simply does not crystallize with annealing at temperatures below 1000°C. We had not communicated to him the giant step in improving consistency given by electrical forming under forward bias (Tsu et al., 1994) with the concept and consequences fully explained by Tsu (1998).

Basically, electrical forming selects the most favorable current path, allowing repeatability and stability, while apparently excluding the participation of many other particles (Figure 7.8). Once the consistency is established, the scheme allows us to study the physics of Si-QDs; however, this is far from being a nanoscale quantum device. Going back to the right-hand side of Figure 7.6 at 77 K, between the initial and the second jump in conductance, $G \sim 80\,\mu S$, consistent with the first QD state occupied by two electrons, the fundamental conductance $G_0 = 78\,\mu S$ (Van Wees et al., 1988). Based on this observation, the data for Figure 7.6 indicate that only one QD is involved, suggesting that our current path is really filamentary, rather than uniform across the contacts. This fact may explain why we have repeatable and consistent measurements and why we have not observed spreading due to a distribution of particle size. In the next section, several plots of conductance versus applied voltage will be shown. In particular, before and after electrical forming, our data clearly show the involvement of only a single QD. As a matter of fact, when the data were first shown to Greene, he told Nicollian and I that from all his experience in surface studies, most field emission measurements actually involve a single electron. With this, Nicollian simply said, "Now you know why Keithley licensed my CV measuring instrument package, the most sensitive up-to-date system."

7.3   Slow Oscillations and Hysteresis

Our funding has been improved sufficiently, allowing us to put more students onto this project and leading to the use of a variety of samples whose main feature is nc-Si embedded in an oxide matrix. Actually, inadvertent forming was first discovered accidentally during measurements in the reverse bias. Then we realized that better control can be achieved by passing a current in the forward bias with a current limiter circuit.

Figure 7.9 shows conductance versus increasing negative bias. With reproducible $G-V$ traces obtained routinely, after forming for the first time we were able to examine the data in detail. The first step jump in conductance in Figures 7.7 and 7.10 is usually represented by $\Delta G = 232~\mu S$ or $6G_0$, with $G_0 = 39~\mu S$. Note that the second jump in Figure 7.10 shows a $\Delta G = 155~\mu S$ or $4G_0$. The factor of 6 indicates that an additional six electrons take part in the conduction and could be attributed by the threefold degenerate 2p state of a sphere. But the next jump involving only four electrons is not well understood, unless we go back to the energy calculated in Figure 7.1.

Figure 7.10 shows conductance versus bias (Tsu et al., 1994). The arrows mark V_n and V_{n+1} where the step increase in conductance occurs because of the addition

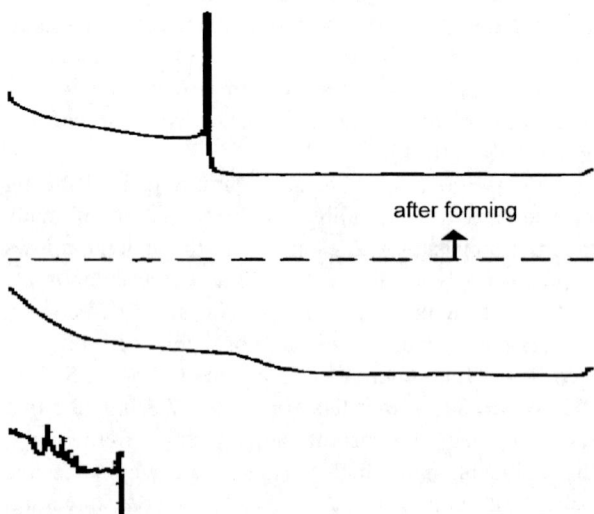

after forming

Figure 7.9 Conductance versus increasing negative bias, shown to the left. Before electrical forming, below the dashed line, $G-V$ is unstable and lacking sharp features. After forming, above the dashed line, $G-V$ is stable and reproducible, and almost always shows at least one sharp peak.
Source: Taken from Bowhill's (1994) M.S. thesis, unpublished.

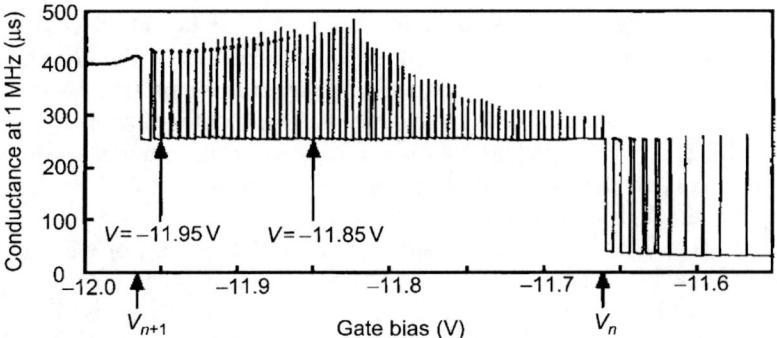

Figure 7.10 Conductance versus the bias at V_n and V_{n+1} having the step jump in conductance and the appearance of an extra conducting channel. At fixed bias −11.95 and −11.85 V, conductance oscillates between the two channels.
Source: Tsu et al. (1994), with permission.

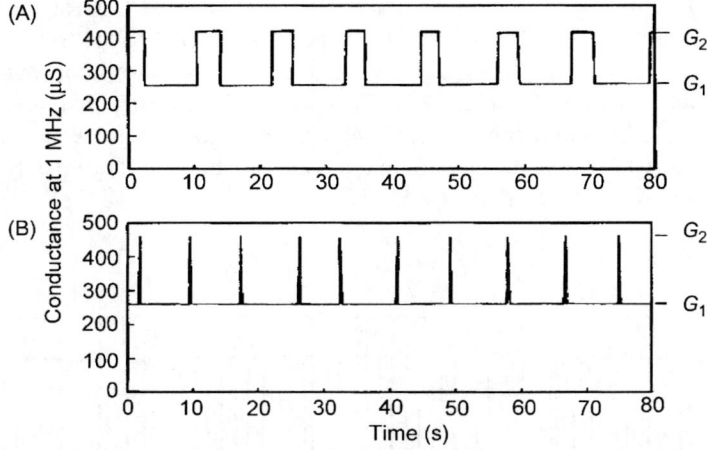

Figure 7.11 Conductance oscillations between G_1 and G_2 at biases of −11.95 V (A), near V_{n+1}, and −11.85 V (B), near V_n.
Source: After Tsu et al. (1994), with permission.

of extra conducting channels. At first we thought the many vertical lines represent many particles. One day Li called me to show some extremely slow oscillations in time at a fixed applied voltage, with periods varying from a fraction of a second to as long as 20 s. Since the circuit impedance does not affect the oscillation, we knew that the oscillation is not induced externally. In other words, these δ-function-like peaks are oscillations in time rather than at different voltage, because if we sweep faster, these δ-function-like peaks are further separated. At a fixed bias, −11.95 and −11.85 V, the conductance oscillates with values approximately between the two channels. Figure 7.11 shows conductance oscillations in time between G_1 and G_2 at two different biases: −11.95 V (A), at a bias close to

V_{n+1}, the beginning of a second channel with G_2 and -11.85 V (B), a bias closer to V_n, where the first channel is conducting with G_1. Note that $\Delta G = G_2 - G_1 = 420 - 260 \, \mu S = 160 \, \mu S \sim G_0$. Again if we press on with only one QD involved, we must accept the fact that the three states shown in Figure 7.1, (012), (013), and (111), with the two lower ones merging into a single state close to a p-like state, so that the additional number of electrons is $6 - 2 = 4$, lead to a step in the conductance given by $\Delta G \sim 4G_0$. This result does not apply to Figure 7.11(B), with $\Delta G \sim 5G_0$, where instead of $\Delta G \sim 4G_0$ or $6G_0$, apparently one electron fails to conduct. This is a good place to emphasize that our work was at first rejected for publication because the reviewer thought these oscillations were telegraph noise, a subject I will go into in some detail later in this section. The nature of these oscillations is very complex, which may not be made clear using the time-independent Schrödinger equation. We think the origin of oscillatory switching lies in the coupling of two nearby QDs, one connected to the contacts while the other is not. The fact remains that the system is extremely complex, calling for more careful measurements, particularly in the nanometer region.

Figure 7.12 shows conductance oscillation measured at 1 MHz with slowing down near the end of a 900 s trace. The switching speed changes from ~ 2 s more than 10 s. In fact the oscillation stops completely after half an hour, ending in the high conductance state, with $\Delta G \sim 155 \, \mu S$ or $4G_0$, indicating that four electrons participated in the conduction process. This experimental fact led us to suggest that the (012) and (013) states in the variational solution shown in Figure 7.1 are degenerate, and this is similar to Figure 7.11.

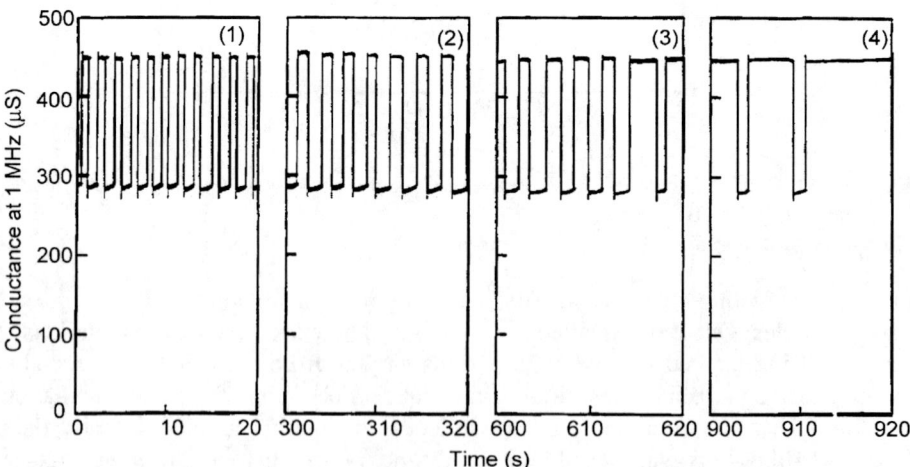

Figure 7.12 Conductance oscillation measured at 1 MHz showing slowing down near the end of a 900 s trace, with switching speed changing from ~ 2 s to more than 10 s, ending in a high conducting state.
Source: From one of the many traces taken in 1993 but never published.

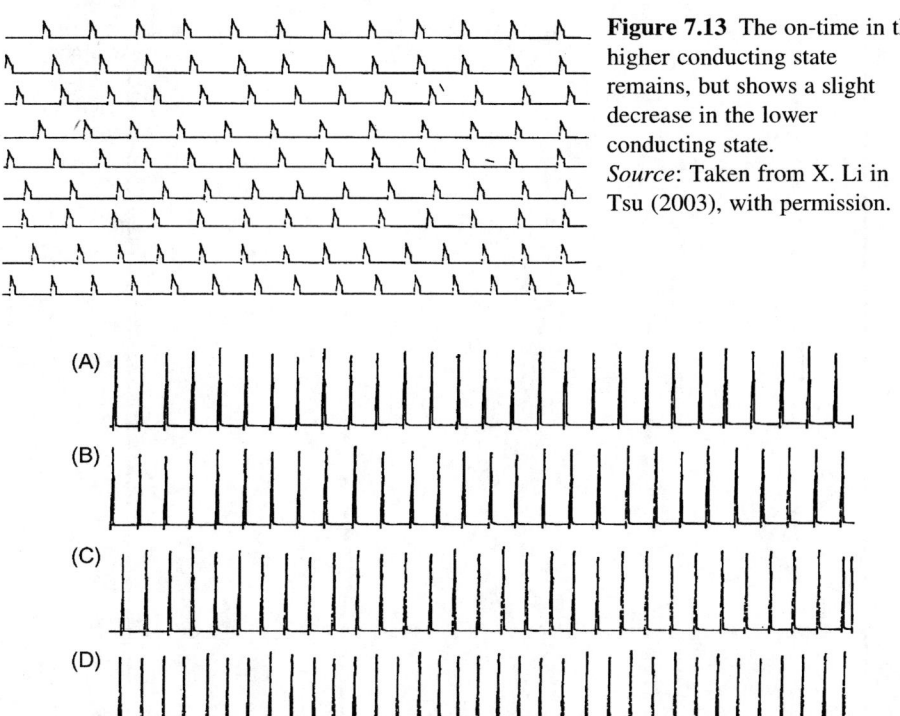

Figure 7.13 The on-time in the higher conducting state remains, but shows a slight decrease in the lower conducting state.
Source: Taken from X. Li in Tsu (2003), with permission.

Figure 7.14 The on-time in the higher conducting state remains unchanged but is slowly decreasing in the lower conducting state.
Source: Taken from X. Li, unpublished.

As far as the switching slows down, it is quite evident from the 600 to 920 s timeframe that the on-time in the higher conducting state is increasing but remains relatively constant in the lower conducting state. Actually, at the beginning, the on-time is quite short. Next, let us show a couple of traces with a somewhat different trend, where the on-time remains relatively constant, but the conductance time in the lower state tends to decrease, as shown in Figures 7.13 and 7.14. There is a small but unidirectional decrease in the period of switching, a 10% reduction during the initial 50—100 periods, which is consistent with a slight reduction of the barrier affecting coupling upon heating.

It is clear that these conductance oscillations, or switching, cannot be explained without bringing in the time dependence of trapping or the residence time of electrons at these levels. They are definitely not telegraph noise. As pointed out in Tsu (1998), if the data were still unstable after electrical forming, the sample was simply discarded. About 20% of the stable and reproducible samples showed oscillations; however, they were different in small details. This tells us that electrical forming is capable of providing a given sample for useful study, but I doubt it would provide a useful sample for device applications. We have taken two devices on the same wafer separated far apart and placed in parallel. Nothing very unusual

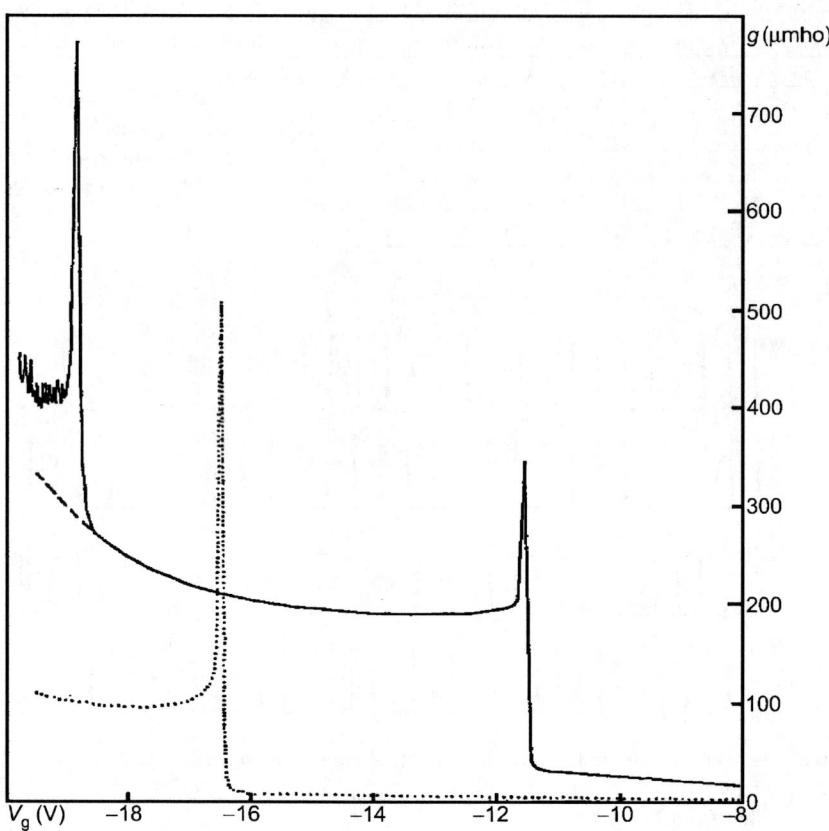

Figure 7.15 The dotted and dashed lines show two adjacent dots connected singly. The solid line shows the two devices connected in parallel. Note jump ΔG (dotted) at ~80 μS and jump ΔG (dashed) at ~160 μS. However, for the parallel connected case, ΔG (solid) ~360 μS, an additional jump of 120 μS, as well as the shift of an additional −2.5 V from −16.5 to −19 V.
Source: Taken from Bowhill's thesis, unpublished.

happens because the current is doubled or the conductance added, whenever the applied voltage is such that both states are occupied. However, when two adjacent devices having slightly different voltages at the jump in conductance are connected singly and in parallel, shown in Figure 7.15, the result is extremely interesting because whether the two adjacent coupled QDs are conducting or not, the potential due to the occupancy of one dot raises the potential of the other, indicating that the two dots are close enough and their potentials are interacting with an additional shift of −2.5 V from −16.5 to −19 V. And there is an additional jump in conductance of 120 μS from ΔG (solid) ~360 μS. This indicates that an extra three more electrons take part, which can only mean that the symmetry has changed from the coupling of two dots.

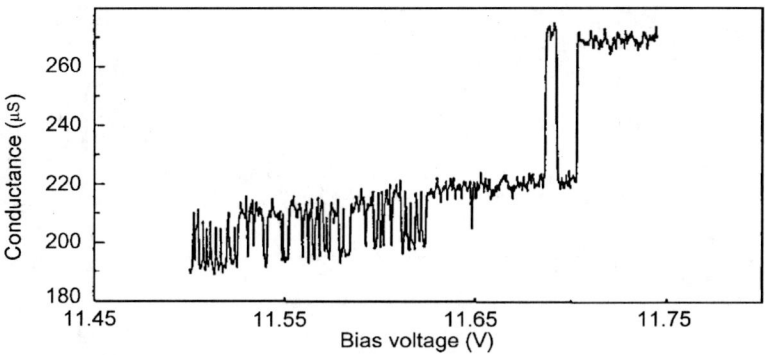

Figure 7.16 Conductance versus bias voltage shows telegraph-like noise. At a fixed voltage, the variation in time is not like all those we have shown as oscillations or oscillatory switching, rather, a typical noise-like spectrum.
Source: Taken from Chen Ding's (1994) thesis, unpublished.

I think by now the reader should be convinced that what we showed was not telegraph noise, or simple breakdown, but involved some complicated quantum conduction. However, since some of our samples do exhibit telegraph-like noise, Figure 7.16 shows a typical case. Broadly speaking, oscillatory conductance involves two QDs or one QD and one defect. But if the defect is coupled to a third or more sites, the complexity alone can lead to more random trapping and re-emission resulting in telegraph-like appearance from ~30% or less of the dots we measured.

Figure 7.17 shows telegraph-like noise spectra of complex current fluctuation prior to oxide breakdown due to discrete multilevel switching (Farmer et al., 1988). Although detailed discussion of slow oscillations is quite complex, it is possible to explain the general mechanisms in terms of a relatively simple model of quantum effects, because, by comparison, the samples we singled out for study are much simpler than those with telegraph-like noise spectra. Since we can produce almost all the reported telegraph noise spectra as well in the type of samples we fabricated using various annealing and electrical forming techniques, I really think the bottom line is that we have used a selective forming process picking the simplest configuration, that is, a current filament consisting of only one silicon dot. As long as we reject the majority of samples as bad samples, our selection process does have merits and is not so different from all the others including passivation of defects, annealing in steam to reduce the interface density of the MOS capacitors, or even something more current, picking out a single thread of a carbon nanotube.

As pointed out by Tsu et al. (1994), whenever eV_1 in Eq. (7.11) is aligned with E_F, the energy $E_1 + e^2/2C$, $E_1 + 2e^2/2C$, includes charging the capacity C, referred to as a Coulomb blockade (Likharev, 1988), which is nothing other than simple electrostatics. When trapping by a defect is present, the first current jump occurs at an applied voltage $V_a = V_1 + Q/C + V_s$, with V_s being the voltage drop due to deep depletion (Tsu, 1993). Suppose an electron is captured resulting in $Q = e(n + 1)$; an

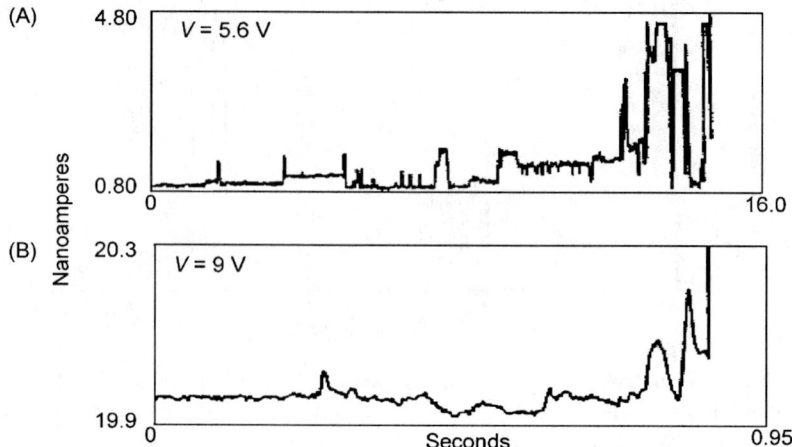

Figure 7.17 Telegraph-like noise spectra of complex current fluctuation prior to oxide breakdown showing discrete multilevel switching.
Source: Taken from Farmer et al. (1988), with permission.

additional voltage of e/C is necessary to maintain resonant tunneling. Because the applied bias is fixed, the conductance will jump down to a lower value. Conversely, whenever an electron is emitted from a trap, the charge Q returns to $Q = en$, and the potential at the QD drops back so that the energy state involved is again aligned with the Fermi level of the contact at V_1, causing the conductance to jump back to a higher value. Therefore, the period is the sum of the electron capture and emission time constants. In this picture, oscillation is the result of a flip-flop between two charge states involving exchange with a defect. Alternately, the two states can be two coupled QDs separated by a barrier resulting in the splitting of energy E_0 into E_+ and E_- with a frequency of oscillation $\omega = (E_+ - E_-)/\hbar = 4E_0 \exp(-\alpha B)/\alpha w\hbar$, where $\alpha = \sqrt{2mU/\hbar^2}$, with U, B, and w, being the barrier height, the width, and the well width, respectively (Tsu, 1993). Assuming that a nonconducting state is weakly coupled to a conducting state *via* an oxide barrier of $U = 3.2$ eV, for a period of 10 s, it is necessary that $B \sim 15$ nm. Since the total layer thickness is $\sim$15 nm, we conclude that the origin of the switching is more likely to be due to a defect state located near a conducting QD rather than to a similar but nonconducting QD state. Conduction peaks have been reported by resonant tunneling *via* bound states of a single donor in a quantum well (Dellow et al., 1992) which is not too different from the subject treated in this section (Figure 7.18).

Typical step-like $G-V$ with hysteresis at 300 K is shown in Figure 7.19, taken from Li (1993). The conductance steps at 40 to 205 to 385 μS give $\Delta G = 40$, 165, and 185 μS. Again the results point to the three higher states merging into a single threefold p-like degenerate state of a spherically symmetric hydrogen-like 1s, 2p state. The two electrons in the 1s state give 78 μS, however, it appears only one electron occupies this 1s state. Occupation of the 2p state, allowing six additional

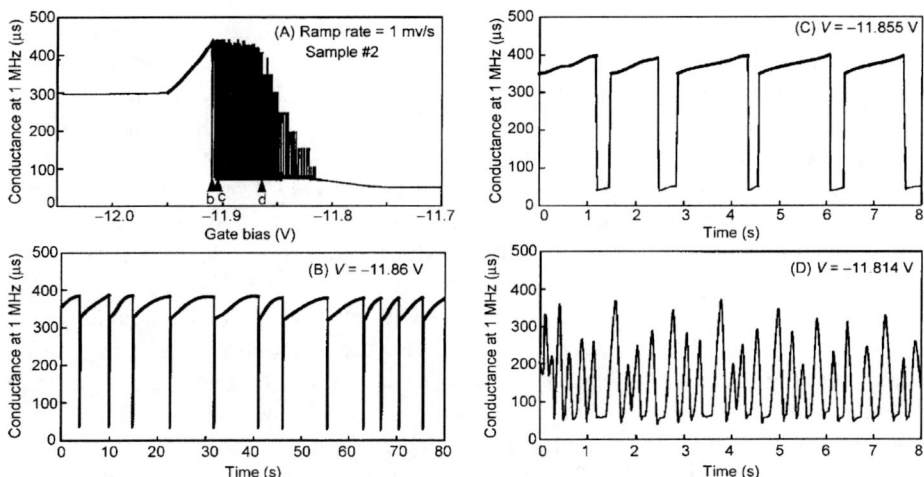

Figure 7.18 (A) Conductance oscillation near and before a peak. At fixed voltages, (B)−(D), oscillation with time is detailed. Oscillation is more complex in (D) where the number of electrons involved in jumping back and forth varies between 4 and 8 while in (B) and (C) the number appears to be fixed at 8.
Source: Taken from X. Li, unpublished.

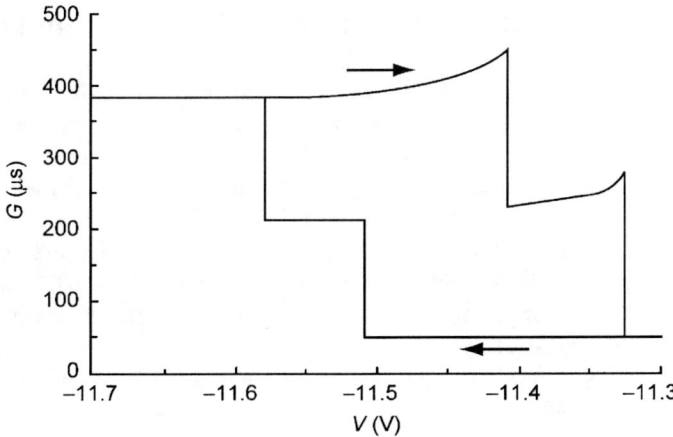

Figure 7.19 Typical step-like $G-V$ with hysteresis at 300 K. Conductances of 40, 205, and 380 μS with $\Delta G = 40$, 165 and 185 μS $\sim G_0$, $4G_0$, and $5G_0$.
Source: After Tsu (2003) and Li's thesis.

electrons, should give ΔG of ~ 240 μS, however, only five appear to be taken up. These facts indicate that the three states in Figure 7.1 are not all merged into a 2p-like state with a total of six electrons. We conclude that perhaps only the first two, (012) and (013), are merged, indicating that it is not a totally spherical symmetric system. Let us look at the energies involved. In Figure 7.19, the voltage at −11.52

and −11.32 V gives ΔV of 0.2 V, and between −11.52 and −11.57 gives a ΔV of 0.05 V as if no charging is involved. In other words, the term $e^2/2C$ does not seem to be involved. The only possibility is that the charges are already present even before the jump in conductance. And the only rationale is that the charges are located at a nearby QD, but for some reason they do not contribute to conductance. Therefore, we have the evidence of trapping from a "nonconducting" QD located near the conducting QD, exchanging electrons and resulting in hysteresis. My previous publication on the subject (e.g., Tsu et al., 1994; Tsu, 1998) stated that we were unable to distinguish between the two possibilities: switching and hysteresis involving coupling of a QD with a nearby defect, or coupling of a QD with a nearby "silent QD." Now I think the ambiguity is at least partially resolved, and we can be reasonably sure that it is the trapping site at a nearby nonconducting QD, rather than a QD coupled to one with a tunneling channel that serves only as a trap for electrons capable of raising and lowering the potential at the conducting QD. What can we say about hysteresis as a result of a different configuration in phase space? The stored charges not only can change the potential but also result in a slight change of bonding configuration giving puckering. The structure is indeed slightly changed, leading to change in the symmetry and selection rules. Occasionally we observed light emission. It was very weak and infrequent, and the origin seems to resemble switching.

7.4   Avalanche Multiplication from Resonant Tunneling

Current−voltage measurements of resonant tunneling through nanoscale silicon QDs connected in parallel reveal large current staircases. The sharpness of the current jumps comes from the action of tunneling through a few groups of silicon particles of various sizes connected in parallel, and the magnitude of the current jumps results from substrate avalanche multiplication of the small injected resonant-tunneling current. Some of the most striking results described apparently originate from a single silicon QD in spite of the fact that many silicon QDs are connected in parallel. However, as pointed out, only a small percentage of samples tested fall into this category. Most of our samples do show fairly consistent conduction peaks especially in cases with lower applied voltage. At higher applied voltage, we consistently observed staircases in $I-V$.

The typical full width at half maximum of the conductance peak is 30 meV. Inhomogeneous broadening, caused by a large distribution of particle sizes, should result in a far broader conductance peak for any appreciable spread in the particle size distribution. We have resolved this paradox, because all the measured linewidths of the conductance peaks show only homogeneous broadening. Localized electrical forming of the sample prior to the observation of resonant tunneling creates a conductive path that selects only a few of the many possible nanocrystallite sizes available in a given device structure. Since only a few nanoparticles are involved in the resonant tunneling, these different groups of particles will cause different peaks, rather than acting to smear a given peak. These major ingredients

lead to a simple model, based on a few particles of different sizes connected in parallel, as shown in Figure 7.20(A) for three sizes, in (B) with different energy states,
and in (C) with three staircases for the current plotted against the applied voltage.
This model also solves the early mystery of why there appear so many jumps or
staircases.

Our applied voltage is divided between the thin oxide layer containing the QDs
V_{ox} and V_S of the substrate depletion layer. Moreover, the electrons tunneling into
the deep depleted region of the substrate result in avalanche multiplication.
Following Nicollian and Brews (1982), we start from the voltage applied to the
device V_{in} as

$$V_{in} = V_{ox} + V_S, \tag{7.12}$$

with

$$V_{ox} = F_{ox}d_{ox}, \tag{7.13}$$

where F_{ox} and d_{ox} are the electric field in the oxide and the oxide thickness,
respectively.

At the oxide/nc-Si interface,

$$\varepsilon_{ox}F_{ox} = \varepsilon_{Si}F_{Si}, \tag{7.14}$$

(A)

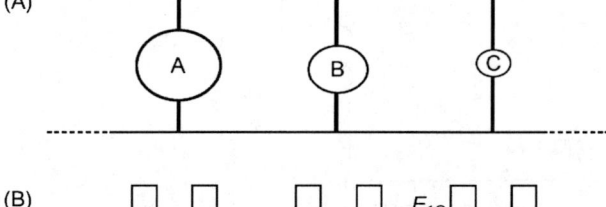

(B)

Figure 7.20 (A) Three nc-Si
particles of different sizes
connected in parallel; (B)
their corresponding energy
states; (C) $I-V$ of the parallel
arrangement showing three
staircases.
Source: After Boeringer and
Tsu (1994), with permission.

(C)

where ε_{ox}, ε_{Si}, and F_{Si} are the effective dielectric constant of the oxide/nc-Si layer, the bulk silicon and the electric field on the bulk silicon side of the interface, respectively. On the bulk silicon side,

$$F_{Si} = \frac{q}{\varepsilon_{Si}} N_D w, \qquad (7.15)$$

in which q and N_D are the electronic charge and the doping density of the substrate, respectively, and the depletion width w is given by $w = (2\varepsilon_{Si}V_S/qN_D)^{1/2}$. Combining these expressions, we arrive at

$$V_{ox} = \sqrt{2q\varepsilon_{Si}N_D}\frac{d_{ox}}{\varepsilon_{ox}}\sqrt{V_S} \equiv C\sqrt{V_S}. \qquad (7.16)$$

As a zeroth order approximation, the oxide/nc-Si layer is modeled as a constant resistor, R_0, with additional resistors, R_{1A}, R_{1B}, and R_{1C} in parallel with the first resistor being successively added as the contact Fermi level sweeps through the energy levels, E_{1A} to E_{1C} of the parallel particles shown in Figure 7.21. The multiplication factor M is given by $M = 1/(1 - (V/V_B)^n$, with V and V_B being the applied and the breakdown voltage, with an empirical parameter n (Miller, 1957; Manduteanu, 1985). Figure 7.21 shows the electrical equivalent circuit model for the tunneling process involved (Boeringer and Tsu, 1995).

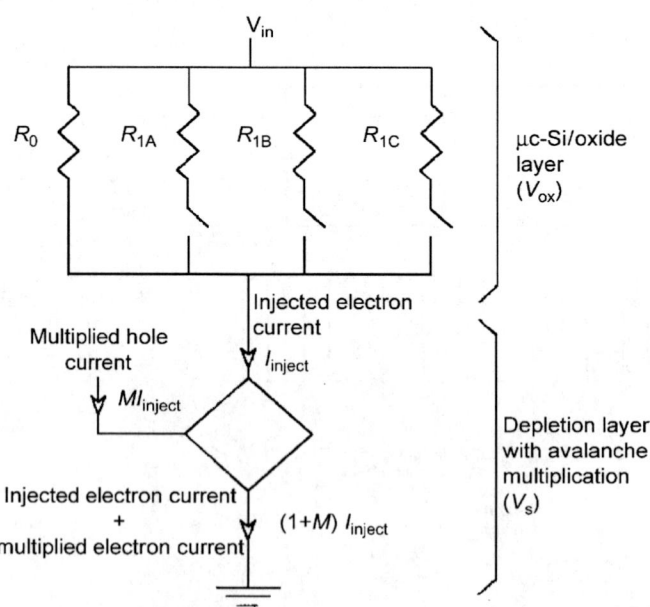

Figure 7.21 Electrical equivalent circuit model showing parallel conduction through parallel particles and avalanche multiplication in the substrate with a multiplication factor M. *Source*: After Boeringer and Tsu (1995), with permission.

For convenience, if we assume a symmetrical double barrier structure, then

$$V_{ox} = 2\frac{E_1}{q} + \frac{q}{C_{eff}},$$
(7.17)

where $E_1 \sim 0.18$ eV is the first energy level in a 5-nm QD (Tsu, 1990) and C_{eff} is given by the increment of the total energy stored to be treated in Chapter 8 (Tsu, 1993). Since the applied voltage is known from the measurement, the constant C can be obtained from Eqs. (7.12) and (7.16) as C is equal to 0.08. The values of the resistors, such as R_{1A}, are determined as the best fit to the measured plot. A remarkably good fit for a typical $I-V$ is shown in Figure 7.22 resulting in the parameters in Table 7.4.

Note that the energies E_1, together with scaling the lowest energy given in Figure 7.1, give us the diameter of the silicon particles in the fourth column of Table 7.4, in general agreement with TEM and Raman measurements (Tsu et al., 1995).

The measured and theoretical $I-V$ curves of a Si-QD device at 77 K have also been fitted by the same procedure. Results show that two sets of staircases are

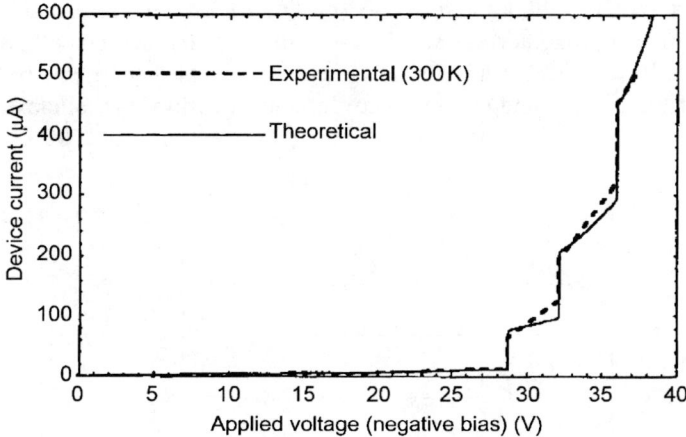

Figure 7.22 Measured and theoretically fitted $I-V$ of a device at 300 K.
Source: After Boeringer and Tsu (1995), with permission.

Table 7.4 Parameters of the Three Particles for the Theoretical Fit to the Experimental Data in Figure 7.22. $R_0 = 58,000 \ \Omega$, n in Eq. (7.12) = 1.8, and $V_B = 45$ V

nc-Si	V_{ox} at E_1 (V)	E_1 (eV)	Calculated Diameter (Å)	R Through E_1 (Ω)
A	0.426	0.170	51.4	12,000
B	0.450	0.180	49.9	8900
C	0.477	0.191	48.5	9000

involved, consistent with the identification of two distinct groups of staircases corresponding to the ground and the first excited states (Boeringer and Tsu, 1994). It should be quite obvious that the bulk of our experimental data were stable and repeatable after we learned how to form the as-fabricated samples electrically. Our basic understanding in terms of tunneling through these crystallized silicon nanoparticles embedded in an oxide matrix may be fitted to our theory using an equivalent circuit with parallel paths. The negative side is the fact that only a very small percentage of the total number of particles covered by our relatively large contact is active. So the dream of a silicon QD field effect transistor (FET) is far from reality.

7.5 Influence of Light and Repeatability under Multiple Scans

My last graduate student, Chen Ding, working on an nc-Si-QD embedded in an amorphous oxide matrix, needed a thesis to graduate and he had just completed our latest $I-V$ and $C-V$ setup. Urged by Nicollian to make one more effort to evaluate whether our idea of making a silicon-based QD FET is viable, I assigned Ding to the task of putting together the new features we have learned since 1989–1990, to fabricate the best devices ready for new simultaneous measurements of the Re- and Im-parts of the current and conductance. Instead of following Poindexter's suggestion that we need to increase the annealing temperature to 1200°C, we built a

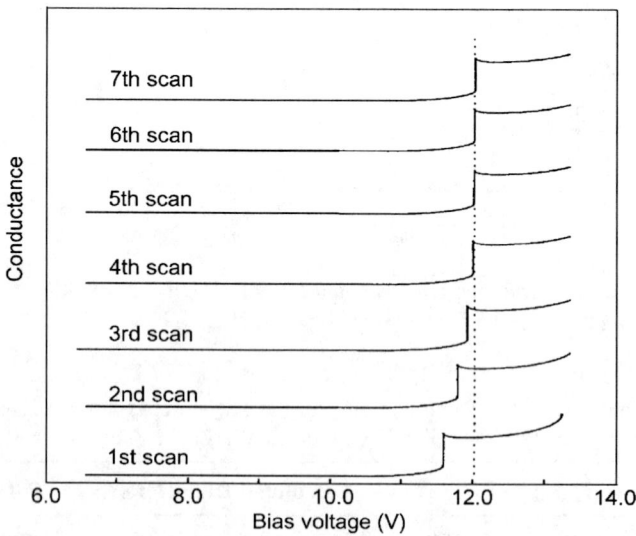

Figure 7.23 Conductance versus bias voltage under multiple scans. There is self-electrical annealing with stable $G-V$ after the fourth scan.
Source: Taken from Ding's thesis (1994).

current limiter for electrical forming under forward bias. The results were striking in some sense and still disappointing when it comes to possible implementation as a real device. Figure 7.23 shows typical conductance with repeated scans. Note that the $G-V$ is totally stabilized from the fourth scan on. This result alerted us that perhaps we should have longer electrical forming is better. Several attempts to form longer did result in totally repeatable $G-V$ right from the first scan. However, we also learned that a better procedure involves first using the electrical forming with forward bias, followed by "forming in the reverse bias," that is, using the electrical repeated scans to stabilize the structure further. This procedure was successful, however, Nicollian said that no one in his right mind would contemplate using such a device. Meanwhile, while working as a consultant to SI Diamond, Inc., I learned, to my surprise, that all electroluminescent (EL) devices require electrical forming. When I told Nicollian about this fact, he replied, "That is why EL devices never made the grade." Within this background, let me show our new results.

Even more amazing is the effect of light shown in Figure 7.24. The magnitude and phase of G is plotted against the bias voltage with light, WL, and without light, NL. Under illumination with a focused microscope light through a filter, typically steps in $G-V$ are transformed to peaks in $G-V$. If there is already a peak, then the peak becomes much sharper and larger under illumination. Using Si, Ge, and GaAs wafers polished on both sides as filters, as well as various Corning filters, it was

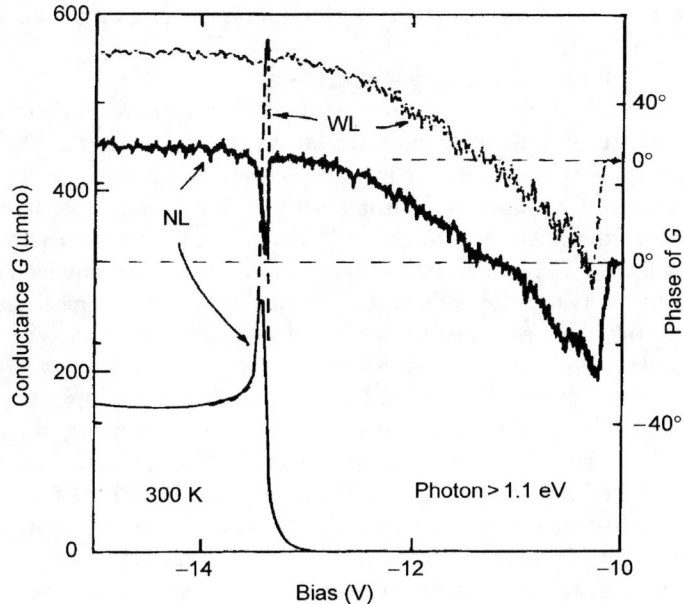

Figure 7.24 Conductance versus bias voltage. The peak in G is almost four times higher with light. On the phase angle Φ versus V plot, the $-25°$ dip near the peak of G disappears under illumination at a photon energy >1.1 eV.
Source: After Ding and Tsu (1995), with permission.

established that light-induced effects disappear when the photon energy falls below the fundamental gap of silicon. In the case of Figure 7.24, the conductance peak is almost four times higher with light. On the phase angle Φ versus V plot, a $-25°$ dip near the peak of G totally disappears under illumination. The disappearance of a substantial phase shift may be interpreted by filling traps from light-generated carriers, resulting in a stronger conductance peak. Plots made by Chen showed three conductance steps with substantial peaks appearing under illumination at the leading edge of each step. Since the generation of e$-$h pairs lowers the substrate resistance, the peaks are shifted toward lower bias voltage under illumination, about a 0.5% reduction. Actually, we should not be surprised by our finding because $C-V$ measurements have been used by Nicollian to identify traps all along (Nicollian and Brews, 1982). There is some difference though. Our procedure does not encompass the Re- and Im-parts of G, rather the amplitude and phase of G.

7.6   Many Body Effects in Coupled Quantum Dots

In two-dimensional systems, DOS is proportional to the conductance G, with $G = gG_0$ where $G_0 = 39\,\mu S$ and the degeneracy factor $g = 1, 2, 3, \ldots$ For a state without any other symmetry-induced degeneracy, $g = 1$ per spin, so that for $+1/2$ and $-1/2$ spins, $g = 2$. I have put together several typical cases representing $g = 2, 4$, and 6 (Tsu, 2008).

After the initial peak, the majority of $\Delta G \sim 160\,\mu S$, with $\Delta g = 4$ for the step, or involving two pairs of $+$ and $-$ spins, with which we are reminded the chemical bonds with two electrons per bond with coordination number 4. There are also $g = 2, 6$, or even 8, with a higher number representing degeneracy when symmetry enters the picture. Occasionally, in some 10% of our sample, we have seen odd numbers, for example, $g = 5$ with $\Delta G = 200\,\mu S$, and usually associated with telegraph-like noise. But some samples show one-dimensional conductance, which may be due to the two coupled QD arranged in line with the current path mimicking a Qwire. *What is never understood or discussed by us is the fact that peaks always precede steps, and why conductance involving higher energy states of the Si-QD shows steps.* Since I have been involved with LaFave in the study of the discrete nature of electrons in capacitance using a dielectric sphere model (LaFave and Tsu 2008), I have acquired a better understanding, or better respect for the electron$-$electron Coulomb interactions. Basically, the coupling of the QDs into a two-dimensional-like system is enhanced by the occupation of two adjacent states be electrons, with interaction terms including the direct Coulomb term, e^2/r_{ij}, as well as all the induced polarization terms on the individual QDs, inside and outside as well as on the interface. *It is the many body effect that creates enhanced coupling in forming the two-dimensional-like system from 0D QD states and results in creating a peak leading the step.* As the applied voltage is increased, electrons tunnel into the QD and occupy the empty states, resulting in all the induced terms as the basis of our calculation for the capacitance (Zhu et al., 2006). The net result is to enhance the interaction between neighboring QDs, essentially by lowering the tunneling barriers with respect to the self-consistent potential of the electrons occupying these states. When a given state is occupied by an electron, the potential energy goes up. *In a nutshell, the many body effect may be simply described by the self-consistent potential from occupation, raising the potential of the individual QD state with respect to the barriers separating these QDs, creating an enhanced coupling between neighboring dots.*

This schematic representation of the enhanced coupling shown in Figure 7.25 is based on the coupling of two Si-QDs, with electron in one and a neighboring QD. Top shows singly occupied individual QDs, middle shows doubly occupied QDs, and bottom shows exchanging occupations leading to oscillations. The exchange should be very fast. When a trap replaces a regular QD, serving as an imposter with very slow emission and capture rates, telegraph-like slow oscillation occurs. If triply occupied, $g = 4, 8, 12$. It is more complicated for the coupling of three. Nevertheless, using the chemical approach to the formation of molecules, it is clear that coupling dictates occupation of states by electrons leading to coupling in the first place. When the coupling spreads in a plane, conductance remains constant.

The spreading of singly and doubly occupied states by electrons are shown in the top and middle cases in Fig. 7.25. Due to Coulomb blockade, a fancy name for the formation of potential barrier, switching between two cases results in oscillation, shown in the bottom of Fig. 7.25. As a reminder, Coulomb blockade is

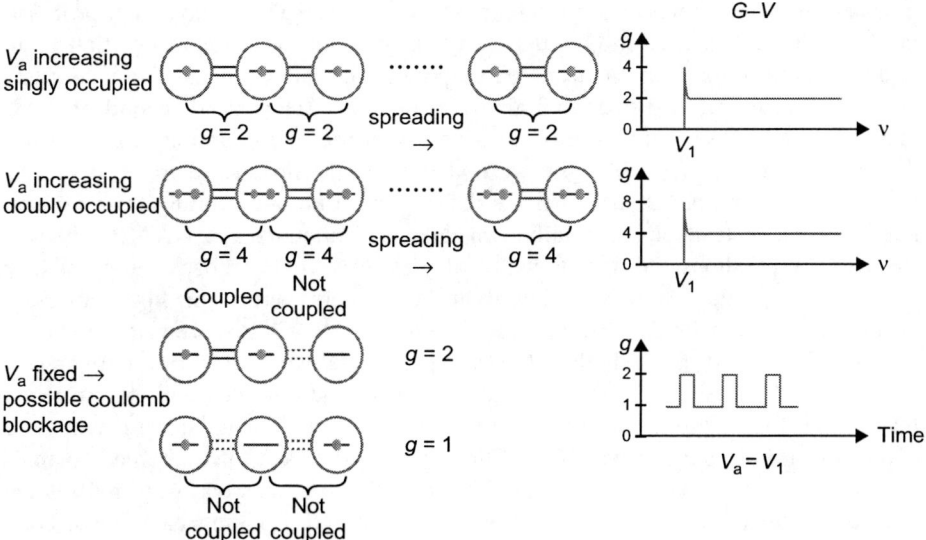

Figure 7.25 A model for the enhanced coupling between QDs from adjacent QDs. The top shows singly occupied individual QDs, the middle shows doubly occupied QDs, and the bottom shows exchanging occupations leading to oscillations, generally fast oscillations. When a trap serves as an imposter of a QD, telegraph-like slow oscillation occurs.
Source: After Tsu (2008), with permission.

nothing but the manifestation of potential due to charges. However, Pauli's exclusion principle is more fundamental than potential barrier, because neutrons have no net charge.

7.7   Summary

To summarize, although occupation of these quantum states results in a more spherically symmetric system, it is not really exact. My estimate is that the two lower excited states, (012) and (013) in Figure 7.1, are merged and the uppermost state (111) is not. Before we discovered the effects of illumination, we thought that trapping is due to a nonconducting QD located next to a conducting QD. The nonconducting QD is charged and raises the potential at the conducting QD so that no extra voltage is needed to move the charges into the conducting dot. The change in conductance is very close to what one would expect from nG_0, with G_0 given by the fundamental unit of conductance $\sim$39 μS, and n, the electron occupation number, consistent with Pauli's exclusion principle. Whenever the first jump in conductance involves two electrons, our data is more predictable, whereas when the first jump involves only one electron, it is more common to observe telegraph-like $G-V$. This fact indicates that a defect traps an electron and holds onto it by drastically changing the potential and blocking another electron. This type of process is precisely what generates random telegraph-like signals. The reason such processes take place far more often in QD tunneling than quantum well tunneling, first of all, is that a large number of electrons participate in the latter process from the transverse degree of freedom overwhelming the trapping effects. And second, the state in a nanoscale QD is quite similar to a defect so that the coupling is large. After we discovered the effects of light, we realized that the process seems to be caused by a variety of trappings, although the data did not rule out trapping from nonconducting dots, instead of some unknown defects. I attended the NATO Advanced Research Workshops on the Physics of Few-Electron Nanostructures held in Noordwijk in 1992. I told Nicollian about how others were using lithographically produced devices with much more consistent data. He did not seem to be swayed by this. He asked me whether these experiments were at room temperature. I replied, "Of course not." He then said that we should continue with our work but that we need to fine-tune our process. Meanwhile I was having success with semiconductor atomic superlattices (see Chapter 6) with results that seemed to make them far more promising as a potential device. Nicollian passed away in 1995 and our pursuit of the Si-based QD device came to an end. Looking back, what could be simpler than forming Si particles by annealing from the amorphous phase? Obviously if I had known all of the complications, most probably I would not have devoted so much effort to it. Nevertheless, the understanding needed to unravel all of the experimental data does present an intellectual challenge that requires more than solid-state physics and atomic physics in what should be properly referred to as the physics of nanoelectronics, as well as good engineering.

The complexity of understanding in what is involved is mind boggling. In semiconductor physics, we are used to a few things, such as effective masses, even effective mass tensors and energy bands that are enumerated in reciprocal space with a designation following group symmetry, etc. The subject is far from child's play; nevertheless, we barely need to keep check of induced charges because surface effects are mostly negligible. In QD, there may be equal or more atoms located on the surface and energy states must take into account these induced charges. Furthermore, one needs to distinguish whether the interacting charges are from another source, such as from the matrix or from the same charge on the dot, because the former case requires the inclusion of Heisenberg's exchange term, while the latter involves only the direct Coulomb term. Since the position of these QDs is not precisely controlled, their couplings are not uniform. It is dangerous to assign broadening to account for a distribution of sizes and separations, because there are "feedback" effects. For example, trapping of an electron from a QD by a nearby impurity results in a change of coordinates to minimize the strain energy. In a simple picture, there is a relaxation resulting in a change of configuration. This is simply a mini local phase transition, which results in hysteresis. We are not saying that none of these phenomena ever happen in our electronic devices. The difference is in sheer numbers. In a low-power MOSFET, there are perhaps far more than 10^3 electrons. Losing a few to a trap, or even some being controlled by a trap is hardly serious. For a typical QD tunneling no more than a few electrons are involved in the process. Here, losing a couple of electrons is serious business. Along this line of argument, I predict the real issue with the implementation of nanoelectronics is not so much as how well we can "clean up" the act, rather, in what role we can expect a device to function. S.S. Chen, now at Florida State University, told me once that the reason "fuzzy logic" was hot was because at one time there were great expectations awaiting its development. *I dare predict that QD nanoelectronics will never be developed into wide usage in computer technology; rather, it will be used in some areas yet to be discovered.*

The device world has been fascinated by the Coulomb blockade leading to the single electron transistor (Fulton and Dolan, 1987), which is seemingly the ultimate electronic device. The fact is that trapping is a very difficult "roadblock" to overcome. With millions of years of evolution, living organisms have developed a system that might be loosely described as "single-ion transistor."

References

Abramowitz, M., Stegem, I.A. (Eds.), 1964. NBS Applied Math Series 55. US Department of Commerce.

Allred, D.D., Gonzalez-Hernandez, J., Nguyen, V., 1988. US Patent 4,792,501, 20 December 1988.

Ando, T., Fowler, A., Stern, F., 1982. Rev. Mod. Phys. 54, 437.

Boeringer, D., Tsu, R., 1995. Phys. Rev. B 51, 13337.

Chou, S.Y., Gordon, A.E., 1992. Appl. Phys. Lett. 60, 1827.

Dellow, M.W., Beton, P.H., Langerak, C.J.G.M., Foster, T.J., Main, P.C., Eaves, L., et al., 1992. Phys. Rev. Lett. 68, 1754.

Ding, C., Tsu, R., 1995. Mater. Res. Soc. Symp. Proc. 378, 757–760.

Efros, Al., Efros, A.L., 1982. Sov. Phys. Semicond. 16, 772.

Farmer, K.R., Saletti, R., Buhrman, R.A., 1988. Appl. Phys. Lett. 52, 1749.

Fortunato, E., Martins, R., Ferreira, I., Santos, M., Maçarico, A., Guimarães, L., 1989. J. Non-Cryst. Solids 115, 120.

Fulton, T.A., Dolan, G.J., 1987. Phys. Rev. Lett. 59, 109.

Hernandez, J.G., Martin, D.S., Chao, S.S., Tsu, R., 1984. Appl. Phys. Lett. 44, 672.

Ioriatti, L., Tsu, R., 1986. Surf. Sci. 174, 420.

LaFave, T.J., Tsu, R., 2008. Microelectron. J. 39, 617.

Li, X.-L., 1993. M.S. Thesis, Department of Electrical Engineering, UNC-Charlotte.

Likharev, K.K., 1988. IBM J. Res. Dev. 32, 114.

Lu, Z.H., Lockwood, D.J., Barlbeau, J.M., 1995. Nature 378, 258.

Manduteanu, G.V., 1985. IEEE Trans. Electron Devices, ED 32, 2492.

Miller, S.L., 1957. Phys. Rev. 105, 1246.

Nicollian, E.H., Brews, J.R., 1982. MOS (Metal Oxide Semiconductor) Physics and Technology, Wiley-Interscience, New York,

Nicollian, E.H., Tsu, R., 1993. J. Appl. Phys. 74, 4020.

Stern, F., Howard, W.E., 1967. Phys. Rev. Lett. 163, 816.

Takagi, H., Ogawa, H., Yamazaki, Y., Ishizaki, A., Nakagiri, T., 1990. Appl. Phys. Lett. 56, 2379.

Tsu, R., 2008. Microelectron. J. 39, 335.

Tsu, R., 1985. Mater. Res. Soc. Symp. Proc. 38, 383.

Tsu, R., 1990. SPIE 1361, 313.

Tsu, R., 1993. Physica B 189, 235.

Tsu, R., 1998. Int. J. High Speed Electron. Syst. 9, 145.

Tsu, R., 2000. Appl. Phys. A. 71, 391.

Tsu, R., 2003. Microelectron. J. 34, 329.

Tsu, R., Döhler, G.H., 1975. Phys. Rev. B. 12, 680.

Tsu, R., Hernandez, J.G., Chao, S.S., Lee, S.C., Tanaka, K., 1982. Appl. Phys. Lett. 40, 534.

Tsu, R., Hernandez, J.G., Chow, S.S., Martin, D., 1986. Appl. Phys. Lett. 48, 647.

Tsu, R., Nicollian, E.H., Reiaman, A., 1989. Appl. Phys. Lett. 55, 1897.

Tsu, R., Ye, Q.-Y., Nicollian, E.H., 1990. SPIE 1361, 232.

Tsu, R., Li, X.-L., Nicollian, E.H., 1994. Appl. Phys. Lett. 65, 842.

Tsu, R., Morais, J., Bowhill, A., 1995. Mater. Res. Soc. Symp. Proc. 358, 825.

Van Wees, B.J., van Houton, H., Beenakker, C.W.J., Williamson, J.G., Kouwenhoven, L.P., van der Marel, D., et al., 1988. Phys. Rev. Lett. 60, 848.

Ye, Q.-Y., Tsu, R., Nicollian, E.H., 1991. Phys. Rev. B 44, 1806.

Zhu, J., LaFave, T.J., Tsu, R., 2006. Microelectron. J. 37, 1296.

8 Capacitance, Dielectric Constant, and Doping Quantum Dots

8.1 Capacitance of Silicon Quantum Dots

We show in the last chapter that charge accumulation plays a major role in the energy states of the quantum-confined nanocrystalline silicon (nc-Si). We are all familiar with the striking difference between the atomic spectra of the hydrogen and helium atoms caused by the presence of an additional electron. The physics is even more complicated when the difference in the static dielectric constant of silicon ($\sim$12) and a-SiO$_2$ ($\sim$4) results in induced polarization. A single electron inside a silicon sphere can interact with its induced polarization in the oxide. With two electrons inside the silicon, electrons and induced polarization interact, resulting in a complicated picture. Adding the first electron results in a hydrogen-like state with the interaction terms described. However, adding a second electron requires a solution somewhat similar to the helium-like state, which has interactions between the two electrons as well as with all the induced charges. The use of perturbation limits our results essentially to the ground state (Babic et al., 1992). Because of complicated interactions, expressing the extra energy due to the addition of an electron in terms of capacitance is not a constant representable by geometry as in classical theory. Only for a large quantum dot does the use of constant capacitance represent a fair approximation (Likharev, 1991). The single most important fact is that the energy of the quantum system is much larger than the electrostatic energy due to the charge of the electron. The ground state energy difference between zero and one electron defines the effective capacitance C_1, and similarly the ground state energy between one and two electrons defines the effective capacitance C_2, etc. In principle, this process can go on to C_n in terms of the energy difference between n and $n + 1$ electrons. However, our approach cannot be readily generalized to more than two electrons as presented by Macucci et al. (1993, 1995), where the important aforementioned induced terms are not included. In spite of the fact that our results use a perturbative calculation, because we took into account the induced charges, a detailed account should be of considerable interest.

We embedded a spherical silicon dot of radius a in an a-SiO$_2$ matrix. Instead of taking the actual barrier height of 3.2 eV between Si and a-SiO$_2$, an infinite barrier height is assumed. The consequence is that for a small radius, except for the ground

Superlattice to Nanoelectronics. DOI: 10.1016/B978-0-08-096813-1.00008-4

state and possibly few low-lying states, higher states are not confined. This is why we restricted our calculation to the ground state even if an exact, instead of perturbative, method is used. Let us point out how complex the problem would be using a finite barrier height. The tailing of the wave function into the matrix necessitates replacement of the dielectric discontinuity by a smooth function in order to avoid the singularity of the associated polarization energy (Stern, 1978). Note that this approximation does not apply for GaAs/AlGaAs dots. For $a > 1$ nm, the effective mass approximation and the static dielectric constant are applicable. However, instead of taking the m_t and m_l as treated in the previous chapter, an isotropic mass of $0.26\,m_e$ and the relative permitivity of Si, $\varepsilon_1 = 12$ and that of a-SiO$_2$, $\varepsilon_2 = 4$ are used. Before we delve into the calculation, let me point out that the charge of an electron is assumed to be infinitesimally divisible in classical theory, which does not really apply at all. And taking the discreteness of electronic charge, even in a classical description, is very complex.

8.1.1 Electrostatics

The calculation of the electrostatic energy terms follows the work by Brus (1983, 1984) and Bottcher (1973). Note that the usual simplification using the image method does not apply because of the curved boundary and the fact that the dielectric discontinuity is not a sheet of infinite conductivity. Thus, Green's function must be used. Green's function inside the sphere is

$$G_{\text{in}}(\mathbf{r}, \mathbf{r}') = \frac{1}{4\pi\varepsilon_0\varepsilon_1|\mathbf{r} - \mathbf{r}'|} + \sum_l A_l r^l P_l(\cos\gamma), \tag{8.1}$$

and outside the sphere

$$G_{\text{out}}(\mathbf{r}, \mathbf{r}') = \sum_l B_l r^{-(l+1)} P_l(\cos\gamma), \tag{8.2}$$

in which $\mathbf{r}$, $\mathbf{r}'$ are the position vectors of the field point and the charge point, respectively, and γ is the angle between these vectors, measured from the origin at the center of the sphere. The coefficients A_l and B_l are determined by the electrostatic boundary conditions at the Si/a-SiO$_2$ interface. With the use of infinite barrier height, the wave function is zero at the surface of the Si sphere, B_l values are not needed for the evaluation of the matrix elements and

$$A_l(r') = \frac{(\varepsilon_1 - \varepsilon_2)(l + 1)r'l}{4\pi\varepsilon_0\varepsilon_1[\varepsilon_2 + l(\varepsilon_1 + \varepsilon_2)]a^{2l+1}}. \tag{8.3}$$

In the case of one electron, an electron induces the bound surface charge density which generates electrostatic potential at the electron's position. Energy associated with this term must include a factor of $\frac{1}{2}$, since it is a self-interaction term. Thus, this energy of self-polarization becomes

$$\phi_s(\mathbf{r}) = \frac{1}{2} \sum_l \frac{q^2(\varepsilon_1 - \varepsilon_2)(l+1)r^{2l}}{4\pi\varepsilon_0\varepsilon_1[\varepsilon_2 + l(\varepsilon_1 + \varepsilon_2)]a^{2l+1}}, \tag{8.4}$$

where the electronic charge is q. In the two-electron case, there are four terms: self-polarization terms for each electrons, Coulomb interaction, interaction between the two induced polarization, as well as the induced polarizations of electron 1 with electron 2 and the induced polarization of electron 2 with electron 1. Therefore the Coulomb term has the form

$$\phi_c(\mathbf{r}_1, \mathbf{r}_2) = \frac{q^2}{4\pi\varepsilon_0\varepsilon_1|r_1 - r_2|}, \tag{8.5}$$

and from Eqs. (8.1) and (8.3), the polarization term is

$$\phi_p(\mathbf{r}_1, \mathbf{r}_2) = \sum_l \frac{q^2(\varepsilon_1 - \varepsilon_2)r_1^l r_2^l P_l(\cos\gamma)}{4\pi\varepsilon_0\varepsilon_1[\varepsilon_2 + l(\varepsilon_1 + \varepsilon_2)]a^{2l+1}}. \tag{8.6}$$

By minimizing the sum of these energies, $E_S + E_C + E_P$, for self-polarization, Coulomb, and polarization for a dielectric sphere consisting of N electrons is presented in Section 8.4, some extremely unexpected features appeared.

8.1.2 Quantum Mechanical Calculation

The Hamiltonian for the one-electron case consists of the kinetic energy for the infinite barrier potential: $V(r) = 0$, $r < a$; ∞, $r > a$, and the self-polarization energy. An exact analytical treatment of the Schrödinger equation is too complex; we resort to the perturbation theory. The spherical Bessel functions are the solutions of the zeroth order Hamiltonian that includes the kinetic energy and infinite barrier potential terms. The lowest eigenfunction

$$\psi_0(r) = Nj_0(\pi r/a)Y_{00}(\Omega), \tag{8.7}$$

in which $N = a^{-3/2}I_0^{-1}$, where $I_0^2 = 0.0506606$. The self-polarization energy is defined by

$$\begin{aligned} E_S &= \langle \psi_0(\mathbf{r})\phi_S(\mathbf{r})\psi_0(\mathbf{r}) \rangle \\ &= N^2 \int j_0^2(\pi r/a)Y_{00}^2(\Omega) \left(\sum_l \frac{q^2(\varepsilon_1 - \varepsilon_2)(l+1)r^{2l}}{8\pi\varepsilon_0\varepsilon_1[\varepsilon_2 + l(\varepsilon_1 + \varepsilon_2)]a^{2l+1}} \right) r^2 \, dr \, d\Omega \end{aligned} \tag{8.8}$$

and contains dimensionless series to be summed numerically,

$$\sum_l \frac{l+1}{\varepsilon_1 + l(\varepsilon_1 + \varepsilon_2)} \int_0^1 x^{2l} j_0^2(\pi x) x^2 \, dx, \tag{8.9}$$

equal to 0.01516. The final form of the self-polarization energy for the ground state is thus

$$E_S = \frac{q^2(\varepsilon_1 - \varepsilon_2)}{8\pi\varepsilon_0\varepsilon_1 a} 0.299, \tag{8.10}$$

which scales as the inverse of the radius a proportional to $\varepsilon_1 - \varepsilon_2$. Note that if the dielectric constant of the matrix is higher than the quantum dot, the self-energy term changes sign. The total ground state energy is then

$$E_1 = 144.6/a^2 + 1.44/a, \tag{8.11}$$

where the units of the energy and the radius are eV and Å, respectively.

The two-electron Hamiltonian includes one-electron terms as before and the Coulomb and polarization terms of the two-electron interaction

$$H = \frac{-\hbar^2}{2m}(\nabla_1^2 + \nabla_2^2) + V(\mathbf{r}_1) + V(\mathbf{r}_2) + \phi_s(\mathbf{r}_1) + \phi_s(\mathbf{r}_2) + \phi_c(\mathbf{r}_1, \mathbf{r}_2) + \phi_p(\mathbf{r}_1, \mathbf{r}_2).$$
$$\tag{8.12}$$

We treat the kinetic energy for the infinite barrier as the zeroth order, and all other terms by first-order perturbation theory. The lowest-order spherical Bessel function is taken as the wave function for each electron in the ground state. An anti-symmetrization is achieved through spin components. The one-electron terms of the two-electron ground state energy are the same as in the one-electron case. The Coulomb matrix element is evaluated in a similar manner as the perturbation treatment of the helium ground state (Bransden and Joachain, 1983),

$$E_C = \left\langle \psi_0(r_1)\psi_0(r_2) \frac{q^2}{4\pi\varepsilon_0\varepsilon_1|r_1 - r_2|} \psi_0(r_1)\psi_0(r_2) \right\rangle, \tag{8.13}$$

which reduces to

$$E_C = \frac{q^2}{4\pi\varepsilon_0\varepsilon_1 a} I_0^{-4} \int_0^1 \int_0^1 j_0^2(\pi x_1) j_0^2(\pi x_2) \times \frac{1}{|x|} x_1^2 x_2^2 \, dx_1 \, dx_2. \tag{8.14}$$

The dimensionless double integral which is computed numerically is equal to 0.00458545. The polarization matrix element

$$E_P = \left\langle \psi_0(r_1)\psi_0(r_2) \sum_l \frac{q^2(\varepsilon_1 - \varepsilon_2)(l+1)r_1^l r_2^l P_l(\cos\gamma)}{4\pi\varepsilon_0\varepsilon_1[\varepsilon_2 + l(\varepsilon_1 + \varepsilon_2)]a^{2l+1}} \psi_0(r_1)\psi_0(r_2) \right\rangle. \quad (8.15)$$

Because of the orthogonality relations for the spherical harmonics, all terms except $l = 0$ vanish. The polarization energy is

$$E_P = \frac{q^2(\varepsilon_1 - \varepsilon_2)}{4\pi\varepsilon_0\varepsilon_1\varepsilon_2 a}. \quad (8.16)$$

It is interesting to note that both Coulomb and polarization energies contain only the $l = 0$ term, while the self-polarization energy contains contributions from all Legendre polynomials. The two-electron ground state energy can be written as

$$E_2 = 289.3/a^2 + 7.42/a. \quad (8.17)$$

The kinetic energy term becomes equal to the other components of the total energy at a radius of 39 Å. For a larger radius than 39 Å, one should use a self-consistent calculation.

8.1.3 Classical Calculation

The behavior of the system for a very large spherical well approaches its classical limit. The length scale approaches the coherence length, assumed to be 100 Å. At this radius, the kinetic energy estimated by the use of the uncertainty principle is ~ 1 meV, which is negligible compared with the electrostatic terms. For a single electron, the polarization has a minimum value with the electron at the center, and excluding the self-energy of the electron, the self-polarization energy is

$$E_1^C = \frac{1}{2}\left(\frac{1}{\varepsilon_2} - \frac{1}{\varepsilon_1}\right) \int_a^\infty \frac{D^2}{\varepsilon_0} d^3r = \frac{1}{2}\frac{\varepsilon_1 - \varepsilon_2}{\varepsilon_1\varepsilon_2} \int_a^\infty 4\pi q^2 \frac{r^2}{16\pi^2\varepsilon_0 r^4} dr, \quad (8.18)$$

which is equal to

$$E_1^C = \frac{1}{2}\left(\frac{\varepsilon_1}{\varepsilon_2} - 1\right)\frac{q^2}{4\pi\varepsilon_0\varepsilon_1 a}. \quad (8.19)$$

Mathematically for two electrons, the problem is harder classically than with quantum mechanics, because we first need to find the positions of the two electrons

Table 8.1 Classically Calculated One- and Two-Electron Electrostatic Energies (in eV)

a (Å)	10	20	30	40	60	80	100	120
E_1^C	0.12	0.06	0.04	0.03	0.02	0.015	0.012	0.01
E_2^C	0.60	0.30	0.20	0.15	0.10	0.075	0.06	0.05
Δ	0.48	0.24	0.16	0.12	0.08	0.06	0.048	0.04

inside a spherical well in the ground state by minimization of the energy that is made up when the repulsive Coulomb and polarization terms push them to the boundary and the self-polarization term pushes them away from the boundary toward each other. Since the positions are symmetrical, we can just take b as the position from the center determined by the minimization of the total electrostatic energy. In terms of $x = b/a$,

$$E_2^C = \frac{q^2}{4\pi\varepsilon_0\varepsilon_1 a}\left(\frac{1}{2x} + (\varepsilon_1 - \varepsilon_2)\sum_l \frac{x^{2l}(l+1)[1 + (-1)^l]}{\varepsilon_2 + l(\varepsilon_1 + \varepsilon_2)}\right). \tag{8.20}$$

The minimum, $E_2^C(\min) = 5.0284$, is found numerically at $x = 0.594$. Table 8.1 gives the calculated classical electrostatic energies for radius a in Å. Evidently it is the discrete nature of the electronic charge that necessitates this procedure.

Before we discuss the significance of our classical calculation, a minimization of the polarization energy to find the most probable position of the two electrons is necessary owing to the discrete nature of the electronic charge. For an infinitely divisible charge density, the simple classical result in terms of the Poisson's equation for charges inside a dielectric sphere in SI units gives

$$V(r < a) = \frac{q}{4\pi\varepsilon_0}\left[1 + 0.5(\varepsilon_0/\varepsilon_1)[1 - (r/a)^2]\right], \tag{8.21a}$$

and

$$V(r > a) = \frac{q}{4\pi\varepsilon_0 r}, \tag{8.21b}$$

applicable for uniformly distributed charges inside a dielectric sphere of ε_1 immersed in ε_0. It should be clear that the complication comes from the discreteness of the electronic charges and is unrelated to quantum mechanical considerations. In other words, capacitance should always be defined in terms of the extra energy stored when an extra electron is added.

Table 8.2 gives the calculated one- and two-electron ground state energies from quantum mechanics. The superscripts k, s, c, p on the energy E refer to kinetic, self-polarization, Coulomb, and polarization interaction terms, respectively. Those values beyond $a = 40$ Å, listed in *italic*, for providing a general trend, indicate that values are very approximate because a self-consistent calculation would be needed.

Table 8.2 One- and Two-Electron Ground State Energies (in eV) from Quantum Mechanics

a (Å)	10	20	30	40	60	80	100	120
E_1^k	1.446	0.362	0.161	0.091	0.040	0.023	0.015	0.010
E_1^s	0.144	0.072	0.048	0.036	0.024	0.016	0.014	0.012
E_1^q	1.59	0.434	0.209	0.121	0.064	0.039	0.028	0.022
E_2^k	2.893	0.723	0.321	0.182	0.080	0.045	0.029	0.20
E_2^s	0.288	0.144	0.096	0.072	0.048	0.036	0.029	0.020
E_2^c	0.214	0.107	0.071	0.054	0.036	0.027	0.021	0.018
E_2^p	0.24	0.12	0.08	0.06	0.04	0.030	0.024	0.020
E_2^q	3.065	1.094	0.568	0.368	0.204	0.138	0.103	0.082
Δ	2.05	0.66	0.36	0.25	0.14	0.100	0.75	0.060

Table 8.3 Capacitances for $n = 0$ and 1 with Q for Quantum and C for Classical Cases

a (Å)	10	20	30	40	60	80	100	500
C_1^C	6.67(−4)	1.33(−3)	2.00(−3)	2.67(−3)	4.00(−3)	5.33(−3)	6.67(−3)	3.33(−2)
C_2^C	1.67(−4)	3.33(−4)	4.97(−4)	6.67(−4)	1.00(−3)	1.33(−3)	1.67(−3)	8.35(−3)
C_1^Q	5.03(−5)	1.84(−4)	3.83(−4)	6.35(−3)	1.25(−3)	1.97(−3)	2.77(−3)	2.30(−2)
C_2^Q	3.90(−5)	1.21(−4)	2.22(−4)	3.20(−4)	5.73(−4)	8.20(−4)	1.10(−3)	6.35(−3)
C_1^Q/C_1^C	0.075	0.138	0.192	0.238	0.313	0.370	0.416	0.690
C_2^Q/C_2^C	0.023	0.036	0.045	0.048	0.058	0.062	0.065	0.076

We define the quantum capacitance and the classical capacitance by

$$E_{n+1}^Q - E_n^Q = \frac{1}{2}\frac{q^2}{C_{n+1}^Q} \quad \text{and} \quad E_{n+1}^C - E_n^C = \frac{1}{2}\frac{q^2}{C_{n+1}^C}, \tag{8.22}$$

where the superscripts Q and C are for quantum and classical cases, respectively. Our results are limited to $n = 0$ and $n = 1$ because of the use of perturbation calculations. However, for a finite barrier height, only a few electrons can be confined in a quantum dot. Table 8.3 gives the values of C_{n+1}^C and C_{n+1}^Q for $n = 0$, 1 with a in (Å) and capacitance C in (fF) with the notation $(-m) = 10^{-m}$.

8.1.4 Summary of Our Calculation

For the convenience of the reader, all the calculated results are summarized as follows.

Quantum mechanical regime (1 nm $< a <$ 4 nm, extrapolated to $a >$ 4 nm):

1. One-electron ground state energy: $E_0^1 = 144.6/a^2 + 1.44/a$
2. One-electron lowest excited energy: $E_1^1 = 295.9/a^2 + 1.55/a$
3. $\Delta E1 = 1.51.3/a2 + 0.11/a$

4. Two-electron ground state energy: $E_0^2 = 289.3/a^2 + 7.42/a$
5. Two-electron lowest excited singlet state energy: $E_0^2 = 440.5/a^2 + 7.89/a$
6. $\Delta E^2 = 151.2/a^2 + 0.47/a$

Classical regime ($a > 10$ nm):

1. One electron located at the center of a sphere of radius a: $E_c^1 = 1.2/a$
2. Two electrons located at $(r = b, \phi = 0)$ and $(r = b, \phi = \pi)$
3. Total electrostatic energy minimized with respect to b gives $b = 0.594$ and $E_c^2 = 6.0/a$
4. $\Delta E^2 = 4.8/a$

Figure 8.1 shows the calculated capacitances. Even though quantum capacitance approaches the value for the classical case, it is still somewhat below at 100 nm. At 2 nm, the quantum capacitance C_Q^2 is only 3.6% and C_Q^1 is only 14% of their corresponding classical values. On the other hand, what is most unexpected is that for the classical C_C^2, it is consistent for all a to be 25% of C_C^1. Obviously when the number of electrons approaches infinity, the effect due to the discrete nature of the electronic charge should disappear.

The calculated energy for a sphere with one electron is somewhat higher than the value given in Chapter 7 because of the self-polarization term included in this calculation, which is very important as far as the capacitance is concerned. In resonant tunneling, the voltage required to align the Fermi level of the contact with the quantum energy depends on the energy calculated self-consistently, which is equivalent to the inclusion of this capacitance. For one electron, the self-polarization term comes from the coupled Poisson's equation in a self-consistent

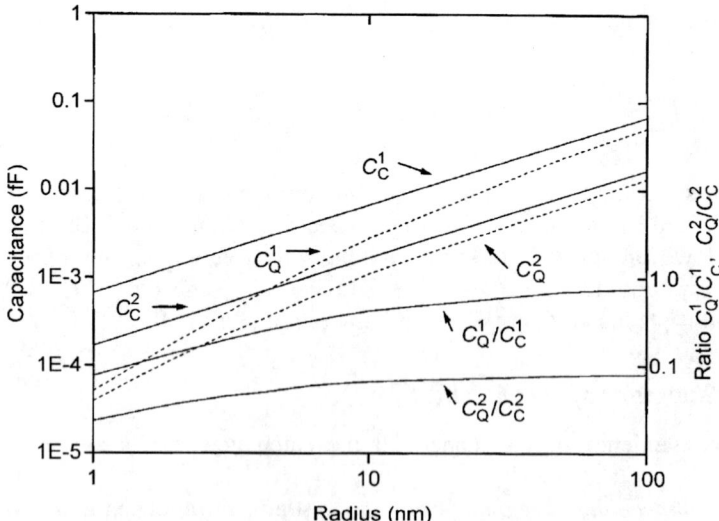

Figure 8.1 Calculated capacitances, where subscript C stands for classical, subscript Q stands for quantum mechanical, and superscripts 1 and 2 are for adding the first and the second electron, respectively.

calculation. However, with two electrons, we have shown that in addition to this self-polarization term, we need to include the polarization and Coulomb terms. For example, at $a = 10$ Å, in addition to the kinetic energy at 2.893 eV, the self-polarization energy is 0.288 eV and the two extra terms, the interaction term from the Coulomb and polarization, provide another 0.454 eV, raising the kinetic energy term by 25%. At $a = 20$ Å, the kinetic energy term is raised by 50% and at $a = 40$ Å, it is raised by 100%. Therefore, as the particle size increases, eventually the kinetic energy is inconsequential such that in a Coulomb blockade (van Houten and Beenaker, 1989), only the capacitance represented by the classical electrostatic stored energy plays a role. As we see, the reverse is true for small-size nanoparticles, where the dominance of the kinetic energy eventually takes over. However, in real structures, this simple picture also breaks down because the kinetic energy is "pinned" by the finite barrier height so that the dominant factor in most cases is still due to the charging effect, and therefore significantly affected by the space charge stored in the states of a quantum dot. We are currently extending our calculation to more than two electrons, up to four in all, when it is a formidable task even to calculate the case of three electrons classically. Our aim is to verify the results of Macucci et al. (1993, 1995), which give higher values for odd numbers and lower values for even additions of electrons. It is interesting to note, as we have shown, that this phenomenon is embodied even in classical electrostatics. In the next section, we shall deal with the reduction of the dielectric constant of a quantum dot, which we have not included in this section for the obvious reason that we did this work before we realized that the dielectric constant depends on the size of the quantum dot.

8.1.5 Comparison with Other Approaches

In attempting to describe a quantum dot as a system of many-electron artificial atoms, Bednarek et al. (1999) used Hartree–Fock for a spherical quantum dot embedded in an insulating matrix with a spherical confinement potential of radius a and a finite barrier height V_0. First of all, with a finite barrier height, whenever the radius a is too small for a given V_0, the bound state does not appear, while the use of an infinite barrier allows bound states for an arbitrarily small radius of the quantum dot. Therefore, their solution in principle should represent an improvement for general applications. Taking the chemical potential μ_n in units of Ry, the Rydberg for silicon, defined by $Ry \equiv \hbar^2/2m^*(a_B^*)^2 = 24.5$ meV with a Bohr radius $a_B^* \equiv a_0 \varepsilon_1 / m^* = 2.4$ Å, from their figure 5 plotted against the radius in units of a_B^*, I tried to convert their calculated results to compare with our results. (For silicon, $\varepsilon_1 = 12$ and $m_l = 0.92 m_e$, and $m_t = 0.19 m_e$ for an approximate isotropic effective mass $m^*/m_e = 0.26$ and a Bohr radius $a_0 = 0.529$ Å.) I found that for $\Delta E_{n,n+1} \equiv \mu_{n+1} - \mu_n$, their values lie between two groups: $\Delta E_{n,n+1}$ for $n = 2, 8, \ldots$ which is $\sim 40\%$ lower than our calculated ΔE^2 and for $n = 1, 3, 7, 8, \ldots$ which is about a factor of three lower. Obviously it is due to the variation caused by following their "periodic table." At this point I would like to express my view on their calculation, as well as the more fundamental aspect of invoking an artificial atom model. First of all an atom is neutral because as electrons are added to the atoms going from hydrogen to Pb, for example, the atoms remain neutral because the extra electrons are precisely balanced by the extra protons inside the nucleus. Therefore, we really should not take too seriously any parallelism between atoms and quantum dots. We can call a quantum dot an artificial atom if we like, but the physics is very different, particularly as we have shown that the difference in dielectric constants between the dot and the matrix produce very large additional energies. Not only the direct Coulomb interaction and the Heisenberg exchange term should be included between the added electrons and the electron already present, these interactions should be taken into account between the induced "image charge" of one with that of the other, including the polarization charges of the two. The reason why their results do not agree with ours in the region where $V_0 = 50Ry$, which should allow a fair comparison, comes from the fact that they assumed no dielectric discontinuity between the quantum dot and the matrix and we found that it is essential to take that into account. I also take issue with their inclusion of many electrons similar to Macucci et al. (1993). A realistic self-consistent calculation including the contacts would have distorted the potential profile rendering a meaningless model of an artificial atom.

I want to discuss some preliminary classical calculations with discrete electronic charge.[1] We have gone from $n = 2$ as far as $n = 20$ electrons, following *Platonic Solid Geometry*, for $n = 4$, 6, 8, 12, 20, corresponding to going from a tetrahedron to a dodecahedron. Electrons are placed at the vertices located at equal distances from the origin. We found that the minimized self-polarization energy is mostly below the classical value until $n = 12$ is passed. In the preceding paragraph, it seemed that there is a difference between even and odd numbers of electrons in the quantum model. Now we know that this is classical but with discrete values for the electronic charge. The classical calculation allows us to include trapped charges at the defect sites inside the oxide gate capacitor, which is really quite relevant. We hope that our results will enrich the understanding of possible instability in the gate oxide in the form of random telegraph noise.

8.2 Dielectric Constant of a Silicon Quantum Dot

As the physical size approaches several nanometers, reduction in the static dielectric constant ε becomes significant. A simple one-oscillator model, an extension of the Penn model, taking into account the quantum-confined silicon sphere of radius a and wire, was first introduced at the 1992 Materials Research Society Boston Conference (Tsu et al., 1993). This work presented a brief derivation of the unpublished work on $\varepsilon(a)$ by Tsu and Ioriatti at the time, leading to the drastic increase in the donor and exciton binding energies and also a model for the self-limiting effect

---

[1] The platonic solid consists of five shapes and was named after the ancient Greek philosopher Plato, who speculated that these five solids were the shapes of the fundamental components of the physical universe.

on electrochemically etched porous silicon caused by the reduction in the dielectric constant. The calculated size-dependent ε is very close to $\varepsilon(\mathbf{q})$ calculated by Walter and Cohen (1970) taking $\mathbf{q} = \pi/a$. The important point is that the expression for the size-dependent dielectric constant involves no adjustable parameters. Therefore, the expression can apply to any solids. It is noteworthy that $\varepsilon(a)$ is more suitable for calculations of donor and exciton binding energies in a finite quantum-confined nanoparticle when the full electrostatic boundary value problem must be tackled. Optical reflectivity measurements show that the refractive index is significantly reduced in porous silicon, PSi, beyond what can be accounted for from porosity, $\sim 70\%$ (Harvey et al., 1992). Using the Bruggeman effective medium approximation it was found, at 5145 Å, that the $n(\text{Si})$ of an Ar laser is 4.2, so that an $n(\text{PSi})$ of 1.8 cannot be all due to voids (Aspnes, 1981). An additional 20% reduction may be due to a quantum size effect as was understood from the reduction calculated for a superlattice (Tsu and Ioriatti, 1985). Figure 8.2 gives the reflectance at two polarizations at 6328 Å. The best fit gives $n = 1.48$. It was concluded then that the additional significant reduction comes from the size-dependent ε.

Between the time Ioriatti and I first submitted our manuscript in 1993 and when it finally appeared in print in 1997 (Tsu et al., 1997), Babic and I had been busy applying $\varepsilon(a)$ to the calculation of the donor and exciton binding energies, as well as developing a model explaining the self-limiting process in electrochemically etched porous silicon. Meanwhile, several calculations appeared between 1994 and 1996 (Wang and Zunger, 1994, 1996a,b; Lannoo et al., 1995), with results almost identical to our simple calculation. A good review of the size-dependent $\varepsilon(a)$ appeared (Yoffe, 2001); however, his statement—that the model developed by Tsu et al. makes many more assumptions relative to other more elaborate calculations—needs to be clarified. We shall see that in our model, within the assumptions that were made such as uniformity of the medium, no local field correction and no wave function extension beyond the quantum dot, there are actually no parameters other than the number of valence electrons, the bulk dielectric constant of silicon, the effective

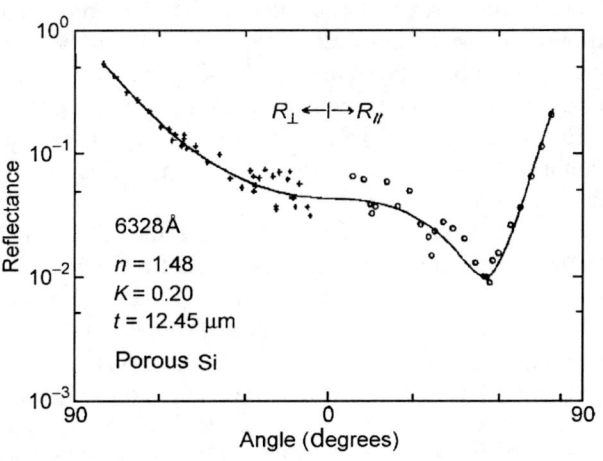

Figure 8.2 Reflectance from porous Si at two polarizations. The best fit gives $n \sim 1.5$, using the measured porosity of 80%. *Source*: After Tsu et al. (1993), with permission.

masses, and the energy denominator, which is taken to be 4 eV. I learned about this 4 eV value some years ago from Morrel Cohen who told me that the important energy is the atomic silicon transition between the ground state and the lowest excited states, a point also used in conjunction with the HOMO–LUMO calculation in Chapter 7.

Reduction of the static dielectric constant becomes significant as the size of the quantum-confined physical systems, such as quantum dots and wires, approaches the nanometer region. A reduced static dielectric constant increases Coulomb interaction energy between electrons, holes, and ionized shallow impurities in quantum-confined structures. A size-dependent static dielectric constant is especially suitable for situations that involve dielectric discontinuity and require a full electrostatic boundary value problem to be solved as in doping quantum dot (Tsu and Babic, 1994a–c), finding the exciton binding energy in Si QD (Babic and Tsu, 1997), and making a model for the self-limiting mechanism when etching PSi (Babic et al., 1992; Tsu and Babic, 1993). For those who might question the simple model in favor of a full pseudopotential computation (Wang and Zunger, 1994), or a semiempirical LCAO computation (Lannoo et al., 1995), we emphasize that our simple model leading to the derivation (Tsu et al., 1997) contains no adjustable parameters and gives better physical insight. In what follows, more detail about the model and the thought that went into it are given.

Strictly speaking, the dielectric function ε is only definable in an unbounded region of space. In Maxwell's equation, it is simply the constitutive parameter of the medium. The wave vector–dependent dielectric function $\varepsilon(q)$ has been derived for cubic semiconductors such as Si, Ge, and GaAs (Penn, 1962; Walter and Cohen, 1970; Baroni and Resta, 1986). This $\varepsilon(q)$ has had many applications, in particular, in the calculation of screened shallow impurity potentials (Morita and Nara, 1966). Theoretical treatment of the dielectric constant in quantum well systems (Kahen et al., 1985; Tsu and Ioriatti, 1985) shows that a significant reduction of ε occurs when the width of the quantum well is reduced to several nanometers or less. However, application of the rigorous $\varepsilon(q)$ in calculations of the donor or the exciton binding energy in quantum dots/wires that have electrostatic boundary conditions to contend with represents a formidable task. These calculations become much more manageable if, instead of $\varepsilon(q)$, a constant but size-dependent effective dielectric constant $\varepsilon(a)$ is used. While the concept of a constant size-dependent effective dielectric constant for a finite body is not rigorous, it represents an approximation that is very suitable for calculations that involve the electrostatic boundary value problem at dielectric discontinuities.

The single-oscillator model is based on a modification of the model by Penn (1962), taking into account the discrete eigenstates of quantum-confined nanoparticles while keeping the oscillator strength fixed and equal to its bulk value. This last assumption has been referred to by Yoffe (2001) as a possible weakness of the theory. The initial version of this work (Tsu et al., 1993) contains ΔE that is a factor of 2 too large, but which subsequently has been corrected when applied to the exciton recombination and binding energies in silicon nanocrystallites and to the donor binding energy (Tsu and Babic, 1993). Wang and Zunger (1994) extended our initial formulation of the size-dependent static dielectric constant $\varepsilon(a)$ of silicon quantum dots using an empirical pseudopotential calculation. Lannoo et al. (1995) applied a semiempirical LCAO technique to calculate the static dielectric constant for silicon quantum dots related to porous silicon. Unfortunately they also referred to the initial version of our work with the factor of 2 error (Tsu et al., 1993) while in fact, the corrected version had already been applied to the calculation of the donor binding energy a year earlier by Tsu and Babic (1994).

8.2.1 Size-Dependent $\varepsilon(a)$

The response of a medium to an applied potential ϕ_0, with ϕ_i representing the induced potential, such that the total potential, usually referred to as the self-consistent potential, ϕ, can be formulated with the use of the quantum mechanical analog of the classical Liouville equation,

$$i\hbar \frac{\partial \rho}{\partial t} = [H, \rho], \qquad (8.23)$$

in which $H = H_0 + H_1(\mathbf{r}, t)$ and $\rho = \rho_0 + \rho_1(\mathbf{r}, t)$, where H_0 is the one-electron Hamiltonian which characterizes the quantum dot, including the nuclei and the boundary conditions, $H_1 = e\phi \exp(i\mathbf{q} \cdot \mathbf{r} - \omega t)$, and ρ_1, the induced number density fluctuation due to an applied potential ϕ_0. With the time dependence represented by $e^{i\omega t}$, the induced electrostatic potential $\phi_i(\mathbf{r})$ is

$$\phi_i(\mathbf{r}) = e \int \frac{\rho_1(\mathbf{r}')d^3\mathbf{r}'}{|\mathbf{r} - \mathbf{r}'|}, \quad \text{with } \phi = \phi_0 + \phi_i. \qquad (8.24)$$

Letting the eigen energy and the eigenstates of H_0 be E_α and $|\alpha>$, then

$$H_0|\alpha\rangle = E_\alpha|\alpha\rangle, \qquad (8.25)$$

and the number density fluctuation

$$\rho_1(\mathbf{r}) = e \sum_{\alpha,\beta}{}' \frac{f_\alpha - f_\beta}{E_\alpha - E_\beta} \psi_\alpha(\mathbf{r})\psi_\beta^*(\mathbf{r}) \int d^3\mathbf{r}' \psi_\alpha(\mathbf{r}')\phi(\mathbf{r}')\psi_\beta(\mathbf{r}'), \qquad (8.26)$$

and the self-consistent potential is represented by the integral equation

$$\phi(\mathbf{r}) = \phi_0(\mathbf{r}) + e^2 \int \frac{d^3\mathbf{r}'}{|\mathbf{r} - \mathbf{r}'|} \int d^3\mathbf{r}'' \chi(\mathbf{r}', \mathbf{r}'')\phi(\mathbf{r}''), \qquad (8.27)$$

in which the susceptibility is given by

$$\chi(\mathbf{r}', \mathbf{r}'') = \sum_{\alpha,\beta}{}'' \frac{f_\alpha - f_\beta}{E_\alpha - E_\beta} \psi_\alpha(\mathbf{r})\psi_\beta^*(\mathbf{r})\psi_\alpha(\mathbf{r}')\psi_\beta(\mathbf{r}'). \qquad (8.28)$$

Not only is it necessary to contend with the integral equation (8.27), the difficulty comes from the nonlocal susceptibility in Eq. (8.28). The problem is drastically simplified and universally defined for unbounded, spatially uniform systems by

$$\chi(\mathbf{r}', \mathbf{r}'') = \chi(\mathbf{r}' - \mathbf{r}''). \tag{8.29}$$

Using the convolution theorem of Fourier integrals where $\phi_0(\mathbf{r})$ varies slowly compared to the lattice constant, the $\mathbf{q}$-dependent dielectric function becomes

$$\varepsilon(\mathbf{q}) = 1 - \frac{4\pi}{\mathbf{q}^2} \chi(\mathbf{q}). \tag{8.30}$$

The reason we go through the usual procedures elaborated by Ehrenreich and Cohen (1959), and discussed in some detail by Harrison (1970), in arriving at Eq. (8.30) will be made clear in order to offer an appreciation of the conditions under which a size-dependent dielectric constant may be defined. It is not because we are dealing with an integral equation (8.27) for the self-consistent potential, because step-by-step iterations can always be used. The most fundamental issue is the assumption in Eq. (8.29), which is not true under the boundary conditions, even if only the electrostatic boundary conditions are used. For a finite structure, $\chi(\mathbf{r}', \mathbf{r}'') \neq \chi(\mathbf{r}' - \mathbf{r}'')$, there is no simple basis set for which the integral equation can be solved for a general $\phi_0(\mathbf{r})$. Thus in principle, a universal scalar dielectric function cannot be defined, although a response function does exist once a special set of input/output arrangements have been specified. The approach of Wang and Zunger (1994) involves the use of the pseudopotential calculation for the absorption and obtaining a dielectric function with the Kramers–Kronig relation. Therefore, the susceptibility in Eq. (8.29) is implicitly assumed. As far as neglecting the local field correction is concerned, it should be much less than the maximum reduction of 10% estimated by Harrison (1970), because in our simple approach, the dielectric constant of the bulk silicon is used which, for the most part, has already accounted for most of the corrections.

Extending the Lindhard formula for the Hartree dielectric function of a free-electron gas to semiconductors (Ziman, 1988), we obtain

$$\varepsilon(\mathbf{q}, \omega) = 1 + \frac{4\pi e^2}{q^2} \sum_{\mathbf{k}, \mathbf{g}} \frac{\left| \langle \mathbf{k} | e^{i\mathbf{q} \cdot \mathbf{r}} | \mathbf{k} + \mathbf{q} + \mathbf{g} \rangle \right|^2 [f_0(\mathbf{k}) - f_0(\mathbf{k} + \mathbf{q} + \mathbf{g})]}{E(\mathbf{k} + \mathbf{q} + \mathbf{g}) - E(\mathbf{k}) - \hbar\omega + i\hbar\Gamma}. \tag{8.31}$$

Using the sum rule for oscillator strength (Thomas–Reiche–Kuhn) (Merzbacher, 1961),

$$\sum_{\beta} (E_\alpha - E_\beta) \left| \langle \alpha | e^{i\mathbf{q} \cdot \mathbf{r}} | \beta \rangle \right|^2 = \frac{\hbar^2 q^2}{2m}, \tag{8.32}$$

and with series expansion keeping only the terms linear in q, Eq. (8.29) becomes

$$\varepsilon(\mathbf{q}, 0) = 1 + \frac{(\hbar\omega_\mathrm{p})^2}{E_\mathrm{g}^2}, \tag{8.33}$$

in which $\omega_p^2 = 4\pi n e^2 / m$ and $E_g \cong E(\mathbf{k} + \mathbf{q} + \mathbf{g}) - E(\mathbf{k})$. It should be noted that the mass is the free-electron mass.

For various situations of interest such as shallow impurities, excitons, and optical absorptions, the important Fourier components involve those in the vicinity of $\mathbf{q} = 0$. Therefore, one is left with the calculation of $\chi(\mathbf{q} = 0)$.

The dielectric constant is a measure of virtual optical transitions. Quantum confinement increases separation of the discrete states resulting in an increase in the energy denominator and a subsequent reduction in ε. First, let us discuss the main feature of the Penn model (Penn, 1962), with the aid of Figure 8.3 showing the electron energy as a function of $\mathbf{k}$ for isotropic three-dimensional nearly free-electron systems. The inset shows ε_2 versus photon energy for Si, giving a justification for setting $E_g = 4$ eV (the average energy separation between the ground state and the first excited state of atomic silicon, $3S^2 3P^2 - 3S^2 3P^1 4S^1 \sim 4.1$ eV; see Figure 6.12), for the single-oscillator model of the static dielectric constant for bulk silicon, where the bulk value ε_B is given by

$$\varepsilon_B = 1 + \left(\frac{\hbar\omega_p}{E_g}\right)^2 . \tag{8.34}$$

The isotropic model fills up an almost free isotropic energy band with all the valence electrons up to an energy E_F, and then a gap of $E_g = 4$ eV is centered at E_F. Round dots in Figure 8.3 indicate the discrete energies and momenta for a sphere of radius a, given by

$$E_{n\ell} = \frac{\hbar^2 k_{n\ell}^2}{2m} , \tag{8.35}$$

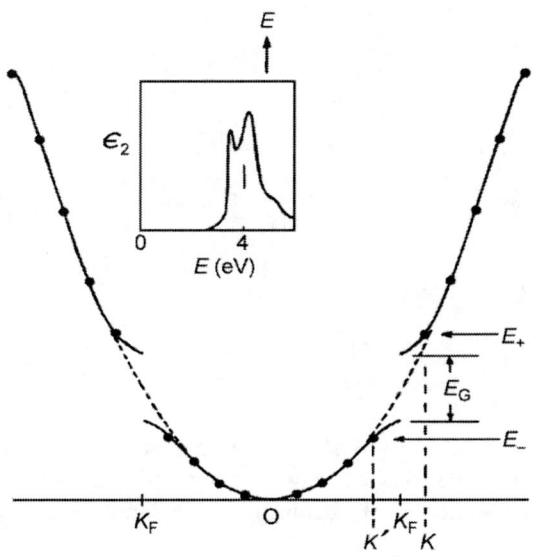

Figure 8.3 Electron energy versus k for an isotropic three-dimensional nearly free-electron model. The inset shows ε_2 versus photon energy, giving $E_g \sim 4$ eV for Si. Round dots indicate the discrete energies and momenta. $E_+ - E_-$ is the new gap for the Si sphere of radius a.
Source: After Tsu et al. (1993), with permission.

where $k_{n\ell} = \alpha_{n\ell}/a$, in which $\alpha_{n\ell}$ are the nth roots of the spherical Bessel function $j_\ell(ka) = 0$. As the radius is decreased, the roots of the spherical Bessel functions are separated further and further apart. Eventually, the energy separation can exceed the gap E_g. Obviously we assume that the wave function is zero at the surface of the sphere, representing an approximation, quite a good one, for the boundary condition. It is important to recognize that the fundamental $\Gamma - \Delta$ gap at 1.2 eV plays no role in the dielectric function. Let us examine in detail why with k_F in the middle of k' and k gives rise to the minimum separation of $\pi/2$. Since $k_F a \gg 1$, the asymptotic expression of $j_\ell(ka) = 0$ at $r = a$,

$$j_\ell(ka) \sim \frac{\sin(kr - \ell\pi/2)}{kr} = 0$$

results in $ka = (n + \ell/2)\pi$, so that the least separation is at $n = n'$ and $\ell' = \ell \pm 1$, giving $k_{n\ell} - k_{n'\ell'} = \pi/2a$ and $k_{n\ell} + k_{n'\ell'} = (2n + \ell - 1/2)\pi/a = 2k_F$, and the corresponding least separation of the energy

$$\Delta E = \frac{\hbar}{2m}(k^2 - k'^2) = \frac{\hbar^2}{2m}(k + k')(k - k') = \frac{\pi E_F}{k_F a}, \tag{8.36}$$

in which $E_F = \hbar^2 k_F^2/2m$. The energy separation of Eq. (8.36) is half the value of the energy separation of the first version of the model where $j_\ell(ka)$ for $\ell > 0$ was erroneously excluded in Tsu et al. (1993) and Tsu (1993).

Now the energies of the states at E_j and E_{j+1}, in the presence of the periodic potential which results in a gap E_g in the bulk, may be found from the coupling of E_j and E_{j+1} at k_j and k_{j+1} (j is a shorthand notation involving both the n and ℓ indices). The new energy of an adjacent pair is given by

$$\det\begin{pmatrix} E - E_1 & E_g/2 \\ E_g/2 & E - E_2 \end{pmatrix} = 0,$$

so that

$$E_\pm = E_F \pm \sqrt{(\Delta E)^2 + E_g^2}. \tag{8.37}$$

The size-dependent dielectric constant from $\varepsilon(a)$ is now

$$\varepsilon(a) = 1 + \frac{(\hbar\omega_p/E_g)^2}{((E_+ - E_-)/E_g)^2} = 1 + \frac{\varepsilon_B - 1}{1 + (\Delta E/E_g)^2}. \tag{8.38}$$

Taking the parameters for Si, $\varepsilon_B = 12$, $E_g = 4$ eV, and filling the energy bands up to E_F with $4 \times 5 \times 10^{22}$ valence electrons per cm^3, giving $E_F = 12.6$ eV and $k_F = 1.81$ Å^{-1} with $m = m_e$, the computed $\varepsilon(a)$ according to the modified Penn model is shown in Figure 8.4.

For comparison with other results, $\varepsilon_B = 11.3$ is also included, together with the plots of the size-dependent screening dielectric constant from Walter and Cohen (1970) and the size-dependent dielectric constant from Wang and Zunger (1994) as well as Lannoo et al. (1995). The crosses shown in Figure 8.4 represent the bulk wave vector−dependent dielectric constant $\varepsilon(q)$ from Walter−Cohen that was converted into $\varepsilon(a)$ by putting $q = 2\pi/d$, where $d = 2a$. This comparison was suggested to me by Marvin Cohen after I showed him our first version. The basis in equating $q = 2\pi/d = \pi/a$ is simply the requirement dictated by the Fourier transform that the confinement in the configuration space of d corresponds to $q = 2\pi/d$ in momentum space, the Heisenberg uncertainty principle. There appears to be good agreement between the simple modified Penn model and the other much more sophisticated calculations. The reduction of $\varepsilon(a)$ is not really significant before the radius of the sphere approaches approximately 15 Å but it becomes really significant for spheres with radii comparable to the lattice constant of Si, $a_0 = 5.43$ Å. The difference between our $\varepsilon(a)$ and other three calculations results from our use of $\varepsilon_B = 12$ for the Si bulk dielectric constant instead of $\varepsilon_B = 11.3$ or 11.4, as in the other works. Using $\varepsilon_B = 11.3$, our results are the same as the results of Walter−Cohen, and very close to the results of Wang and Zunger as well as Allan et al. (1995) and Lannoo et al. (1995). Finally, it should be noted that as the sphere radius a is reduced below 1 nm, approaching the atomic Si, as mentioned before, $E_g \geq 4.1$ eV will further reduce $\varepsilon(a)$. It is remarkable that $\varepsilon(a)$, given by such simple theory, compares so closely to the results of the far more sophisticated calculations.

At the early stage of structural investigation of porous silicon, quantum wire was considered to be a model for porous silicon. Therefore, we would like to include the modified Penn model for quantum wire of radius a. Everything for the sphere applies to the wire except that the spherical Bessel junction $j_\ell(ka) = 0$ is

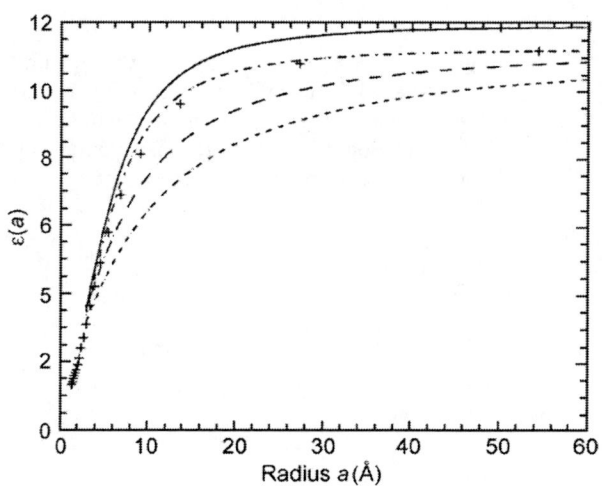

Figure 8.4 Size-dependent static dielectric constant $\varepsilon(a)$ versus the radius a in angstroms for silicon: solid line—modified Penn model with $\varepsilon_B = 12$; dash-dot line—same as before but with $\varepsilon_B = 11.3$; crosses—converted from Walter−Cohen with $q = \pi/a$; long-dash line—from Lannoo et al.; and short dash line—from Wang−Zunger. *Source*: After Tsu et al. (1997), with permission.

replaced by the cylindrical Bessel junction $J_\ell(ka) = 0$, then the density of states of a quantum wire is

$$n(E) = \frac{1}{4\pi^2} \left(\frac{2m}{\hbar^2}\right)^{3/2} \sum_{n\ell} (E - E_{n\ell})^{-1/2}. \tag{8.39}$$

As it turns out, owing to the isotropic electron mass, the computed constant as a function of the wire size has the same appearance, although the positions of the energies $E_{n\ell}$ are different for the wire and sphere cases.

Since dielectric function, similar to elastic constant, is considered to be a constitutive parameter, which applies to an unbounded region of space, the concept of a size-dependent dielectric function requires further discussion.

Figure 8.5 shows the plasmon dispersion for $q < q_s$. Classically and at high temperatures, the Fermi velocity should be replaced by thermal velocity, where screening is referred to as the Debye screening. The greater the electron density, the larger is the plasma frequency and the larger is q_s. Since q is inverse to length, as the region shrinks, only electrons with large q participate as independent particles. However, all electrons in a quantum dot are phase coherent, with q determined by the size, not by the density. For small quantum dots, q is very large. By virtue of the phase coherency, the boundary condition dictates the interactions. Even a single electron can interact with its induced charge distribution at the boundary. Here is the very fundamental issue—Do we separate the medium from the geometrical boundary?

Traditionally, we formulate the dynamics of a situation by global material parameters such as density, band structure, effective mass, dielectric function, and even elastic constant and melting point, and geometrical factors are to be accounted for by boundary conditions and integration on a given surface or volume. The complication arises not only owing to the mixing of material parameters and boundary

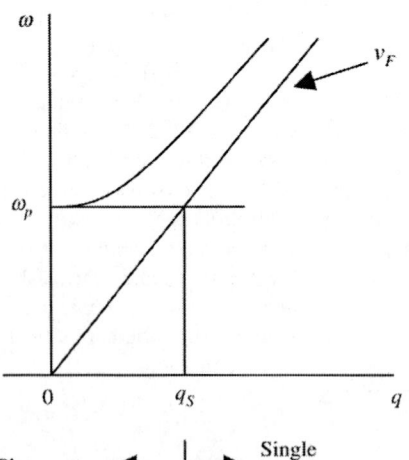

Figure 8.5 Plasmon dispersion for $q < q_s$, with $q_s = \omega_p/\nu_F$, where many-body effects give rise to plasmon, while for $q < q_s$, a single electron description dominates. A large plasma frequency, large q_s, gives a smaller screening length.

conditions, but because both are in fact losing their identities. Therefore, only a microscopic description, in terms of individual particles, atoms, and molecules can provide a unique solution of the dynamical situation. My preoccupation with the dielectric constant, capacitance, and so on, in this chapter, therefore should be viewed as a way to present a description capable of providing the physics and engineering for the design and optimization of a device. Our first attempt to publish our work, presented during the spring of 1993, ended in disaster from my personal point of view. Without a forum like the Annual Material Society Conference, I would not have been able to present the partial results in December 1992 in the first place. The following work of Wang and Zunger and of Lannoo et al. would not have appeared and the review by Yoffe (2001) would certainly have been different from what he wrote. The crucial point is that most of us, including myself, without solving the complicated problems in their entirety, were simply not equipped or willing to go far enough. Morrel Cohen once told me this. I simply asked how far is far enough? For example, it is certainly true that a global description is not adequate for many problems related to quantum dots; nevertheless, there is a definite solution for the response once the input/output is specified. In short, the usual expressions for the dielectric functions are global, but the dielectric function of a quantum dot is really a response function; local, not global. In this case, it seems that we are forced to take each problem on its own. Actually, this is not a recent phenomenon; fundamental models are always created for a simplified situation with the solution, it is to be hoped, capable of describing the problem at hand by a wider range of possible applications. In closing this section, I want to offer my view. We should continue to develop models even though they may only be valid in a limited realm of applicability, because, by doing so, others can join in the effort in searching for a more universally acceptable solution.

8.3 Doping a Silicon Quantum Dot

Doping is important in semiconductor devices. In n-type semiconductors, if the binding energy of the shallow donor is sufficiently low, then the electron is appreciably ionized into the conduction band at the operating temperature, giving rise to conduction under an applied voltage. In p-type semiconductors, the dopant is a shallow acceptor and the charge carrier is a hole. What happens when the dopant lies deep, that is, when the binding energy if much greater than $k_B T$, the electron is not ionized into the conduction band so that conduction cannot take place. Let us use silicon as an example for the rest of our discussion. The binding energy of phosphorus dopant is 47 meV which is $\sim 2k_B T$ at room temperature. Therefore, at room temperature, the P-site has a charge state of +1 because one electron is lost from P-site to the conduction band. Suppose there are 5×10^{16} cm^{-3} of P-doping. The ratio of [P]/[Si] is 10^{-6}, or one in every 100 Si-sites. In each Si unit cell there are eight Si-atoms, so that the probability of a P-atom is 1/12 of the Si unit cells. Contrary to the common intuition that one needs a coherent length covering many atoms in order to form overlapping energy states into an energy band, in most cases only few are needed. In the very first publication on the man-made superlattice, Esaki and I used Heisenberg's uncertainty principle to argue that all we need is three coupled quantum wells to mimic a superlattice. It is even less for the three-dimensional case. When Tanaka and I were calculating the density of states of amorphous silicon (Tanaka and Tsu, 1981), we discovered that by using 11 tight-binding parameters it is possible to generate the band structure of silicon. And 11 interactions do not even represent all of the third neighbor distances. I want to establish, for the argument needed for the case of quantum dots, the coherent length needed to form overlapping states allowing the transfer of electrons from one dot to the next. Obviously the minimum is two, because with two coupled dots, an electron can move from one to the next, then on to the third, and so on. For man-made superlattices, we need at least a mean free path $\ell > 2d$, with d being the period of the superlattice. For negative differential conductance (NDC) and Bloch oscillation, one needs an even longer ℓ, discussed in detail in Chapter 1. If coupling involves only two dots, then the two original uncoupled states at a common E are transformed into two, $E_{\pm}$ with $\Delta = E_+ - E_-$. If all the dots are in contact, an electron injected into a given dot results in conduction under an applied voltage per dot of $V_d < \Delta$, otherwise something else is needed, usually phonons. What happens if a given dot is doped? As long as all the dots are in contact, or more precisely coupled under an applied voltage, the electron in the given dot can move to the next, resulting in conduction. When they are not all in contact and not all coupled together, the electron will be confined in the cluster, transforming the cluster into a charged capacitor, raising the potential and blocking further tunneling into the cluster. Suppose now that we can drastically increase the dopant density such that there is a unit probability of any dot being occupied by a dopant. If all the dots are in contact, the conduction will increase n-fold. What happens if they are only connected into unconnected clusters? High dopant density means that many dots in a cluster are doped, allowing more electrons to occupy the states in the cluster while remaining neutral, thus reducing charging of the clusters. Suppose we keep the dopant density fixed, while reducing the dot size. The extent of the wave function of a dopant is basically the Bohr radius in a solid. For Si, $r_B \sim \varepsilon/m^*$ is about 25 Å. Once an electron is ionized into the conduction band, it is localized within a mean free path ℓ that is usually longer than 25 Å. Thus, we say that an electron in a localized state is transformed into a nonlocal state, the band state. In a quantum dot of radius 10 Å, taking into account the reduction of ε, r_B is about the size of dot. But fundamentally the extent of the wave function is obviously the confining potential barrier. To achieve unit probability of occupation by a dopant in each dot with a radius of 10 Å requires a doping density $>5 \times 10^{19}$ cm^{-3}, approaching the solid solubility of phosphorus in silicon. That is aside from the fact, as we shall see in this section, that the binding energy for a radius of 10 Å is about 1.5 eV, making ionization into the ground state of the confined silicon quantum dot impossible. Let us go on to the derivation of the binding energy of shallow dopants before we finish discussing all the consequences.

Fundamentally, quantum confinement pushes up the allowed energies resulting in an increase in the binding energy of shallow impurities such as in the cases of quantum wells (Bastard, 1981) and the superlattice (Ioriatti and Tsu, 1986). In a quantum dot of radius a, the reduction of the size-dependent static dielectric constant $\varepsilon(a)$ results in a significant increase in the binding energy of shallow impurities (Tsu and Babic, 1993, 1994a−c). Since the formation of PSi by electrochemical etching depends on the current, a significant increase in the binding energy can cut off extrinsic conduction leading to a self-limiting process during the formation of the porous silicon (see the preliminary discussion in Tsu and Babic, 1993). With a more detailed study, however, the reduction in $\varepsilon(a)$ contributes to only a portion of the increase of E_b, with the bulk of the increase due to the induced polarization charges at the boundary of the dielectric discontinuity (Tsu and Babic, 1994a−c). The physical picture is as follows: (1) the reduction of the static dielectric constant plays a role in increasing E_b of a donor or accepter via reduction of dielectric screening; (2) a more significant term is due to the induced charges at the dielectric interface between the quantum dot and the matrix in which the dot is embedded. With ε_1 and ε_2 denoting the dielectric constant of the particle and the matrix, for $\varepsilon_1 > \varepsilon_2$, the induced charge on the donor is of the same sign resulting in an attractive interaction with the electron of the dot, pushing deeper the ground state energy of the donor and resulting in an appreciable increase in E_b. For $\varepsilon_1 < \varepsilon_2$, the opposite is true, E_b is reduced allowing possible extrinsic conduction even at room temperature. Discussion of the totally different behavior of PSi in air and water was pointed out by Lehmann and Vial at the Grenoble Workshop. Tsu and Babic (1993) suggested that the different behavior of PSi in an aqueous solution and in air may be attributed to the difference in the binding energies. In short, matching the dielectric constant of the quantum dot and the matrix can considerably reduce E_b, thus allowing doping, a vital point to recognize in considering the optoelectronic role of quantum dots Tsu et al. (1994).

The validity of the effective mass approximation for a quantum dot depends on the range of the Bloch function for the dot, which must be less than the width of the Brillouin zone. For GaAs wells, the effective mass approximation is valid down to a well width of $\sim$20 Å (Preister et al., 1983). Owing to the nearly identical crystal structures of Si and GaAs, and thus the nearly identical sizes of the Brillouin zones, the effective mass approximation should provide adequate results for a Si quantum dot down to a radius of 10 Å. Interestingly, it was the inclusion of $\varepsilon(q)$ in the calculation of E_b that resulted in fair agreement with the experimental value of 47 meV for P in Si (Pantelides, 1978). Using $\varepsilon(a)$ derived in the last section for ε_1, Tsu and Babic (1994a−c) derived E_b for a Si dot embedded in various matrixes. The Hamiltonian may be written as

$$H = -\frac{\hbar^2}{2m_e}\nabla^2 + V(r) + \phi_c(r) + \phi_p(r) + \phi_s(r), \qquad (8.40)$$

where

$$V(r) = \begin{cases} 0, & r < a, \\ \infty, & r \geq a, \end{cases}$$

and the direct Coulomb potential ϕ_c between the donor and the electron is

$$\phi_c = -\frac{q^2}{4\pi\varepsilon_0\varepsilon_1 r}, \tag{8.41}$$

and the self-polarization between the electron and its induced charges, following Babic et al. (1992), is

$$\phi_s = \frac{1}{2}\sum_l \frac{q^2(\varepsilon_1 - \varepsilon_2)(l+1)r^{2l}}{4\pi\varepsilon_0\varepsilon_1[\varepsilon_2 + l(\varepsilon_1 + \varepsilon_2)]a^{2l+1}}, \tag{8.42}$$

and taking the s state for the spherically symmetric ground state, the polarization term between the electron-induced polarization of the donor (Babic et al., 1992) is

$$\phi_p = -\frac{q^2(\varepsilon_1 - \varepsilon_2)}{4\pi\varepsilon_0\varepsilon_1\varepsilon_2 a}. \tag{8.43}$$

Note that we have excluded the self-polarization term between the donor and its induced polarization, because this interaction contributes to the donor formation energy when the donor is introduced into the quantum dot. The ground state energy of the donor, E_0 is obtained by a minimization of E_0 with respect to the parameter c in the trial function

$$\psi(r) = \left[1 - \left(\frac{r}{a}\right)^2\right]e^{-r/c}. \tag{8.44}$$

Note that this trial wave function satisfies the boundary condition of $\psi\,(r = a) = 0$.

For a given radius a, $\varepsilon_1 = \varepsilon(a)$ taken from Figure 8.6 is used to obtain the ground state numerically. How we define the binding energy and approximate matrix needs some discussion.

Figure 8.7(A) shows how we define the binding energy $E_b = E_1 - E_0$, where E_1 is the lowest allowed state in a neighboring particle without a positively charged donor, but includes the self-polarization and the ground state energy E_0 of the donor. This definition makes sense only when the donor density is such that majority of silicon particles contain no donor. In Figure 8.7(B), (a) shows the actual situation where a donor at the center of a sphere is surrounded by other spheres without donors and (b) shows our simplified model where the sphere with the donor is immersed in a uniform matrix of ε_2. The induced charge at the dielectric interface should be reduced in (a).

Figure 8.8 shows the calculated donor binding energy E_b versus the Si sphere of radius a for few matrices. Instead of taking ε_2 for water as 80, we were advised by L.M. Peter at the Les Houches Winter School 1994 that 6 should be used because within a thin layer of water in contact with silicon, referred to as the primary salvation sheet, the dipoles are bound and resist orientation by an

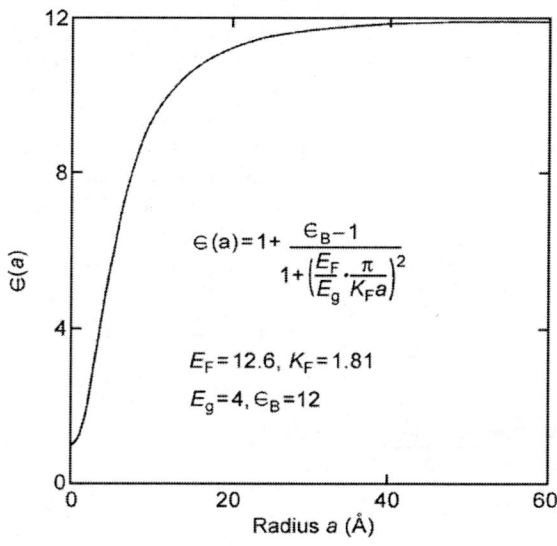

Figure 8.6 Size-dependent ε versus the radius a of a Si sphere, with E_F in eV and k_F in $\mathring{A}^{-1}$.
Source: After Tsu and Babic (1994c), with permission.

$$\epsilon(a) = 1 + \frac{\epsilon_B - 1}{1 + \left(\frac{E_F}{E_g} \cdot \frac{\pi}{K_F a}\right)^2}$$

$E_F = 12.6$, $K_F = 1.81$

$E_g = 4$, $\epsilon_B = 12$

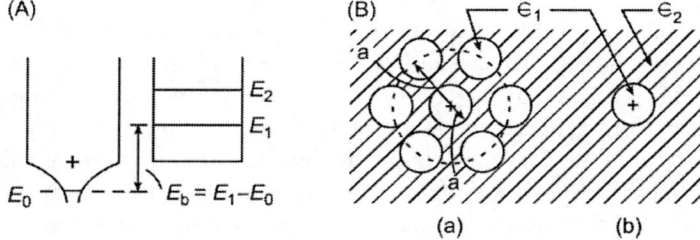

Figure 8.7 (A) Definition of the binding energy E_b in terms of the difference between the lowest allowed state in a neighboring particle without the donor and the ground state energy E_0 of a donor. (B) (a) Actual situation where a donor at the center of a sphere is surrounded by other spheres and (b) our simplified model where the sphere with the donor is immersed in a uniform matrix.

external field, resulting in a much lower value for the dielectric constant of water. A detailed account is given by Bockris and Reddy (1973). Note that without the dielectric difference, $\varepsilon_1 = \varepsilon_2$ at 20 $\mathring{A}$ and $E_b \sim 0.17$ eV, a huge difference from the case where $\varepsilon_2 = 1$ for a PSi particle in air, where the PSi particle is essentially a good insulator in air.

Porous silicon is usually formed by anodic etching of p-type silicon. Although our calculations have been for n-type silicon, the conclusions are applicable to the p-type. The reason for this is the dominance of the electrostatic energy terms, which are the same for either donors or acceptors. Dealing with donors has allowed simplification of the kinetic energy term compared to the case of acceptors where one has to treat light and heavy hole degeneracy and use the much more complicated Luttinger−Kohn Hamiltonian (Luttinger and Kohn, 1955). The dramatic

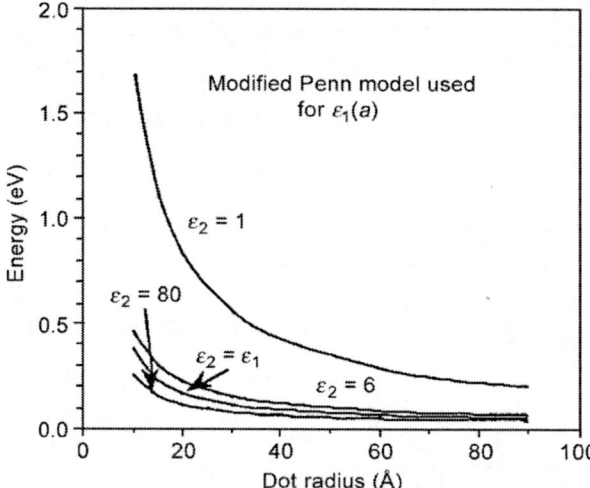

Figure 8.8 The donor binding energy E_b versus the Si sphere for a few matrices. Instead of taking ε_2 for water as 80, we were advised to use 6 instead. See text for the reason. *Source*: After Tsu and Babic (1994c), with permission.

increase of the accepter binding energy due to dielectric mismatch and quantum confinement offers an explanation for the self-limiting etching of PSi. At the beginning of etching, acceptor binding energy is low, with the same value as in the bulk. A positive voltage applied to the p-type silicon produces an accumulation of mobile holes at the silicon electrolyte interface enabling etching. Figure 8.9(A) shows the energy bands for silicon and the redox states in the solution. As etching progresses, the dimensions of unetched silicon are reduced and the binding energy of acceptors increases sharply. The concentration of free holes decreases, making silicon appear intrinsic. Figure 8.9(B) shows the energy bands in nanosilicon with respect to the redox states in the solution. Without the accumulation of holes at the interface, electrochemical etching cannot proceed. Although some holes can tunnel from the acceptors to the solution, this does not constitute etching because no silicon bond at the interface is involved.

8.4 Capacitance: Spatial Symmetry of Discrete Charge Dielectric

Capacitance is a measure of the ability to store electrons and is conventionally considered to be a constant dependent upon the shape of metal contacts and the dimensions of the system. In general, equal-potentials of dielectric systems without metal contacts depend on the spatial distribution of discrete electrons. The capacitance is defined in terms of the total interaction energy of N-electrons confined in a dielectric sphere. The distribution of N-electrons is obtained from minimization of the total Coulomb and polarization interaction energy as well as the formation energy, the work done on the system. Our discrete charge dielectric model, DCD, gives rise to an electrostatic capacitance agreeing with $N = 1$ and $N = \infty$. The fundamental physics involves the change of symmetry created by the introduction of each additional electron into the confining volume space. For example, a single electron confined

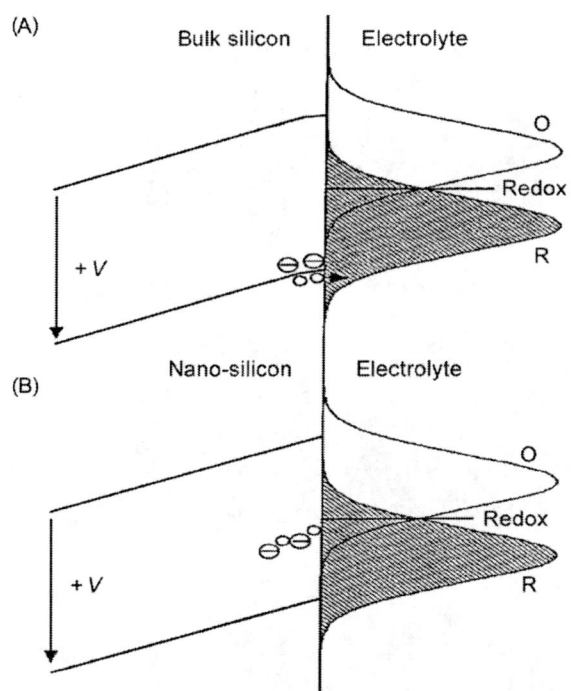

(A)

Bulk silicon   Electrolyte

O

Redox

R

+ V

(B)

Nano-silicon   Electrolyte

O

Redox

R

+ V

Figure 8.9 Energy band diagram of silicon with respect to the redox states in solution under an applied positive anodic voltage for (A) bulk silicon when etching can proceed and (B) silicon nanoparticles with no etching. *Source*: After Tsu and Babic (1994c), with permission.

inside a dielectric sphere is located at the center. As we have seen in Section 8.1.3 for two electrons, Coulomb repulsion pushes them away from each other until the induced charge on the surface pushes them toward the interior. This is because whenever the dielectric constant inside is greater than outside, the induced charge forms repulsive interaction, the fundamental cause of dielectric confinement. Before we go further, it is instructive to show the computed locations of N-electrons upto 12, shown in Figure 8.10, originally computed by Zhu et al. (2006) and discussed fully in LaFave and Tsu (2009).

Let us point out the important principles involved. The starting point of our computation is Green's function approach with Eqs. (8.1)−(8.6), of the Poisson equation. Let us jump ahead to give a brief tour of what lies ahead. The most important physical principle involved is the symmetry of the N-electron system. Each additional electron totally changes the symmetry as in any phase transition. Therefore, capacitance of few electron system is *monophasic*. The change of symmetry is most drastic whenever an odd number is encountered, in particular, with prime numbers, the change in the total interaction energy is noticeably larger. Let us further jump ahead by revealing something even more startling. The trend looks so much like the periodic table of the chemical elements. At this stage, LaFave told me that the trend is roughly following the ionization energy of the elements, but the values are off. I suggested that one should not be surprised by the fact that we did not even use the Schrodinger's equation, because our theory is quasi-stationary without the kinetic energy; therefore Poisson equation can indeed be used. However, I suggested that

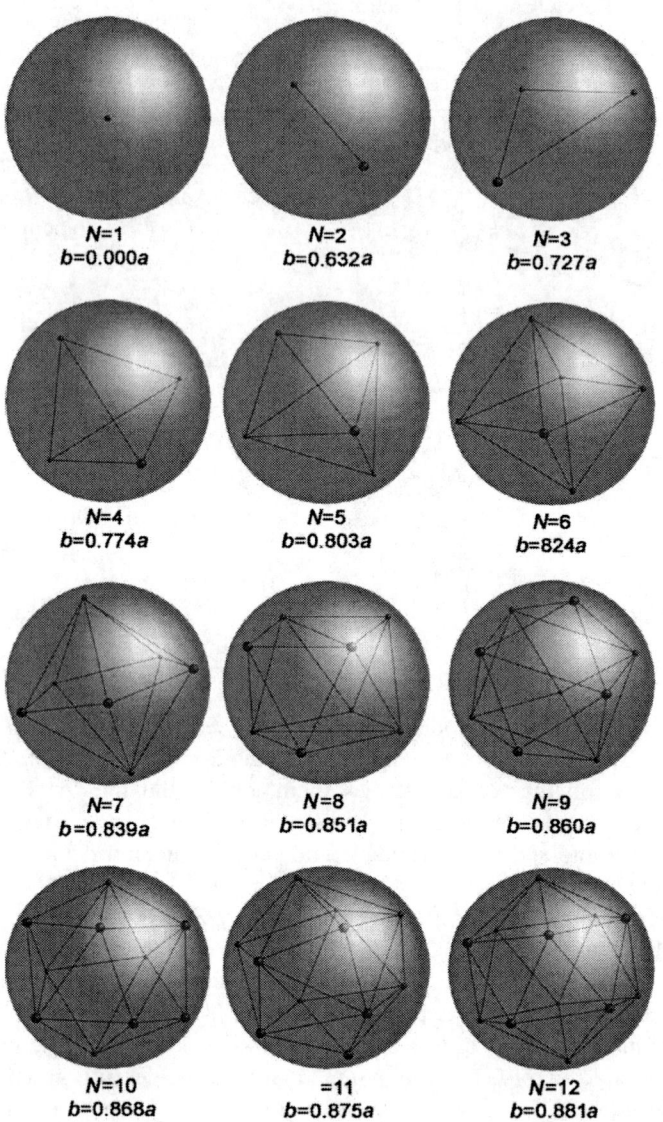

Figure 8.10 N-electrons in a dielectric sphere.
Source: After Zhu (2006), with permission.

we should change the radii with each addition of electron according to the atomic radii. With that factor, the plot of total interaction energy versus N, the number of electrons, is close to the experimentally determined ionization energy of the elements. Furthermore, at N = 11, the so-called sodium anomaly also shows up. At N = 83 for the element Bi is the last stable element in the periodic table before radioactive decay starts due to the inability of the confinement, in this case, inside the nucleus, in comparison to the electrostatic Coulomb repulsive forces. I have pointed out before that we should always go ahead with our instinct, as we have discovered the type-III superlattice for the wrong reason, and now we have discovered the secret of the periodic table for the wrong reason again. Joe Quinn, an accomplished mathematician at UNC−Charlotte, told me that mathematically our procedure generates the essence of the spherical harmonics. Let us show some of the computed results as well as point out the cardinal principles involved. Note that the inter-phasic energy difference is lower for even numbers such as 2, 4, . . . , 10, 12, 14, . . . quite similar to the atomic shell model.

In Figure 8.12, W^+ is defined to compare N-electrons with $N − 1$ electrons with one extra electron added to the center to prevent the addition of an extra charge LaFave and Tsu (2009). By placing at the center we preserve the charge so that the difference may be attributed to the change of symmetry. Therefore, the good agreement indicates that symmetry is indeed the most important parameter in determining the trend of the ionization processes.

We come to the reality that whenever the number of electrons in a given device is down to 100 instead of 1000 as in many so-called nano-devices, with each addition or subtraction of an electron, there results in a change of symmetry. Therefore, as the MOSFET is reduced to nanoscale, we must take account of the role of symmetry change with change of the number of electrons. We shall go into more discussion in the last chapter of this text.

I asked Quiyi Ye, who worked for a number of years developing the smallest source-drain (S-D) MOSFET. I invited her to give a seminar at UNC−Charlotte when she was working on 35 nm S-D MOSFET. I asked her to estimate the smallest number of electrons involved on the gate. She came up with 380 electrons,

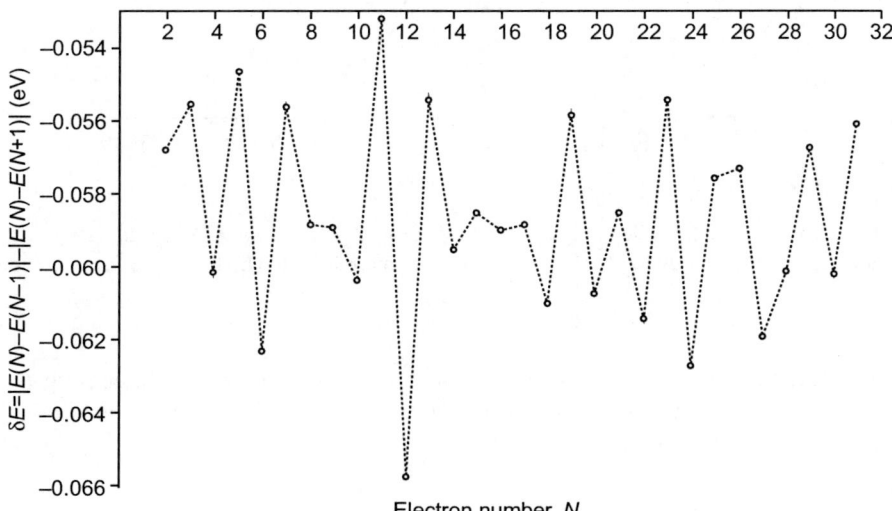

Figure 8.11 Interphasic energy differences exhibit even-number phase energy lying closer to preceeding $[N − 1]$ phases than neighboring $[N + 1]$ phase, indicating greater even-N symmetries. *Source*: After LaFave and Tsu (2008), with permission.

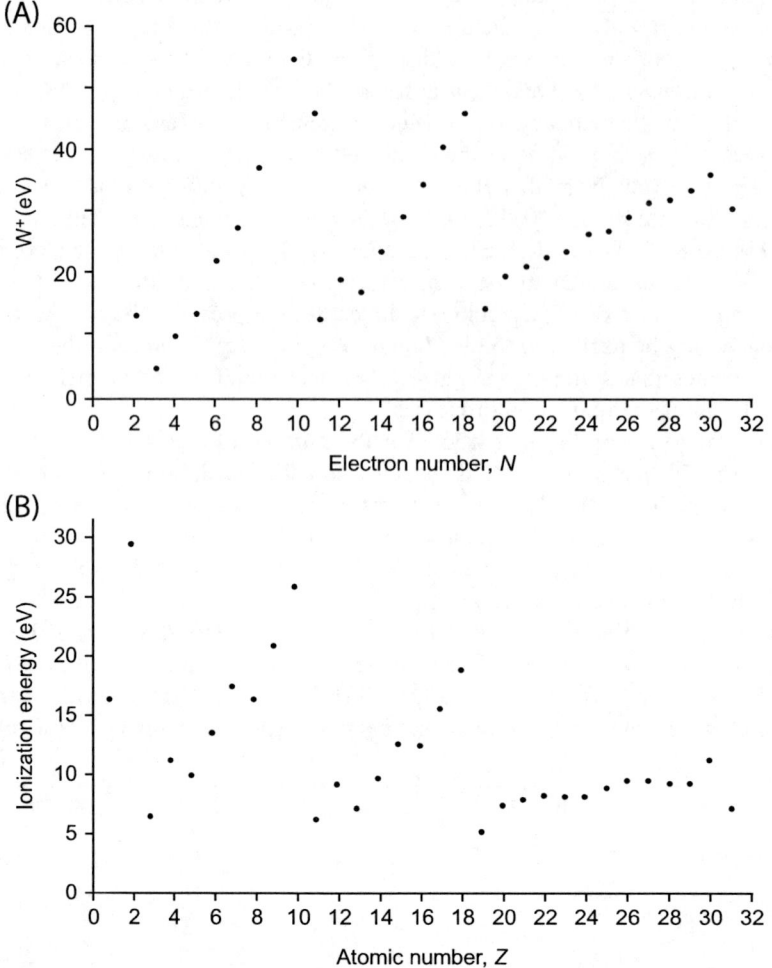

Figure 8.12 (A) Interphasic energy $W^+ = E(N - 1 + 1e \ at \ center) - E(N)$, a quantity most related to symmetry, versus Z using $\varepsilon = \varepsilon_0$ and atomic radii and (B) known ionization energy of the neutral atoms.
Source: After LaFave and Tsu (2009), with permission.

getting closer to our DCD model of capacitance. This nonlinearly increasing function with the number of electrons N certainly requires new consideration when N is further reduced to below 100.

8.5 Summary

We take on three subjects, the quantum capacitance, the size-dependent dielectric function, and the doping of a quantum dot—all are different in the quantum regime.

The main difference in quantum capacitance from its classical definition is the fact that there is no such thing as storage of charge alone and the charge of electron is discrete. We have shown that the classical capacitance taking into account the discreteness of electronic charge is quite different when there are only few electrons. Only when the number of electrons is large does the discreteness of electronic charge cease to exhibit new features. When the size is below a couple of nanometers, quantum mechanically the kinetic energy of the stored electrons becomes dominant over the electrostatic stored energy. Pauli's exclusion principle plays an additional role in quantum capacitance. In our formulation, one electron in the ground state of a sphere is nothing but the ground state of a hydrogen-like state and the ground state of a helium-like state for two electrons. For simplicity, while still retaining the dominant features for describing a quantum capacitor, we assumed that the sphere has a different dielectric constant from the matrix, but the barrier confining the electrons is taken to be infinity. Thus, a step toward a more realistic description should include a finite barrier, which would result in more complications. At a radius of 2 nm, the first electron added to a sphere has a quantum capacitance of 7 times below that of the classical capacitance, and the second electron added has a quantum capacitance of 27 times below its classical counterpart. *Note that even our classical capacitance is not the same as the textbook value because of the discreteness of electronic charge.* The model we used to calculate the classical capacitance, with discrete electronic charge, is based on the minimization of the electrostatic energy. This procedure allows us to include fixed trapped charges and thus may serve to elucidate instability in the gate oxide even without quantum effects.

The size-dependent dielectric function is not defined according to the usual definition, because of the loss of the global nature. We emphasize that even the general $\varepsilon(q)$ is global while $\varepsilon(a)$, with its boundary condition, is closer to the definition of localized response function. Using a single oscillator placed at the Fermi energy of all the valence electrons without any other adjustable parameter, the calculated value for silicon turns out to be almost identical to $\varepsilon(q = 2\pi/a)$, obtained from pseudopotential calculation. This fact leads to the realization that the pseudopotential calculation is perhaps no more accurate than the single oscillator description for the dielectric constant. Part of the reason I gave is the fact that the $\varepsilon(a)$ result does utilize several parameters, such as the value for the bulk dielectric constant, the number of valence electrons, the silicon unit cell size, and the energy E_g to be at the maximum absorption peak for silicon. In fact, what I want to point out is that one oscillator does not mean one parameter—the reality is more like four parameters.

The reduction of the dielectric constant at size approaching couple of nanometers drastically reduces the dopant binding energy from 13.6 eV for the hydrogen atom to ~1 eV. Even if the dopant happens to be inside a quantum dot of couple of nanometers in size, the quantum dot is still intrinsic. How can we ensure that dopants can indeed be inside the quantum dot as well as all the other quantum dots? The simple intuition tells us that uniform doping of all the quantum dot is not possible. If it comes to the point that we must uniformly dope the quantum dots, which is not possible with present techniques, then I think we must do away with doping. On the other hand, it appears that carrier injection should be the main mode of utilizing quantum dots for nanoelectronics. There is one huge difference: injection of electrons adds negative charge, while doping gives charge neutrality.

References

Allan, G., Delerue, C., Lannoo, M., Martin, E., 1995. Phys. Rev. B 52, 11982.
Aspnes, D.E., 1981. Proc. SPIE 276, 188.
Babic, D., Tsu, R., 1997. Superlattices Microstruct. 22, 581.
Babic, D., Tsu, R., Greene, R.F., 1992. Phys. Rev. B 45, 14150.
Baroni, S., Resta, R., 1986. Phys. Rev. B 33, 7017.
Bastard, G., 1981. Phys. Rev. B 24, 4714.
Bednarek, S., Szafran, B., Adamowski, J., 1999. Phys. Rev. B 59, 13036.
Bockris, J., Reddy, A.K.N., 1973. Modern Electrochemistry, Plenum Press, NewYork, p. 156.
Bottcher, C.J.F., 1973. Theory of Electric Polarization, second ed. vol. 1. Elsevier, Amsterdam.
Bransden, B.H., Joachain, C.J., 1983. Physics of Atoms and Molecules, Longman, New York.
Brus, L.E., 1983. J. Chem. Phys. 79, 5566.
Brus, L.E., 1984. J. Chem. Phys. 80, 4403.
Ehrenreich, H., Cohen, M.H., 1959. Phys. Rev. 115, 786.
Harrison, W.A., 1970. Solid State Theory, McGraw-Hill, New York.
Harvey, J.F., Shen, H., Lux, R.A., Dutta, M., Pamulapati, J., Tsu, R., 1992. Mater. Res. Soc. Symp. Proc. 256, 175.
Ioriatti, L., Tsu, R., 1986. Surf. Sci. 174, 420.
Kahan, K.B., Leburton, J.P., Hess, K., 1985. Superlattices Microstruct. 1, 295.
LaFave, T.J., Tsu, R., 2008. Microelectron. J. 39, 617.
LaFave, T.J., Tsu, R., 2009. Microelectron. J. 40, 791.
Lannoo, M., Delerue, C., Allan, G., 1995. Phys. Rev. Lett. 74, 3415.
Likharev, K., 1991. In: Ferry, D. (Ed.), Granular Nanoelectronics. Plenum, New York.
Luttinger, J.M., Kohn, W., 1955. Phys. Rev. 97, 869.
Macucci, M., Hess, K., Iafrate, G.J., 1993. Phys. Rev. B 48, 17354.
Macucci, M., Hess, K., Iafrate, G.J., 1995. J. Appl. Phys. 77, 3267.
Merzbacher, E., 1961. Quantum Mechanics, Wiley, New York.
Morita, A., Nara, H., 1966. J. Phys. Soc. Jpn 21 (Suppl.), 234.
Pantelides, S.T., 1978. Rev. Mod. Phys. 50, 797.
Penn, D.R., 1962. Phys. Rev. 128, 2093.
Preister, C., Allan, G., Lannoo, M., 1983. Phys. Rev. B 28, 7194.
Stern, F., 1978. Phys. Rev. B 17, 5009.
Tanaka, K., Tsu, R., 1981. Phys. Rev. B 24, 2038.
Tsu, R., 1993. Physica B 189, 235.
Tsu, R., Babic, D., 1993. Special Issue of NATO ASI Series for NATO Workshop on Optical Properties of Low Dimensional Silicon Structures, CNET, Grenoble, France, 1–3 March 1993, Kluwer, Dordrecht, p. 203.
Tsu, R., Babic, D., 1994. Appl. Phys. Lett. 64, 1806.
Tsu, R., Babic, D., Feng, Z.C., Tsu, R. (Eds.), 1994. Porous Silicon. World Scientific Publishing, Singapore.

Tsu, R., Babic, D., 1994. In: Vial, J.C., Derrien, J. (Eds.), Porous Silicon Science and Technology, Winter School Les Houches, Febuary 1994. Springer, Berlin, p. 111.

Tsu, R., Ioriatti, L., 1985. Superlattices Microstruct. 1, 295.

Tsu, R., Ioriatti, L., Harvey, J.F., Shen, H., Lux, R.A., 1993. Mater. Res. Soc. Symp. Proc. 283, 437.

Tsu, R., Babic, D., Ioriatti, L., 1997. J. Appl. Phys. 82, 1327.

van Houten, H., Beenaker, C.W.J., 1989. Phys. Rev. Lett. 63, 1893.

Walter, J.P., Cohen, M.L., 1970. Phys. Rev. B 2, 1821.

Wang, L.-W., Zunger, A., 1994. Phys. Rev. Lett. 73, 1039.

Wang, L.-W., Zunger, A., 1996a. Phys. Rev. B 53, 9579.

Wang, L.-W., Zunger, A., 1996b. Phys. Rev. B 54, 11414.

Yoffe, A.D., 2001. Adv. Phys. 50 (1), 168−171.

Zhu, J., Lafave, T.J., Tsu, R., 2006. Microelectron. J. 37, 1293.

Ziman, J.M., 1988. Principle of Solids, second ed. Cambridge University Press, Cambridge.

This page intentionally left blank

9 Porous Silicon

9.1 Porous Silicon: Light-Emitting Silicon

Following the first report on efficient photoluminescence (PL) in electrochemically etched silicon into porous silicon or PSi (Canham, 1990), the field of investigation has become very active owing to possible applications in silicon-based optoelectronics. The electronics industry is almost totally dominated by silicon technology, although without significant interaction with light because of its indirect band-to-band optical transition, unlike GaAs, silicon is unable to play a role in optoelectronic devices. Our main interest is focused on the light-emitting aspect of PSi that has been explored since 1990, although the history of PSi dates back to 1956 (see review by Gösele and Lehmann, 1994). The origin of visible PL at around 1.6 eV and higher was attributed to quantum confinement from the very first paper by Canham (1990) and confirmed by an increased optical absorption edge (Lehmann and Gösele, 1991) and by correlation with Raman shift (Tsu et al., 1992). However, there were equally convincing arguments with respect to polysilane surface centers (Prokes, 1993) as well as siloxene molecules (Stutzmann et al., 1993). Let us comment on the involvement of two equally credible groups with diametrically opposite interpretations backed by experimental data. First of all, PSi is easy to form but harder with which to institute fine control, such as molecular beam epitaxy (MBE), for the fabrication of devices. Second, PSi involves electrochemistry, which is not nearly as developed as precise deposition techniques like chemical vapor deposition (CVD) and MBE. And third, a possibly important fact is that both quantum confinement and "surface crud" contribute to the observed PL. In addition to the reasons mentioned, a high-resolution cross-section transmission electron micrograph (TEM) of PSi clearly identifies crystallites of size ~3 nm (Cole et al., 1992) and there is sufficient proof to invoke the quantum confinement model even if surface complexes do exist. This is the mindset I shall pursue in the following.

The PL in PSi at room temperature is attributed (Babic and Tsu, 1997) to the recombination of excitons confined in silicon nanocrystallites whose diameter is ~3 nm. The transition is mainly a second-order phonon-assisted process of bulk silicon which involves an electron from the bottom of the conduction band and a hole from the top of the valence band. In well-passivated samples, there are no surface trap states in the silicon energy gap that would provide nonradiative recombination paths. Radiative electron–hole recombination is practically the only

Superlattice to Nanoelectronics. DOI: 10.1016/B978-0-08-096813-1.00009-6

process effectively left. This picture of PSi PL is essentially consistent with the picture described in Brus et al. (1995).

However, measurements from Schuppler et al. (1995) suggest significant smaller Si nanocrystallites than those reported. Calculations by Hill and Whaley (1995) appear to support significantly smaller Si-nanoparticles being responsible for the visible PL. A nonradiative channel competing with the radiative exciton recombination is possibly provided by electron tunneling out of the quantum dot (QD) (Vial et al., 1992). Too little passivation leads to many surface states leading to temperature dependent non-radiative decay rates.

Figure 9.1 shows the calculated exciton recombination energy plotted against the radius of a silicon sphere for several values of the dielectric constant of the matrix ε_2 with the dielectric constant of the Si sphere ε_1 given by Figure 8.6 of Chapter 8 (Babic and Tsu, 1997). Note that the recombination energy is not sensitive to ε_1 and has a value of ~1.9 eV for a dot of diameter 3 nm, due to near mutual cancellation of the electron and hole polarization self-energies and their polarization interaction energy for the excitons involved. Our calculation, based on effective mass model, agrees well with the tight-binding model used by Martin et al. (1994).

It is instructive to compare and contrast excitonic radiative recombination in silicon nanocrystallites and in bulk Si, both at room temperature. In the bulk at room temperature, it is more likely that the exciton will be broken up by a phonon than it will encounter the right phonon-assisted radiative recombination. This is due to the low exciton binding energy in bulk Si of 14.7 meV (Shaklee and Nahory, 1970) compared to the energy of the phonon necessary to conserve momentum in the radiative recombination. Exciton breakup is facilitated by the quasi-continuum of available states in both valence and conduction bands. The e–h pair liberated from exciton by a phonon flies apart through the quasi-continuum, thus totally disabling radiative recombination. In Si QDs, the exciton binding energy is much

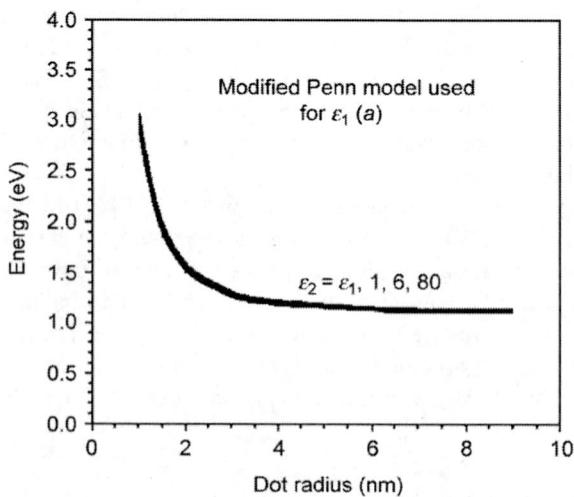

Figure 9.1 Calculated exciton recombination energy versus Si dot radius for several values of the dielectric constant of the matrix ε_2.
Source: After Babic and Tsu (1997), with permission.

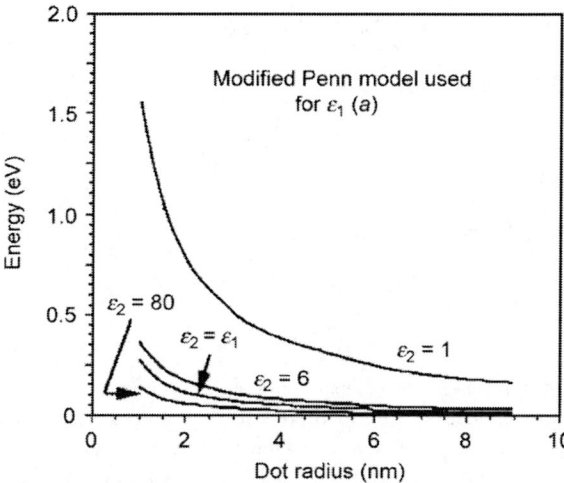

Figure 9.2 Exciton binding energy versus the radius of the dot for several values of the dielectric constant of the matrix. *Source*: After Babic and Tsu (1997), with permission.

larger than the energy of the phonon necessary for radiative recombination. The quasi-continuum of valence and conduction band states is modified to a discrete set of energy levels. A thermal phonon cannot break up the exciton inside the QDs, allowing time to wait for the right phonon to come along for radiative recombination.

Figure 9.2 shows exciton binding energy plotted against the radius of the dot for several values of the dielectric constant of the matrix. Applied to a typical PSi crystallite of radius 1.5 nm, the exciton binding energy ranges from 82 meV ($\varepsilon_2 = 80$) to 1.03 eV ($\varepsilon_2 = 1$) and 0.23 eV with $\varepsilon_2 = 6$ for water (see the discussion following the citation of Figure 8.8). For $\varepsilon_2 = \varepsilon_1$, where only the Coulomb term remains, the exciton binding energy is 0.16 eV, still more than an order of magnitude higher than the bulk value of 14.7 meV. This major increase in the Coulomb interaction part of the exciton binding energy is caused by the increased overlap of the electron and hole wave functions. The difference between the exciton binding energy and the Coulomb energy is due to the self-polarization energy of the electron and of the hole. As shown in Eq. (8.4) and discussion, the self-polarization energy scales as $(\varepsilon_2 - \varepsilon_1)/\varepsilon_2$ and accounts for most of the exciton binding energy, therefore, the size-dependent dielectric constant (Figure 8.4) must be used for the calculation. Regardless of which value is used for the matrix, the exciton binding energy is much greater than $k_B T$ at room temperature. Therefore, excitons confined to QDs are well bound electrostatically irrespective of the surrounding matrix. For comparison with other calculations, see further discussions in Babic and Tsu (1997).

Quantum confinement changes the positions of the Γ valence band maximum and Δ conduction band minimum levels. Note that the conduction band maximum at Γ, experimentally determined to be $\sim$2.9 eV (Chalikowsky et al., 1989) should also be modified by confinement, as shown in the schematic sketch in Figure 9.3. It is reasonable to assume that the curvature of the bands at Γ for both conduction and valence bands is nearly the same; therefore, no stable excitons can exist in bulk at Γ ($E-k$ bends down instead of bending up). *Via* quantum confinement,

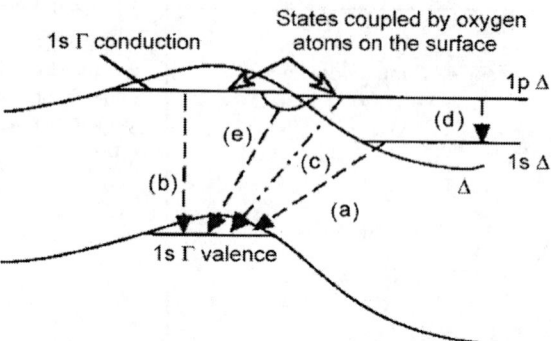

Figure 9.3 Schematic diagram showing bulk silicon valence and conduction bands, confinement-induced quantum states with electrons confined both at the Γ and Δ points of the Brillouin zone. The 1p excited state at Δ maybe coupled to the 1s state at Γ by the surface oxygen complex. The possible transitions (a) red peak at 1.96 eV, (b) weak and fast blue peak at 2.75 eV, (c) suppressed transition at 2.66 eV, (d) infrared peak at 0.76 eV, and (e) blue peak at 2.6 eV are observed in oxidized samples.
Source: After Babic and Tsu (1997), with permission.

because the quasi-continuum is discretized, excitons are stable as usual. The electrostatic interaction between the electron and the hole are similar to the already calculated lowest indirect exciton with its envelope wave functions. Thus, the estimated exciton recombination energy of the confined direct exciton in Si nanocrystallites is ~2.75 eV for a diameter of 3 nm PSi. Recombination of these direct excitons maybe the origin of the weaker but faster blue PL in well passivated and nonoxidized PSi samples (Petrova-Koch et al., 1993). It is faster but weaker because it competes with an extremely strong nonradiative relaxation channel that relaxes electrons toward the bottom of the conduction band. The crucial part of this nonradiative channel is the first excited p-like electron state in the QD constructed from the bulk Si states around the conduction band minimum at the Δ point. Radiative recombination from the first excited p electron state to the lowest hole state is suppressed by the selection rule involving the envelope wave functions. This p-like electron at Δ and a hole at Γ of the valence band form an indirect exciton whose energy is 2.66 eV for a ~3-nm QD in PSi, which is less than 0.1 eV below the direct excitons, allowing for extremely efficient vibronic—phonon transitions between the direct exciton to the indirect excitons at the Δ point, competing with the recombination of the direct exciton. The energy difference between 2.66 and 1.96 eV is 0.7 eV lying in the infrared, and may very well be the reported infrared PL in PSi (Fauchet et al., 1994). PSi subjected to rapid thermal oxidation shows blue PL at 2.6 eV attributed to surface states involving oxygen complexes (Kanemitsu et al., 1993).

Besides PSi, similar luminescence phenomena were observed in man-made silicon nanocrystalline structures, fabricated by the vacuum deposition technique discussed in Figures 6.4, 6.6, and 6.34 of Chapter 6, as well as using a

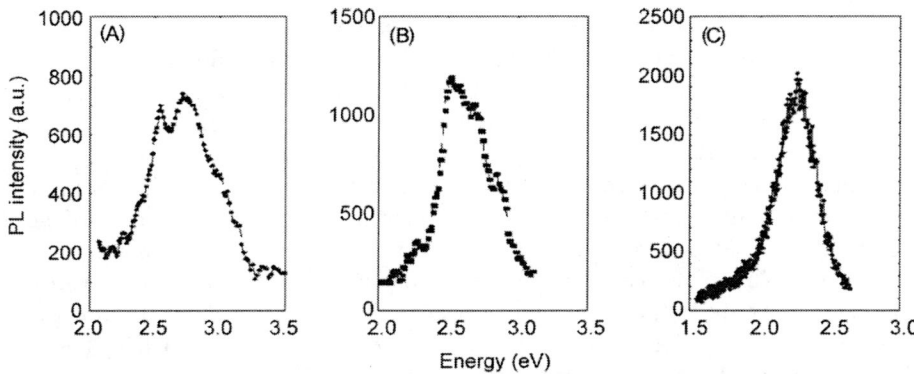

Figure 9.4 PL of silicon clusters in a SiO$_2$ matrix. The intensity of PL is related to the cluster densities found by TEM. Red shift results from (A) to (C) with increasing Si in the oxide matrix.
Source: After Tsu and Zhang (2002), with permission.

high-temperature aerosol method and selecting the size by chromatography (Littau et al., 1993) and a nonthermal microwave gas phase synthesis using a mixture of argon and silane (Zhang et al., 1994).

Elemental semiconductors Ge, Si, and C embedded in a SiO$_2$ matrix exhibited fairly strong and stable PL, with peaks ranging from infrared to blue wavelengths (Zhang et al., 1995) and elaborated in Tsu and Zhang (2002). In the case of Ge, samples for a light-emitting diode (LED) were fabricated from 45-nm-thick poly-crystalline Si films deposited initially as amorphous films by e-beam evaporation onto 70-nm SiO$_2$ films thermally grown on n$^+$ Si substrates, followed by Ge implantation, to create a supersaturated solid solution of Ge in the SiO$_2$ film with an approximately uniform Ge ($\sim$5 nm in diameter) concentration of 5%. The samples were subsequently annealed at 600°C, 1×10^{-6} Torr, for 40 min to induce precipitation. The electroluminescent (EL) spectrum was broad and peaked at 1.2–1.4 eV. In the case of silicon, samples of Si-clusters can be prepared by sputtering SiO$_2$ onto silicon wafers without additional heating. After annealing at 800°C for 20–30 min in N$_2$, the typical PL spectra show the typical quantum size effect in Figure 9.4(A)–(C) with an increasing Si to oxide ratio. A more detailed discussion on the extended X-ray absorption fire structure (EXAFS) characterization and mechanisms of cluster size control maybe found in Zhang et al. (1995). In the case of C-clusters in a SiO$_2$ matrix, blue emission results (Zhang et al., 1996).

In 1971, inspired by P. Sorokin to develop a possible blue or UV laser, I undertook a project putting rare earth elements like Eu and Sm into CaF$_2$ (Tsu and Esaki, 1970a,b). I noticed that luminescent efficiency of the desired lines was quenched after introducing more than 0.5% of rare earth into CaF$_2$. The explanation is quite simple. Whenever the impurity concentration reaches the third-nearest neighbor, as is generally applicable to all solids, the probability of the dopants forming a complex increases rapidly. Not only does the complex have its own signature of optical transitions, the nonradiative channels become more significant because coupling to phonons in the matrix is drastically increased owing to vibronic states of the complex coupling with the phonons of the solid. The experimental data at low temperatures clearly show the multiple vibronic peaks in addition to the "zero-vibronic" peak. The vibronic frequency is not so different from the phonon frequencies because phonon dispersions maybe adequately represented by two so-called Keating constants (Keating, 1966) with the stretching constant overwhelmingly dominating the bending constant. In quantum confinement of nc-Si, the problem is compounded by the indirect transition dominating even at a diameter of 2–3 nm. The game plan is to find the "proper" matrix, which is detached from the nc-Si crystallites, not only with respect to electronic detachment with barriers like those used in the formation of superlattices and quantum wells, but more importantly as a barrier to phonons. The only way is to increase the porosity of the matrix, making the matrix as close to a mere skeleton as possible. Therefore, PSi is really very, very special. The skeleton provides a current path that allows electrical excitations, while blocking phonon couplings. It is safe to state that the study of PSi involves more people than any other endeavor in science and engineering and primarily it is simple to make. However, after 15 years, there is not a single device that has entered into common usage. It is equally safe to state that the problem it faces is almost insurmountable.

9.2   PSi: Other Applications

The biggest obstacle preventing PSi devices from joining the class of commercial products is the lack of mechanical robustness and stability. Tsybeskov and Fauchet (1994) took a step in the right direction by reducing the current during electrochemical etching. We focused our study on the morphology of PSi (Filios et al., 1996) by drastically reducing the current as well as reducing the hydrogen floride (HF) concentration to less than 25%. Typically, $J \leq 10 \, \mathrm{mA \, cm^{-2}}$ with an anodization time of 20–45 min in the dark is considered to have been gently etched. By comparison, $J \geq 30 \, \mathrm{mA \, cm^{-2}}$ is considered to have been heavily etched, measured using PL, Raman, and audio force microscopy (AFM). The results are quite convincingly improved, as shown in Figure 9.5, where the gently etched sample shows up interference fringes in the PL spectrum. The slight blue shift with the heavily etched sample is obviously due to the decrease in size of the nc-Si.

Figure 9.6 shows the AFM comparison between the gently etched (A) and heavily etched (B) samples.

What is most encouraging is the discovery of epitaxial growth of Si on PSi, resulting in a free-standing thin film of epitaxially grown silicon (Solanki et al., 2004). Figure 9.7 shows the AFM of a much improved PSi surface morphology after annealing at 1050°C for 30 min in H_2 (Solanki et al., 2004).

Layer transfer techniques based on PSi have been developed to obtain a high-quality thin film of Si on a foreign substrate aimed at reducing the substrate cost for photovoltaic (PV) purposes (Brendel, 1997). The separation of the PSi layer from the substrate maybe either before or after epitaxial growth of high-quality

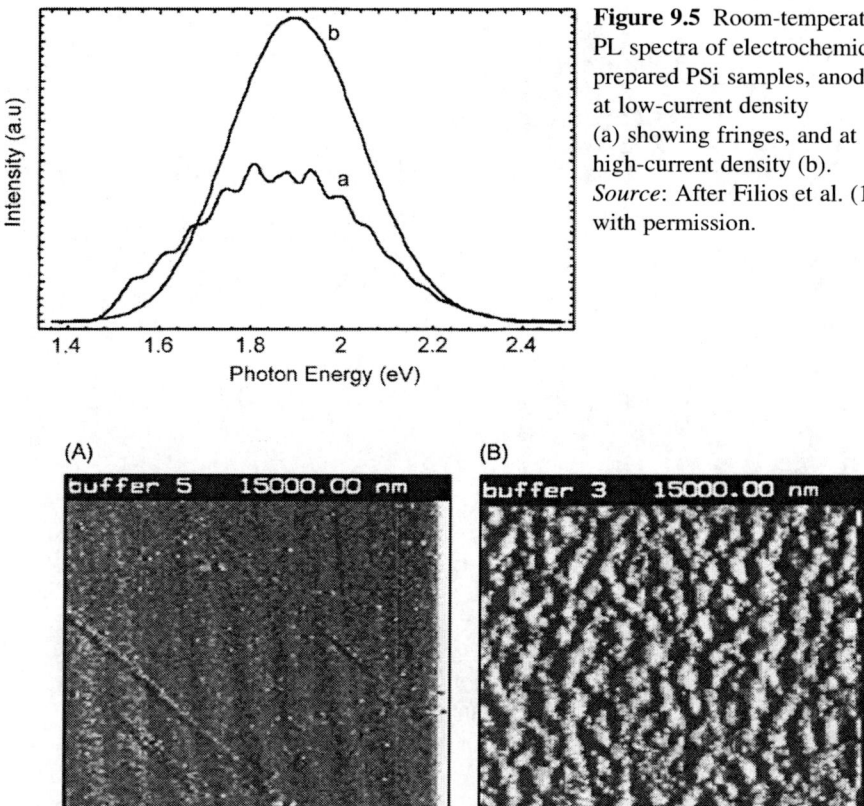

Figure 9.5 Room-temperature PL spectra of electrochemically prepared PSi samples, anodized at low-current density (a) showing fringes, and at high-current density (b). *Source*: After Filios et al. (1996), with permission.

Figure 9.6 AFM images of electrochemically prepared PSi samples, anodized at low- (A) and at high-current density (B).
Source: After Filios et al. (1996), with permission.

silicon (Tayanaka et al., 1998). In PV applications, the process allows the realization of stable and high-efficiency solar cells in a thin film form. After heat treatment at >1050°C in a H_2 atmosphere for 30 min, the structure of PSi is drastically changed in such a way that the top surface of the low-porosity region becomes smooth due to closing of pores, and thus ready for epitaxial growth of monocrystalline silicon (Tayanaka et al., 1998; Rinke et al., 1999). Thus far, the main driving force behind this work is to find a way to reduce the cost of high-quality PV material as well as a better way to produce silicon on insulator (SOI) for a better complementary metal oxide silicon (CMOS). However, at least to my mind, another potentially huge application is the bridging of PSi for its optoelectronic capability, and depositing high-quality monocrystalline epitaxial silicon on top for devices such as CMOS, thus realizing the dream of a silicon-based optoelectronic device, a subject to which many special international conferences have been devoted, e.g., the European MRS Conference in 2002 (Ni et al., 2003).

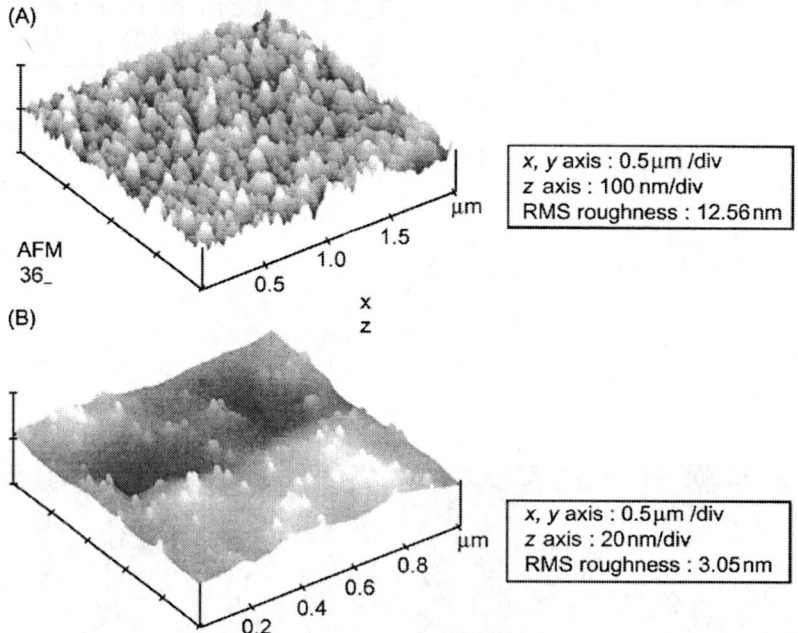

Figure 9.7 AFM of PSi surface anodized in 20% HF in acetic acid at 75 mA cm^{-2}. (A) As prepared, (B) after annealing at 1050°C for 30 min in H$_2$.
Source: After Solanki et al. (2004), with permission.

Another subject discussed during this meeting (Ni et al., 2003), involved the question of the broad linewidth of PSi and QDs in general. I pointed out that the major contributor to the line-broadening process is the coupling to the outside world, I/O for intended coupling, as well as losses for unintended coupling. However, the "crud" on the surface constitutes a major challenge to theoretical formulation, either with diffused boundary conditions, or splitting the surface species into two parts, one lying slightly inside and another slightly outside the QD as done by Stern (1978), to avoid the singularity at the boundary. With all the expectations of QDs in devices, the question of linewidth is an important subject for investigation.

9.3 Summary

Visible luminescence in PSi is basically due to quantum confinement, however, oxygen complexes also contribute significantly. The morphology maybe improved by reducing the etching rate. Silicon can be epitaxially grown on PSi. I argue that the possibility of embedding luminescent QDs of CdSe or PbS, etc., into high-porosity PSi, sealed off by epitaxial silicon growth, offers the best chance for electronic and optoelectronic applications. The possibility of introducing luminescent QDs, even with a periodic pattern, into PSi followed by epitaxial Si growth maybe useful in optoelectronic applications. And replacing luminescent QDs by wide-gap QDs may serve as a gate capacitor to control source—drain current in a field effect transistor. Therefore, the capability of epitaxial growth on PSi has enormous device potential.

After more than 10 years of making a glorious entry to the world of devices, I take this opportunity to comment on the difficulties in making the grade in the world of devices. Point-contact transistors opened the door for solid-state electronic devices, but it was the pn-junction and Field Effect Transistor (FET) making the grade. In 1973, I accompanied *Jacques Maisonrouge*, chairman and chief executive officer of IBM World Trade, to China. He asked me which discipline represented the greatest contribution toward IBM. I replied that electrical engineering was first, followed by physics. He replied while laughing that mechanical engineers were first by a wide margin, because robustness and mechanical stability is the key for any technology. PSi is far from meeting these criteria.

References

Brendel, R., 1997. Proceedings of the 14th European PV Solar Energy Conference, Barcelona, Spain, p. 1354.

Babic, D., Tsu, R., 1997. Superlattices Microstruct. 22, 582.

Brus, L.E., Szajowski, P.F., Wilson, W.L., Harris, T.D., Schuppler, S., Citrin, P.H., 1995. J. Am. Chem. Soc. 117, 2915.

Canham, L.T., 1990. Appl. Phys. Lett. 57, 1046.

Cole, M.W., Harvey, J.F., Lux, R.A., Eckart, D.W., Tsu, R., 1992. Appl. Phys. Lett. 60, 2800.

Chalikowsky, J.R., Wagener, T.J., Weaver, J.H., Jin, A., 1989. Phys. Rev. B 40, 9644.

Fauchet, P.M., et al., 1994. SPIE Proc. 2144.

Filios, A.A., Hefner, S.S., Tsu, R., 1996. J. Vac. Sci. Technol. B 14, 3431.

Gösele, U., Lehmann, V., 1994. In: Feng, Z.C., Tsu, R. (Eds.), Porous Silicon. World Scientific, Singapore, p. 17.

Hill, N.A., Whaley, K.B., 1995. Phys. Rev. Lett. 75, 1130.

Kanemitsu, Y., et al., 1993. Mater. Res. Soc. Symp. Proc. 298, 205.

Keating, P.N., 1966. Phys. Rev. 145, 637.

Lehmann, V., Gösele, U., 1991. Appl. Phys. Lett. 58, 856.

Littau, K.A., Szajowski, P.F., Muller, A.J., Kortan, A.R., Brus, L.E., 1993. J. Chem. Phys. 97, 1224.

Martin, E., Delerue, C., Allan, G., Lannoo, M., 1994. Phys. Rev. B 50, 18258.

Ni, W.X., Grutzmacher, D., Priolo, F., Tsu, R. (Eds.), 2003. European MRS 2002 on silicon-based optoelectronics, Physica E 16, Strasbourg, France.

Prokes, S.M., 1993. J. Appl. Phys. 73, 407.

Petrova-Koch, V., Muschik, T., Kovalev, D.I., Koch, F., Lehmann, V., 1993. Mater. Res. Soc. Symp. Proc. 283, 179.

Rinke, T.J., Bergmann, R.B., Brugemman, R., Werner, J.H., 1999. Solid State Phenom. 67/68, 229.

Schuppler, S., Friedman, S.L., Marcus, M.A., Adler, D.L., Xie, Y.H., Ross, F.M., et al., 1995. Phys. Rev. B 52 4910.

Shaklee, K.L., Nahory, R.E., 1970. Phys. Rev. Lett. 24, 942.

Solanki, C.S., Bilyalov, R.R., Poortmans, J., Beaucarne, G., van Nieuwenhuysen, K., Nijs, J., et al., 2004. Thin Solid Films 451–452, 649.

Stern, F., 1978. Phys. Rev. B. 17, 5009.

Stutzmann, M., Brandt, M.S., Rosenbauer, M., Weber, J., Fuchs, H.D., 1993. Phys. Rev. B 47, 4806.

Tayanaka, H., Yamauchi, K., Matsushita, T., 1998. Proceedings of the 2nd World Conference on PV Solar Energy Conversion, Vienna, Austria, p. 1272.

Tsu, R., Esaki, L., 1970a. Phys. Rev. Lett. 24, 455, ICPS, 1970.

Tsu, R., Esaki, L., 1970b. In: Keller, S.P., Hensel, J.C., Stern, F. (Eds.), Proc.. of ICPS, US Atomic Energy Commission, Cambridge, MA, p. 282.

Tsu, R., Zhang, Q., 2002. In: Koch, C.C. (Ed.), Nanostructured Materials. Noyes Publications, Norwich, p. 527.

Tsu, R., Shen, H., Dutta, M., 1992. Appl. Phys. Lett. 60, 112.

Tsybeskov, K., Fauchet, P.M., 1994. Appl. Phys. Lett. 64, 1983.

Vial, J.C., Bsiesy, A., Gaspard, F., Herino, R., Ligeon, M., Muller, F., et al., 1992. Phys. Rev. B 45, 14171.

Zhang, Q., Bayliss, S.C., Al-Ajili, A., Hutt, D.A., 1994. Nucl. Instrum. Math. Phys. Res. B 97, 329.

Zhang, Q., Bayliss, S.C., Hutt, D.A., 1995. Appl. Phys. Lett. 66, 1977.

Zhang, Q., Bayliss, S.C., Frentrup, W., 1996. Solid State Commun. 99, 883.

10 Some Novel Devices

10.1 Field Emission with Quantum Well and Nanometer Thick Multilayer Structured Cathode

At the First International Symposium on cold cathodes held on October 25–26, 2000, in Phoenix, Arizona, as part of the 198th Meeting of the Electrochemical Society, I was involved in pulling hot electrons from a semiconductor surface originating above the Fermi surface, as cooling mechanism. A double-barrier resonant tunneling (DBRT) structure was placed at the surface with a barrier. A second barrier is formed by the vacuum. Resonant tunneling is capable of selecting electrons from a particular energy range, thus serving as an energy selector, and is used to remove electrons from the surface of a semiconductor into the vacuum. It is referred to by us as inverse Nottingham cooling (Tsu and Greene, 1999; Tsu, 2001a,b; Yu et al., 2002). Basically, in addition to the use of resonant tunneling as an energy filter, the discrete energy level in a quantum well (QW) produces two-dimensional electrons (2D electrons), allowing the electrons in a given state to be at an energy above the bottom of the conduction band, gaining extra energy at the expense of the work function and resulting in lowering of the effective work function for field emission. At that meeting, I learned from Binh how to engineer a thin insulating layer near the surface, named SSE for solid state emission cathode, to trap the space charge resulted in lowering of the work function *via* a triangular-shaped QW when an electric field is applied, similar to surface inversion in a typical MOSFET (Binh and Adessi, 2000). In fact, the ideal is similar to that of Korotkov and Likharev (1999), applied to cooling. I suggested to Binh that we should join forces by using a resonant tunneling structure to serve both as a 2D-electron system as well as to lower the space charge. As shown in Figure 10.1(A), the structure we chose uses a III−*N* system to take advantage of the low work function of high band-gap semiconductors.

The first barrier is n-$Al_{0.5}Ga_{0.5}N$ with a GaN QW. The second barrier is the vacuum. Two QW states, E_1 and E_2, are shown with widths due to unavoidable losses. Figure 10.1(B) shows the schematic of the measuring apparatus. We first obtained a molecular beam epitaxy (MBE) grown sample on sapphire without showing substantial emission. While showing these poor results to A. Khan, he suggested that he could grow a sample with the same geometry on SiC, possibly with an improved structure as well as being capable of high current.

Superlattice to Nanoelectronics. DOI: 10.1016/B978-0-08-096813-1.00010-2

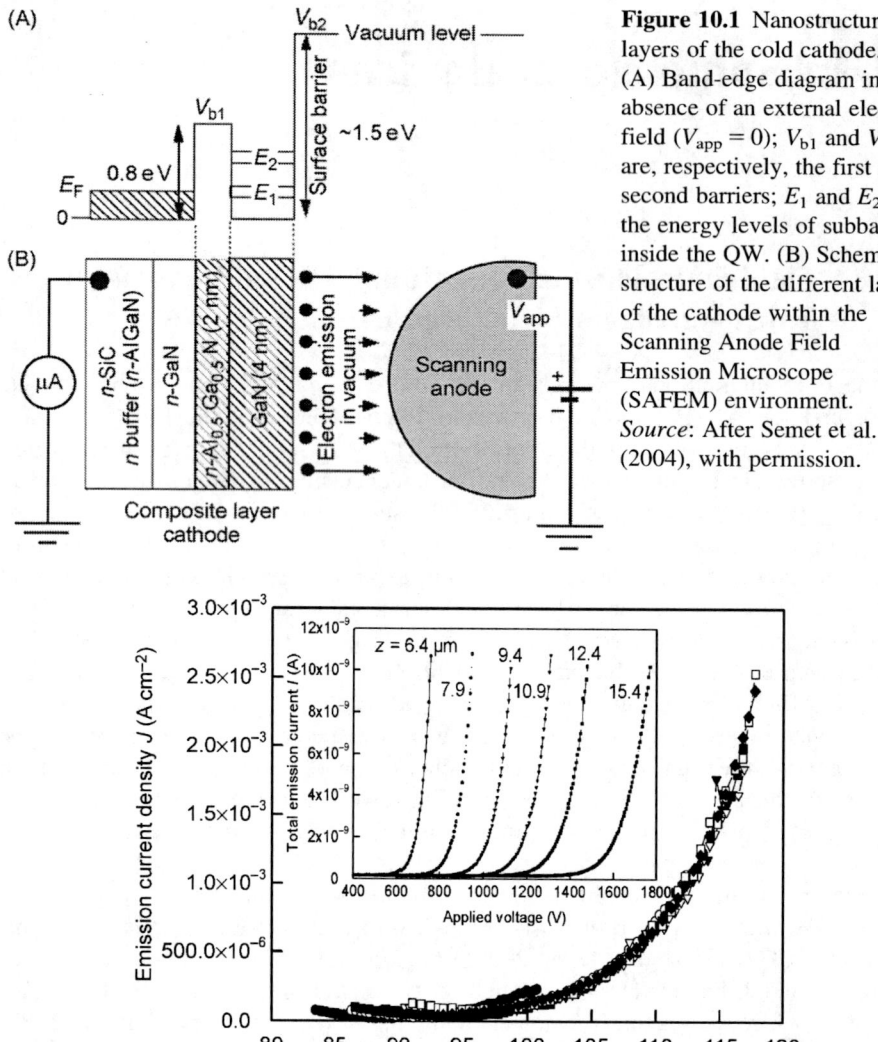

Figure 10.1 Nanostructured layers of the cold cathode. (A) Band-edge diagram in the absence of an external electric field ($V_{app} = 0$); V_{b1} and V_{b2} are, respectively, the first and second barriers; E_1 and E_2 are the energy levels of subbands inside the QW. (B) Schematic structure of the different layers of the cathode within the Scanning Anode Field Emission Microscope (SAFEM) environment. *Source*: After Semet et al. (2004), with permission.

Figure 10.2 Emission current density J_{mes} versus the actual field F at the surface of the nanostructured solid state emitter (SSE) cathode at 160°C. The insert shows the total emission current I versus applied voltage V_{app}, for six different distances z between the anode and the cathode.
Source: After Semet et al. (2004), with permission.

Figure 10.2 shows the emission current density J_{mes} plotted against the actual field F at the surface of the nanostructured cathode at 160°C, i.e., in the presence of a thermionic-like emission. In the insert are the plots of the total emission current I plotted against applied voltage V_{app}, for six different distances z between the

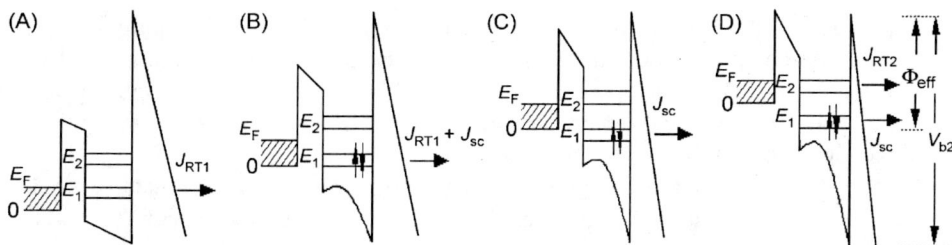

Figure 10.3 Illustration of the different field emission mechanisms by schematic band-edge diagrams of the nanostructured planar cathode with an applied field F at room temperature. (A) Resonant tunneling mechanism only; (B)–(D) evolution with space charge formation inside the GaN layer with, as a consequence, an effective lowering of the surface barrier. In addition to the resonant tunneling and due to the occupation of the quantum state E_1, electrons occupying this state (e.g., whenever the level E_1 moves below 0) can tunnel out of this single barrier *via* the usual F–N tunneling, resulting in J_{SC} (C), and the total current $J_{FN} = (J_{RT} + J_{SC})$. Note that the vacuum field is not drawn to scale in this illustration. *Source*: After Semet et al. (2004), with permission.

anode and the cathode. The convergence of all the I–V data toward a unique J_{mes}–F plot for different z values is a confirmation of the correctness of the emission measurements and analysis of Binh et al. (2001). From these data, a Fowler–Norheim (F–N) behavior for the emission characteristics is obtained from a nearly straight line plot of $\ell n(J_{mes}/F^2)$ against $1/F$.

At first we did not include the self-consistent potential by neglecting the space charge in the subbands. Calculated results indicate that substantial emission current starts only at a field in excess of 5×10^6 V cm^{-1}. However, when we included the space charge, 2D electrons occupation when a given subband is either aligned or moving below the Fermi level of the contact, emission started at an electric field at least 5 times lower than that. Figure 10.3 shows the different field emission mechanisms of the nanostructured planar cathode with an applied field F. Note that: (i) to be able to present these band diagrams within this figure, the field representation is not at the same scale inside the cathode and outside in the vacuum particularly if one considers GaN having $\varepsilon = 8\varepsilon_0$ with an applied field in the range of 50 V μm^{-1}; and (ii) further reduction from the induced image charges due to the space charge in the QW is not shown in this sketch.

In our two-step tunneling model, a larger lowering of the work function due to space charge in the QW is crucial. The idea is that when this 2D-like quantum state is occupied, it results in a space charge in the QW, leading to additional lowering of the effective work function defined by the energy of the source of electron at the vacuum level. A precise quantitative approach requires the use of the Airy function when a voltage is applied to the structure, with self-consistent calculations. However, to estimate the lowering of the work function we used a simple approach, involving first finding the potential due to the space charge inside the 2D QW. Using this calculated lowering of the work function, the usual field emission expression derived from F−N theory is then fitted to the experimentally measured field emission. It is important to note that lowering the work function consists of two parts, the 2D-quantum states that have a "flat" energy profile right up to the surface at any applied voltage and the space charge "bulging" of the potential shown in Figure 10.3.

To estimate the space charge effect, assuming a perfect confinement, we calculate the charge density n at the lowest level E_1 state in the QW of width w,

$$n = e(m^*/\pi\hbar^2)(E_2 - E_1)(2/w) \sin^2(\pi x/w), \tag{10.1}$$

where m^* is the effective mass of the electron with charge e. Solving the Poisson's equation we arrive at the potential energy of the space charge V_{SC}, for $0 \leq x \leq w$,

$$V_{SC} = (\rho_0/4\varepsilon)\{(w/\pi)^2 \sin^2(\pi x/w) + wx - x^2\}, \tag{10.2}$$

where $\rho_0 \equiv e(m^*/\pi\hbar^2)(E_2 - E_1)(2/w)$ and ε is the dielectric constant. The maximum value of V_{SC} is at $x = w/2$, $V_{SC}(w/2) = 0.25(\rho_0/\varepsilon)w^2(\pi^{-2} + 0.25)$ and the average of $V_{SC} \equiv V_{SC}(\text{av}) = 0.62 V_{SC}(w/2)$. Taking the average of the difference $V(w) - V(0)$, the total lowering of the work function $\Delta\Phi$ is

$$\Delta\Phi \equiv V_{b2} - \Phi_{\text{eff}} = V_{SC}(\text{av}) + 0.5(V(w) - V(0)) + E_1. \tag{10.3}$$

The effective barrier Φ_{eff} is the actual barrier at the surface after the lowering and can be determined experimentally from the (J_{mes}, F) plots, i.e., Φ_{FN}. For an estimation of $\Delta\Phi$ we have taken $m^* = 0.22m_0$, $\varepsilon = 8\varepsilon_0$ for GaN, $V_{SC}(w/2) = 0.4$ eV, $0.5 \times (V(w) - V(0)) = 0.62$ eV, and $E_1 = 0.18$ eV. This gives $\Delta\Phi = 1.05$ eV, i.e., $\Phi_{\text{eff}} = 0.45$ eV for $V_{b2} = 1.5$ eV. This calculated value 0.45 eV for the effective surface barrier is very close to the experimental values Φ_{FN} measured from the (J_{mes}, F) plots, which were in the range of 0.25−0.53 eV. Therefore, we conclude that after the occupation of the quantum level for the electron in the state E_1 lying below $E = 0$, the tunneling current $J_{\text{FN}} = J_{\text{RT}} + J_{\text{SC}}$ is given by the F−N tunneling through a single barrier created by the vacuum, with an effective barrier of only a few tenths of an electron volt, Figure 10.3(D). This lowered barrier at the surface controls the variation of the emitted current J_{FN} with field.

Repeat measurements shown in Figure 10.4, Semet et al. (2008), of the DBRT with 4 nm of GaN well and 2-nm $Ga_{0.5}Al_{0.5}N$ barrier with $V_{b2} = 1.5$ eV on GaN/SiC device resulted in the measured resonant states $E_1 = 0.071$ eV, $E_2 = 0.282$ eV, and $E_3 = 0.617$ eV. The work function from Eq. (10.3) due to space charge in the QW is now $\Delta\Phi = 0.94$ eV and $\Phi_{\text{eff}} = 0.56$ eV, which is higher than 0.45 eV, the previous value. It is quite possible that the device has deteriorated from repeated handling over the course of four years.

The typical SSE with TiO_2 on Pt under increasing applied electric field is shown in Figure 10.5. Note that substantial tunneling results at an electric field of 1.4×10^6 V cm^{-1}. On the other hand, emission current density J versus the applied field F for various thickness of the TiO_2 shows that at a common $J = 10^{-2}$ A cm^{-2},

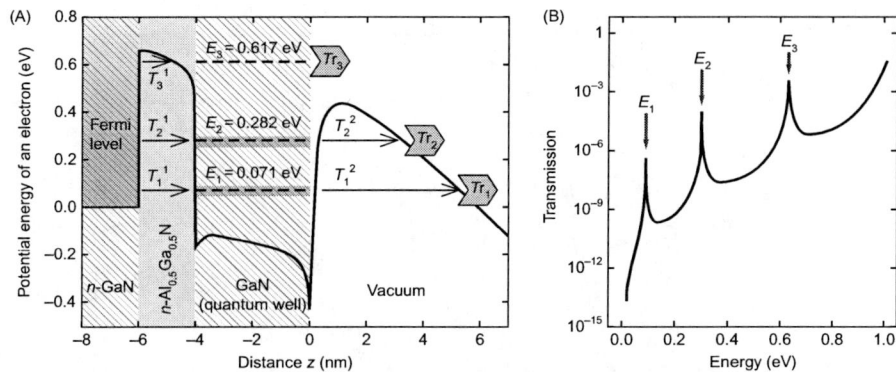

Figure 10.4 (A) Computed band diagram of DBRT on GaN/SiC with the transmission peaks shown in (B) with an electric field of 5×10^5 V cm^{-1}.
Source: After Semet et al. (2008), with permission.

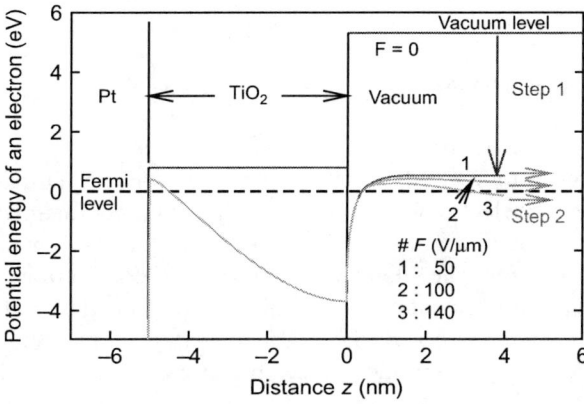

Figure 10.5 Typical SSE with TiO$_2$ on Pt. Applied E field increases from 1: 50 V μm^{-1} to 2:100 V μm^{-1}, to 3: 140 V μm^{-1}, showing increasing electron tunnel from E_F, left, to the vacuum, right. *Source*: After Semet et al. (2008), with permission.

the applied electric fields are ~ 0.5 and 1.6 MV cm^{-1}, respectively (Figure 10.6). Obviously thinner TiO$_2$ is better. At this stage it seems that a simple deposition of a few nanometer thick TiO$_2$ can outperform the complicated DBRT approach. However, we are comparing one device using DBRT structure versus many samples of SSE with TiO$_2$ on Pt. In principle, the DBRT scheme offers a wider range of design possibilities when other specifications are considered such as, ultimately, the stability and robustness under high current emission, and that applies to both cold cathode as well as Nottingham cooling.

This work is really quite significant, not only because the cold cathode *via* field emission is important in its own right, but also, for example, as source of electrons in vacuum electronics, such as the famous traveling wave tube (TWT), which is still widely in use as a high-power wide-band amplifier, a fact not generally known outside of the space electronics field. A robust resonant tunneling structure is not obtainable by surface treatment, for example, by using cesium, an electropositive

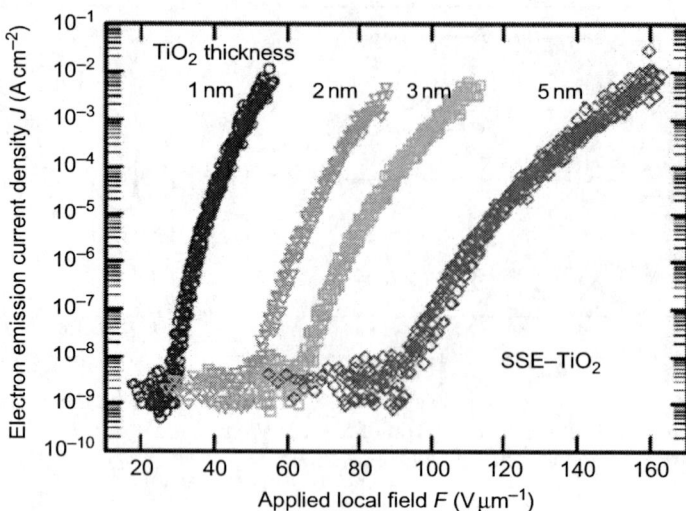

Figure 10.6 Emission current density J versus the applied field F for various thickness of the TiO_2.
Source: After Semet et al. (2008), with permission.

substance, on the surface of silicon to lower the work function. But more than that, it is a typical example of how nanoelectronics consisting of quantum confinement in one direction maybe the only viable quantum device that has the standard input/output (I/O) in a planar configuration. The I/O for quantum dot (QD) electronics has thus far eluded all efforts at implementation. I shall elaborate on this point later, dealing with contacts on a planar surface. In the next section, I shall show how QDs maybe incorporated into optoelectronic devices because of their fantastically low power threshold for a nonlinear response.

10.2 Saturation Intensity of PbS QDs

The saturation intensity for lead sulfide in a titanium dioxide—glycerol matrix (PbS/TiO_2—glycerol) on a glass substrate as a function of quantum cluster concentration and cluster size has been studied (Kang et al., 2004). The saturation intensity in these materials is strongly dependent on the size of the semiconductor nanocrystals, the concentration, or the sample thickness. As discussed in the last chapter on the introduction of luminescent materials such as CdS or PbS into PSi, a sol-gel such as TiO_2—glycerol serves as an excellent matrix for decoupling phonons and other unwanted coupling between the QDs and the matrix. Porous silicon, in addition to providing decoupling, is capable of carrying current by the silicon skeleton. The basic feature in samples is that they are reflective at a certain range of the incoming intensity of the optical field and are transparent over a threshold intensity beyond which the output and input intensities are related linearly. The system studied involves a very dilute distribution of PbS QDs with dimensions of $\sim 3-10$ nm embedded in a matrix of TiO_2−glycerol. We found that the threshold of power separating absorption bleaching, P_{th}, is linearly related to the thickness, with values of P_{th} of a few milliwatts per square centimeter, more than three orders of magnitude below that of QDs with dimensions of about 1 μm, representing a typical solid.

Titanium isopropoxide (99%), glacial acetic acid (99.7%), glycerol (99%), and lead acetate trihydrate (99%) were purchased from Aldrich and used without further processing. Aqueous lead acetate (2.5 mmol) was mixed in a solution of 10 ml ethanol and 8 ml acetic acid. Titanium isopropoxide (2 ml), various amounts of glycerol, and 5 moles of thiourea were subsequently added. The solution was stirred for 24 h and approximately 70% of the solvent was evaporated using a rotary evaporator before spin coating. Thiourea was used as a sulfur source, which prevents the formation of PbS before heating, following the reactions:

$$NH_2CSNH_2 + 2H_2O \rightarrow H_2S(g) + CO_2(g) + 2NH_3(g)Pb^{2+} + H_2S \rightarrow PbS + 2H^+.$$
$$(10.4)$$

The viscous sol was spin-coated to the glass substrate at a spinning rate of 3000 rpm for 30 s. The as-prepared film was transparent, but became dark after annealing for 10 min at 160°C, indicating the formation of nanocrystalline PbS. Four thicknesses of films starting from 6.8 μm were fabricated by successive coating after heat treatment.

Figure 10.7 shows the threshold power $P_{th} = 3.1, 6, 8.5$, and 13.5 mW cm^{-2} for the PbS cluster size of 10 nm plotted against the thickness $d = 6.8, 13.6, 20.4$, and 27.2 μm, respectively. The samples were a thin film of PbS/TiO_2−glycerol with

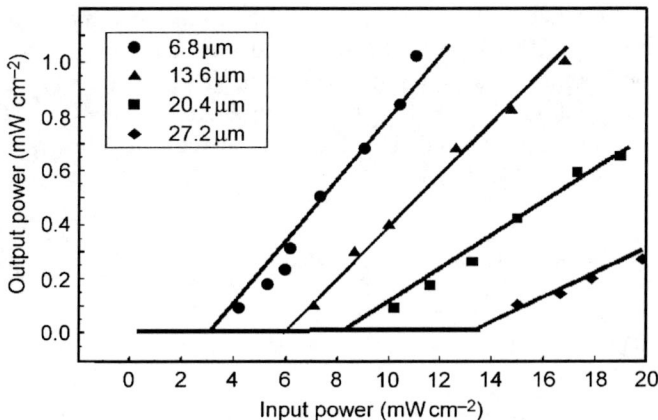

Figure 10.7 The O/I intensity distribution for several thin film PbS/TiO_2 samples on a glass substrate.
Source: After Kang et al. (2004), with permission.

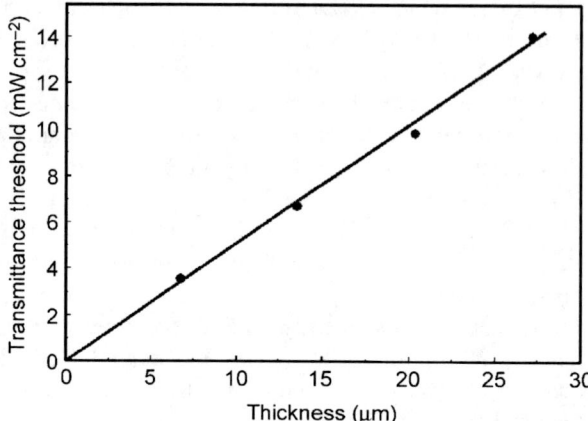

Figure 10.8 Saturation intensity as a function of sample thickness. *Source*: After Kang et al. (2004), with permission.

thicknesses ranging from 6.8 to 27.2 μm on a glass substrate. The key point is that these samples consist of dilute nanoparticles of PbS, having size $a < \Lambda$, coherence length of an electron in the nanoparticles embedded in a TiO_2-glycerol matrix, and essentially with one transition between the ground state E_1 and excited state E_2 in each PbS QD.

Figure 10.8 shows the saturation intensity as a function of sample thickness, allowing a selection of thickness for specific saturation intensity.

Note that the result is linear through the origin. This fact indicates that (a) the total number of transitions is governed by the total number of QDs; (b) various layers of QDs have the nearly same density per unit area; (c) these 10-nm-size QDs are determined by X-ray diffraction, and IR spectra determined by optical means characterize these QDs of PbS.

If distribution of QDs in a layer remains fixed then the total number N_{QD} is proportional to the thickness. This assumption leads to a simple model allowing comparison of various cases where only the excited state can interact with the matrix. Neglecting the spontaneous emission term, with Ω_{12} denoting the transition rate from E_1 to E_2, which is equal to Ω_{21} from E_2 to E_1, and a decay rate Ω_0 due to the interaction between the QD and the matrix, we have the following coupled rate equations, writing $\Omega = \Omega_{12} = \Omega_{21}$,

$$\frac{dn_2}{dt} = (n_1 - n_2)\Omega - n_2\Omega_0 \tag{10.5}$$

$$\frac{dn_1}{dt} = -(n_1 - n_2)\Omega. \tag{10.6}$$

Substituting, $n_2 = C(t)\exp(-\Omega_0 t)$ and $n_1 \approx n_{10}\exp(-\Omega t)$ we arrive at

$$\frac{dC}{dt} = n_{10}\Omega \exp[-(\Omega - \Omega_0)t],$$

resulting in

$$n_2 = n_{10} \frac{\Omega}{\Omega - \Omega_0} \{ 1 - \exp[-(\Omega_0 - \Omega)t] \} \exp(-\Omega t). \tag{10.7}$$

The ratio of

$$\frac{n_2}{n_1} = \frac{\Omega}{\Omega_0 - \Omega} \{ 1 - \exp[-(\Omega_0 - \Omega)t] \}. \tag{10.8}$$

At $\Omega_0 = \Omega$, L'hospital's rule leads to

$$\frac{n_2}{n_1} = \Omega t. \tag{10.9}$$

Saturation occurs at $n_1 = n_2$ thus at a time t such that $\Omega \tau = 1$ defines saturation, where τ is the decay time defined by Ω_0. Therefore, the experimentally determined threshold value P_{th} allows us to find the product $\Omega \tau$ or $\Omega_0 \tau$. It is interesting that as long as we can assume that the channel of relaxation *via* the matrix is only through the excited state E_2, regardless of how large this interaction is, saturation occurs at $\Omega_0 \tau = 1$. The relationship given by Eq. (10.8) maybe further improved using a higher-order iteration, allowing n_1 as well as n_2 to be affected by the relaxation term $n_2 \Omega_0$; however, the account we have presented is sufficient for our present purposes.

Experimentally, the value of P_{th} ranges from 3.1 to 13.5 mW cm^{-2}, a rather moderate power for the threshold, indicating that coupling between the PbS QDs and the TiO$_2$−glycerol matrix is rather low, making it a nearly ideal system for a purely optical switch or other devices. Before we leave the present discussion, we want to point out that the plot of n_2/n_1 versus Ω from Eq. (10.8) represents a well-behaved function with a definite point at $\Omega = \Omega_0$, delineating the onset of saturation. A list of future investigations may involve the temperature effect of the role of phonons and vibronic states and nonradiative decay channels as well as the frequency dependence of this coupling as $E_2 - E_1$ varies with the size of the QD.

For a large particle of PbS, ~ 1 μm in size, which is much greater than $\Lambda \sim$ 10−25 nm, the transmission versus the thickness is highly nonlinear (Kang et al., 2004), because the total transition is summed over the volume of the particle, which is a factor of 10^6 greater than the nanoparticles embedded in a comparable volume. This fact explains the much larger P_{th}. Unlike PbS where electron−hole pairs created by the incident photons of a Nd-YAG laser $\gg 0.4$ eV, the band gap can cascade down *via* phonon interaction in nanoparticles with discrete quantum states, transfer of excitation to phonons is generally not possible.

Although the number of states in a QD is closer to a molecule, there are however multitudes of surface states, mainly due to complexes from bonds between the QD and the matrix. The fraction of complexes increases as the surface-to-volume ratio increases. The linear relationship in Figure 10.8 indicates either a lack of any surface crud, which is quite unlikely, or more likely that the complex involves energies in the UV range, which are not accessible to excitation by the Nd-YAG laser. Therefore, if UV excitation is used, the P_{th} versus thickness maybe better represented by the case without quantized states in the nonlinear threshold—thickness relationship, as for the case of large particle size, leading to a much higher value of P_{th}. Thus, by probing with a different frequency, it should be possible to identify the recombination channels due to the surface complexes.

The difficulty in providing I/O prevents us from taking full advantage of the quantum properties generally expected from nanoscience in relation to electronics and optoelectronics. We can build a quantum system with energy barriers for electronic confinement. However, for phonons, it is a different story. It is almost impossible to isolate a QD from a phonon. Therefore, the use of a sol-gel as well as the possible use of porous silicon mentioned in the previous chapter represents an ideal situation. Sulfur compounds like CdS (Brus, 1994, 1996; Alivisato, 1996) and PbS are also important in this work. The all pervasive degradation, for example, of oil as a lubricant and metal contacts, is caused by oxygen, because oxygen is so plentiful and has such a high electronegativity. I would like to tell the story of how I became convinced that oxygen is the main spoiler. I was taking a London University external B.Sc. during the early 1950s majoring in Physics and Chemistry. My professor showed us how to make small aluminum particles, collected on a filter paper after the final rinse. We were told to put the filter paper in the sun to see what happened. When it dried, the rapid oxidation set the paper on fire! Sulfur is one of the better elements at resisting oxidation, and since the environment is generally not full of sulfur, sulfides are more stable compared to column V elements like P and As, although the nitrides are more stable. Perhaps a lnN QD has a future. Although QDs of CdS, CdSe, and PbS are more stable than most III—Vs, they still need to be encapsulated and protected. Nevertheless, PSi is really my all time favorite, because the possibility of epitaxial silicon growth on top of PSi virtually seals the active nanoparticles from the intrusion of oxygen as well as providing current connectivity. I am not sure whether PSi can ever be developed to the point of general use. At least in principle it possesses the ingredients to serve as a "perfect matrix" for QDs.

10.3   Multipole Electrode Heterojunction Hybrid Structures

Since the introduction of the man-made superlattices and QW structures, the field has taken off with the development of quantum slabs (QS), quantum wires (QWs), QDs, and nanoelectronics in general (Meirav et al., 1990). This rapidly expanding field owes its success to the development of the epitaxially grown heterojunctions and heterostructures originally used to confine carriers in injection lasers (Kroemer, 1963; Alferov, 1965; Hayashi, 1984). Meanwhile, advances in lithography allow potentials to be applied in the nanoscale dimension, mainly with in-plane electrodes, leading to the possibility of quantum confinement without heterostructures (Tsu, 2002, 2003). Actually, quantum states in the inversion layer of field effect

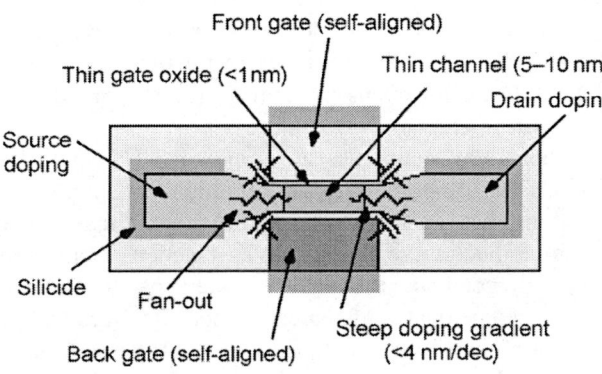

Front gate (self-aligned)

Thin gate oxide (<1nm)

Thin channel (5–10 nm)

Drain doping

Source doping

Silicide

Fan-out

Back gate (self-aligned)

Steep doping gradient (<4 nm/dec)

Figure 10.9 Schematic cross-section of an ideal double-gate FET.
Source: After Wong et al. (1977), with permission.

transistors (FETs), formed by the application of a large gate voltage appeared several years before the introduction of superlattices and QWs (Stern and Howard, 1967). The quantum Hall effect was first discovered in the Si inversion layer (von Klitzing, 1980). Apart from the devices, many fundamental phenomena connected with quantum conductance effects will not be elaborated upon, however, we should point out that most experiments on the Aharonov—Bohm effect and conductance oscillation in a magnetic field, and so on, have already embodied the use of hybrid electrodes (see, e.g., Washburn and Webb, 1992). In this section on the multipole electrode heterojunction hybrid structure (MEHHS), the hybrid structures of heterojunctions and applied potentials *via* multipole electrodes for a much wider variety of structures for future quantum devices are discussed (Tsu, 2002, 2003). The technology required to fabricate these electrodes, to some degree is routinely used in the double-gate devices targeted for improving efficiency of complementary metal oxide silicon (CMOS) devices (Wong et al., 1997).

Figure 10.9 shows a typical double-gate FET. Note that the double-gate structures could very well be the gates needed to confine electrons provided the dimension is below the inelastic mean free path of the channel. We shall see later how similar this structure is to the schemes involving multipole electrodes.

Resonant tunneling *via* man-made double-barrier heterostructures, the resonant tunneling diode (RTD), was introduced 10 years before working devices appeared (Sollner et al., 1983). Quantum cascade lasers (QCLs), developed at BTL under the direction of F. Capasso, incorporating the principles of superlattices appeared almost 25 years after the introduction of the man-made superlattices (Faist et al., 1994). The main advantage of using double-gate CMOS and DG-FETs is to control the electric fields in the active region of extremely short channel devices. More importantly, doping becomes unnecessary, as in RTDs. In spite of the fact that the DG-FET originates from the work done by engineers in silicon technology, the adoption into the mainstream of ULICs may very well be at least 10 years away. On the other hand, it must be recognized that with the ever-decreasing size of integrated circuits (IC) structures, the goal stressed here for the use of heterojunction multipole electrodes for quantum devices may very well be reached *via* silicon technology. In short, research in silicon technology and quantum devices is converging toward each other, but full implementation maybe many years away. A few specific examples are detailed here to stimulate a rapid adoption of a hybrid system for the formation of quasi-discrete states, the resonant states, for quantum devices. From a broad view point, the Si/SiO_2 inversion is a hybrid system.

Let us go into some detail about why multipoles are important. The multipole expansion of an electric field has a very important feature; the higher the multipole, the faster the field falls off with distance. In fact this is essentially the reason why heterojunctions, being neutral, can be much more abrupt than pn-junctions. Sometimes we consider quantum confinement by geometrical boundaries as a separate means of achieving quantum states; however, geometrical confinement cannot take place without band-edge alignment of the heterostructures. For example, a microwave resonator is formed from a section of the waveguide with a geometrical constriction. The constriction is precisely the result of discontinuities, such as the difference in dielectric constants or the difference between an insulator and conductor. Although the origin of confinement maybe ultimately traced to the potential, for the purpose of describing and characterizing the fundamental mechanisms for engineering designs, it is very useful to distinguish between confinement by heterostructures and confinement by multipole electrodes, as well as a hybrid scheme using both.

10.3.1   Examples of Heterojunction Multipole Electrode Hybrid Structures

Next we shall discuss the hybrid system with multipole electrodes applied to a QS formed by heterojunctions, for a much wider variety of structures for use in future quantum devices (Tsu, 2002, 2003). Meanwhile advances in lithography allow potentials to be applied in nanoscale dimensions leading to the possibility of quantum confinement without heterostructures (Song et al., 1998). In principle, multipole electrodes can provide confinement as well as control of symmetry for specific device functions (Tsu and Datta, 2002). Such potentials maybe designed (i) for arbitrary geometry, (ii) to produce softer scattering, and (iii) to be dynamic, e.g., to turn the device on and off. An example of a hybrid system with multipole electrodes applied to a QS formed by heterojunctions is shown in Figure 10.10, taken from Tsu (2002). The difference between the electrodes in Figure 10.10 and the double gates in Figure 10.9 is the multipole nature of the ($\pm$) pair of electrodes. Note that the potentials at the top, $t = 0$, and the bottom, $t = d$, are ideal for confining an electron, but at the center, $t = d/2$, the confinement is much weaker.

The thickness of the QS d should be less than the inelastic scattering length of the material forming the QS, usually no more than a few tens of nanometers. The maximum separation of the ($+$ $+$) pair and ($-$ $-$) pair should be determined by the breakdown field of the material, generally no more than $\sim 10^7$ V cm^{-1}. Figure 10.11 shows a case where the thickness of the slab is much greater than the separation of the ($+$ $+$) and ($-$ $-$) pairs of electrodes. Therefore, for $d \sim 10$ nm, the bottom electrodes should not be placed more than few d lengths away. The requirements that the top $+1$ V and -1 V electrodes must be sufficiently separated, limited by the breakdown voltage, together with the minimum distance

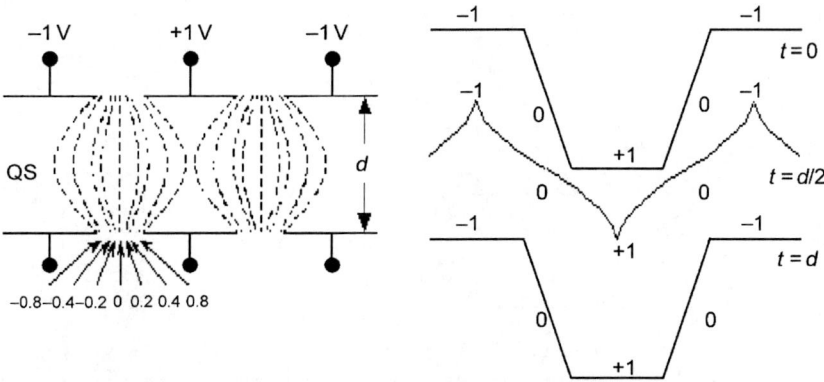

Figure 10.10 Left: the potential inside a QS of thickness d with multipole electrodes on top and bottom of the slab to effect quantum confinement; right: potential versus distance at three levels: $t = 0$, $t = d/2$, and $t = d$. Note that the confining potentials are reduced in the middle of the slab.
Source: After Tsu (2002), with permission.

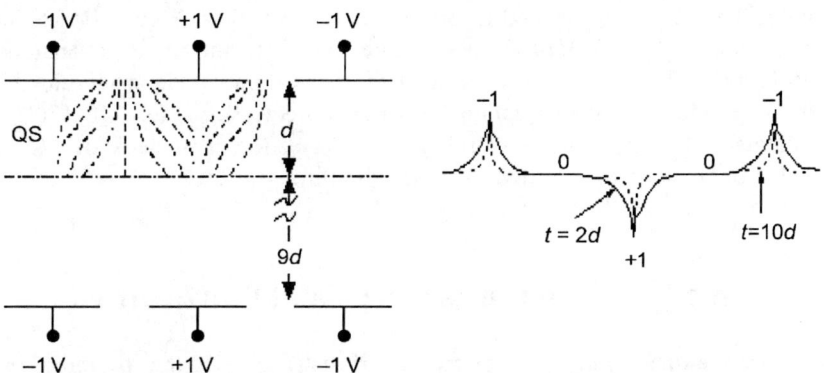

Figure 10.11 Left: the potential inside a QS with multipole electrodes placed at the top and bottom; right: potential versus distance at the bottom of the QS, $t = d$, and "delta-function like" potential spikes versus distance when the electrodes are separated by $10d$. Note that in this case, confinement at $t = d$ almost disappears because these spikes are not really delta functions with the peak position fixed at $+1$ V and -1 V.
Source: After Tsu (2002), with permission.

separating the top and bottom electrodes present a severe technical challenge though not an impossible one. The right side of Figure 10.11 shows the case where the bottom electrodes are placed $10d$ below the top electrodes, the confinement at $t = d$ almost disappears because these potential spikes are not delta functions. The multipole electrodes in Figures 10.10 and 10.11 are similar to most buried gates currently being investigated for highly efficient CMOS-FET. The fabrication is complex but possible. Wong (2002) details an excellent account for the self-aligned

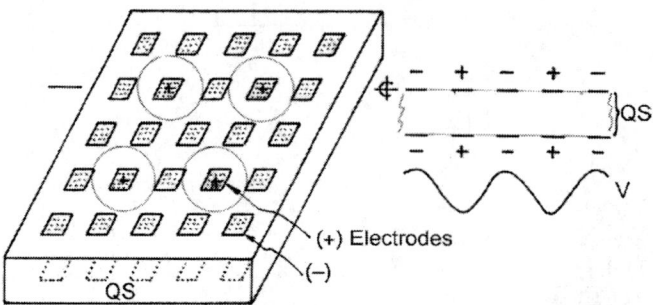

Figure 10.12 A schematic drawing of a 3D-hybrid structure QS with a 2D array of electrodes on top and bottom. The positive and negative electrodes are arranged to produce confinement as QDs. The 3D potential profile is shown on the right, with electrons confined in the minima.

double-gate fabrication process. The major difference between what is presented here and the DG-FET is the multipole nature of the electrodes. It is quite possible that this hybrid confinement scheme would be realized by pursuing high-end silicon technology rather than by those investigating nanoelectronic devices. This last view can be traced to my experience at BTL and IBM, where I witnessed how great a program and effort are needed to bring an idea to reality. Figure 10.12 shows a schematic drawing for a 3D-hybrid structure QS. With the rapid advancement in the resolution of lithography, such structures maybe fabricated, serving as the precursor to a wide variety of quantum dot-field effect transistors (QD-FETs). The main theme is that by using electrodes in a plane, the bottleneck of I/O maybe avoided. Connections to the electrodes maybe through the use of vias connecting contacts in various 2D-like structures.

10.4   Some Fundamental Issues: Mainly Difficulties

In bulk crystalline solids, symmetry manifests itself in the periodic part of the Bloch function, which depends on the wave vector **k** of the Brillouin zone. The symmetry of the man-made quantum state is determined by the symmetry of the heterostructure and the applied potentials. Like atomic and molecular physics, for two electrons occupying a given state, a symmetric state goes with the anti-symmetric spin function, the singlet, while the antisymmetric state goes with the three symmetric spin states, the triplets. The applied potential can change the symmetry resulting in rearrangements of the spin parts. Moreover, determining the boundary conditions due to band-edge off-set, together with the dielectric discontinuity due to the difference between the dielectric functions of the regions of interest, for the QD or QW, constitute a formidable task in computational physics. For example, the dielectric discontinuity was fully included in the treatment of the doping of a QD and the capacitance of a QD in Chapter 8. However, in these treatments it was assumed that the electron wave function goes to zero at the boundary. We know that is not the case because the RTD will never work with this type of simplified boundary condition. We simply point out the involvement. If an electron induces a potential at the dielectric mismatch, there will be an inside and outside effect (outside refers to the matrix), resulting in extra Coulomb energy. However, if the induced electron wave function originates from another site in the matrix, then the Heisenberg exchange term should be included. All these features are quite negligible in QW structures, but are magnified by a large factor in QD structures simply because of the small volume of the active region. When the size is sufficiently reduced, a Poisson distribution replaces Gaussian statistics so that fluctuations from the mean by inadvertent defects become unavoidable and switching results. This is a classical problem associated with random processes, which prevents the implementation of redundancy and robustness. The familiar concept of resistance capacitance (RC) time constant is still applicable, except that capacitance is not simply given by the geometry, and resistance is not given by the lossy part of the transport, but rather includes the fundamental wave impedance of the electrons, which will be treated in the next chapter. The quantum mechanical definition of capacitance in terms of stored energy includes kinetic as well as electrostatic energy. Therefore, capacitance is much reduced in QDs, not only due to the reduction in dielectric constant, but also due to the increased kinetic energy inside the QD. Lastly, we should remember that atomic physics is based on spherically symmetric potentials. With arbitrary symmetry, one needs to examine each individual case (Tsu and Datta, 2002). Nevertheless, as with all theoretical and computational issues, we resort to simple solvable geometry to obtain guidelines for engineering design.

I/O is the most difficult problem for these nanoscale devices (Tsu, 2001a,b). First of all, when one speaks of a voltage applied at a particular region, terminal or contact, an equal potential surface is implied. As size decreases to the nanoscale region, generally only metals may qualify. One cannot rely on doping to reduce the Schottky barrier. For example, solid solubility limits the maximum doping concentration, usually by about 0.1%. Silicon, for example, has 5×10^{22} atoms per cm^3. In a volume space defined by 10 nm linear dimension, there are 50 dopants at the 0.1% doping level. However, for a doping density of 10^{18} cm^{-3}, there is only one dopant in the volume. The doping density fluctuates wildly so that the device cannot function. Therefore, it is safe to say that metallic contact or semimetallic contact must be used to ensure I/O to the active quantum device, if for nothing more than the demands of electrostatics.

Many dramatic results have been claimed in the work of single electron transistor, for example by Reed et al. (1988). First, these results take advantage of the relatively larger size operating at very low temperatures or use scanning tunneling microscopy (STM) probes for smaller devices operating at higher temperatures. Both cases represent benchtop experiments, a longway from being able to apply to the world of electronic devices. Meanwhile there is a steady stream of results and claims involving nanoscale particles in optical applications (Brus et al., 1996). However, these are also far from any real devices. First of all, photons, with their large wavelength, invariably contact many particles in parallel. A parallel system like optical "blinking" is intrinsically unstable. Let us use a very simple argument against the system in parallel. A resistor with two contacts is basically a parallel system. Then why is it stable? It is stable because the resistor is operating in the linear range, the Ohm's law regime, so that the conductance is proportional to the area of the contact. More than 10 years ago, Tsu and Nicollian took up the investigation of tunneling *via* nanoscale silicon particles, ~3 nm in size, between two large contacts (see Chapter 7). Many strange phenomena occurred. First, there was the traditional tunneling *via* resonant state of the Si QDs resulting in a delta-function-like conductance structure, which is to be expected. The charging and discharging of the QD, the so-called Coulomb blockade is also to be expected. Even the hysteresis curve is to be expected from inadvertent trapping of electrons. These trapping sites do not even need to be located inside the QD as long as the potential of the charged traps can affect the energy of the QDs. But, what is most disturbing is the conductance oscillations in time, or more precisely, on/off switching of the conductance caused by the traps. Therefore, the old rule always applies, cleaning up the defects and impurities. Only, because the active number of electrons is many orders of magnitude reduced, the cleaning up is far more stringent and may take a generation or more. More on this will be given in the next chapter.

The MEHHS is ready for possible application. The system that is ready may very well be the electron quantum waveguide, where fabrication techniques for heterostructures are sufficiently mature and high-resolution lithography is sufficiently advanced to allow the formation of this hybrid system.

10.5 Comments on Quantum Computing

In recent years, the research community has feverishly pursued the lure of quantum computing (QC). Basically, the binary bits are replaced by Q-bits, a sort of quaternary bit, which includes putting the phase of the state into the logical gates of a classical computing scheme (Deutsch, 1985; Mullins, 2001). Let me comment on QC using familiar examples. Memory is totally classical, in other words no phase in is involved, because once a bit is stored, a "0" or a "1", a day later it is still a "0" or a "1". This is a similar situation to when a policeman catches a speeder and enters the record of conviction. However, he is likely to have used a phase sensitive scheme, the Magic-Tee (also known as the Michealson interferometer) to catch the speeder. Therefore, QC utilizes the wave nature of electron to serve as the phase sensitive part of the scheme in order to accomplish certain task. I think many people fail to catch on to this idea and this gives rise to great argument. There is no fundamental problem for me when someone plans to use the wave nature for the phase sensitive part of the mission, commonly referred to as processing. However, there are problems in the real system with losses and leakage, no different the reason I use a non-Hermitian operator for an open system. Clearly the idea involving using the Magic-Tee to catch the speeder is an open system! Mathematically a complete set can represent any function. However, because of the need for I/O, devices are not totally isolated but partially open, and one is forced to deal with resonance states instead of eigenstates of the system. Together with unavoidable losses and finite bandwidth, serious problems exist to this day in the implementation of the concept. In real terms, it would be remarkable if someone could assemble 10 terms in a series expansion with any degree of precision, for example, with five significant figures for each term. One can digitalize a signal, but one cannot digitalize an electron wave function because the wave function is normalized. Nevertheless, we mention QC devices because they may become important someday, and more importantly, the use of an electron quantum waveguide with MEHHS maybe a vehicle for achieving the dream of QC. My reading on all this is different. I think QC will never come into common usage, not even with cryptology. However, the concept may find some use in a niche yet to be discovered.

10.6   Recent Activities in Superlattice

10.6.1   THz Sound in Stark Ladder Superlatices

I have discussed the formation of Stark Ladder in Section 1.6 under application of a constant electric field to a periodic system. Basically with a constant field, a constant potential energy difference exists between adjacent cells, a ladder structure for the energy state appears. Recently Beardsley et al. (2010) has succeeded to create coherent ultrafast acoustic phonons using a doped semiconductor superlattice under applied electric field. If the superlattice is biased such that the energy difference between adjacent periods adjacent period exceeds the width of the minibands, localization occurs as discussed in Sections 1.6.1−1.6.3, and electron transition occuring between adjacent wells is phonon assisted as shown by Tsu and Döhler (1975). The theoretical treatment by Glavin et al. (2002) and experimental verification by Kini et al. (2005) definitely established the coherent nature of the phonon involved. Recently, a 50-period GaAs/AlAs SL grown on a 0.4-mm semi-insulating GaAs substrate showed coherent amplification of THz sound (Beardsley et al., 2010).

The device was structured with up-to-date design following previous results of Kini et al. (2005), using 5.9 nm GaAs well with 3.9 nmAlAs barrier, uniformly doped with Si to $n \sim 2 \times 10^{16}$ cm^{-3}. Contact has n^{+} with a 20 nm undoped GaAs spacer to avoid mobility degradation. The top contact is ring shaped to allow optical excitation for experiments. In the dark, the current starts at 50 mV, representing the separation between the contact potential of the emitter and the lowest level of the QW state. As shown in Figure 10.13(B), both Bragg reflection marked B, due to the slight mismatch of the elastic constants of GaAs and AlAs, as well as the coherent folded phonon (FP) with energy close to the gaps in the Γ-phonon dispersion, were excited using the femtosecond pump−probe technique. The optical wavelength of the pump is determined by the eh-resonance of the SL, being ~ 1.6 eV. (Trigo et al., 2006). Note that there is an unspecified gap in the middle of the device shown in Figure 10.13. I think the gap is used as matching for the phonon resonance. In fact I always thought that such a gap may even be used in QCL for optimization.

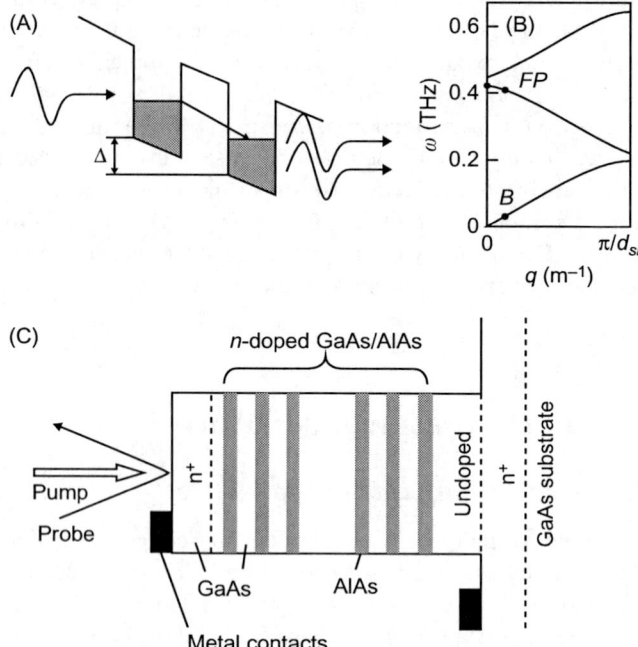

Figure 10.13 (A) Phonon-assisted tunneling between adjacent QWs in an SL by stimulated emission; (B) folded LA in a GaAs/AIAs SL with a period $d = 10$ nm with B: Brillouin scattering and FP: folded phonon; (C) device structure.
Source: After Beardsley et al. (2010), with permission.

Application of an electric field to a weakly coupled semiconductor superlattice gives rise to an increase in the coherent folder phonon, FP, generated by a femtosecond optical pulse. The condition is whenever the start energy $eFd >$ energy of the phonon, in this case , the FP phonon. I used to think that the QCL represents ultimate complexity in the design and fabrication related to research and development involving superlattice. Now I think this work represents a step jump in the sophistication and careful design of the superlattice structure.

10.6.2 Quantum Cascade Laser

Not long after the publication of the first edition of *Superlattice to Nanoelectronics*, I decided if I had the chance to publish a second edition, I would include a section on quantum cascade laser, QCL, developed at BTL under Federico Capasso's direction. But why did I leave it out in the first edition? To answer this I need to explain my state of mind. If you notice that the approach in my book was to emphasize principles, because I thought that once complicated engineering details were included, there would be no line to draw between one device and another. The reader should have noticed that my book is somewhat different from traditional technical books, which concentrate on theory and experimental details. I placed more emphasis on the understanding of the principles involved. When I was at ECD in early 1980s, I often had lunch with Marty Steel, who ran the design group at Ford Motors. Once he asked me how many years it takes to produce an expert in a typical high-tech company. I said, "It takes at least ten or fifteen years." He replied, "Not so! Five years to make an expert, ten years to make a dinosaur!" On another occasion during the APS March Meeting not long before the success of the QCL, someone introduced me to Federico Capasso. He said to me, "Ray, I read quite a few of your papers. You name one and I shall tell you what you said." I was going to flunk him by mentioning a complicated paper he could not possibly have heard of. But I mentioned one he knew, and he became a hero. It entered my mind that Capasso is not only capable of attacking complicated problems, he also has the self-confidence to draw big applause. In fact, this story sums up what it is all about with QCL. I am not discussing the details of how to make a good QCL, because it is not so different from making a good MOSFET. Engineering progress is made in small steps, one at a time, after a start with a big bang or a little puff, but often, these little steps may last more than a century. Shortly after the first publication in *Science* by Faist et al. (1994), I summarized their ingenious success story in *Nature* (Tsu, 1994). Parts of my account are summarized here.

The QCL is a III—V semiconductor multiple QW structure with layers of gallium indium arsenide and aluminum indium arsenide. Figure 10.14 shows one of the 25 active regions, showing injector from the right and collector from the left,

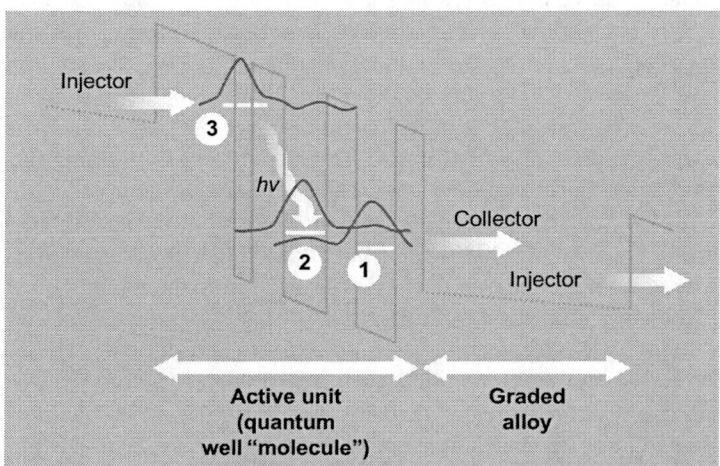

Figure 10.14 Conduction-band energy of one of the 25 active regions in the QCL by Faist et al. (1994) in *Science,* showing the active region consisting of three coupled QWs, undoped. The collector—injector as right and left contacts for the active region are doped to $\sim 10^{17}$ cm^{-3} and graded to be almost flat at the operating voltage. The overall device contact is heavily doped but contain a lightly doped buffer to prevent impurity scattering from reaching the active region.
Source: After Tsu (1994), with permission.

with three double-barrier quantum wells, DBQW, serving the role of a three-level laser. Level 3 with a narrow well is the upper level with electrons injected from the left contact. Level 2 is separated from level 3 by a relatively narrow well with the bottom level designed for substantial coupling with photons, such that $E_3 - E_2 = \hbar\omega$, the designed frequency for the QCL. Another QW with a level E_1 is also closely coupled to the lower level E_2. To prevent accumulation of electrons at the bottom level, E_1 is designed to couple strongly with the collector by having a relatively thin barrier between the collector and the well 1. In fact the design of the graded alloy collector—injector is really very sophisticated, because it is sawtooth without voltage, but almost flat across at the operating voltage. This design is to ensure that at operating point, little voltage appears across the graded region to minimize power consumed in the collector—injector section. In fact it is not just a matter of efficiency; without such design, electrons would be accelerated to the next cell, smearing the Fermi function so that the QCL would not operate at all. The first successful design produced 30 mW peak power at $\sim$70 THz.

Both schemes, involving cascading with a superlattice, by Kazarinov and Suris (1971) as well as a patent by Esaki and Tsu (1977), in fact would not work because of domain formation. The success at BTL was due to breaking the long staircase into segments for control. Therefore, we can see that knowledge, once known all seems so obvious. It was not done then, otherwise both the Russians and we would have onto it too. Nevertheless, there is a huge lesson to be learned about the direction of the future in superlattice research. Separating the whole superlattice into pieces was precisely what Esaki and I did with the DBRT. But we did it for the wrong reason, as discussed in Chapter 2, to counter the objection of Ian Gunn that our negative resistance was not due to Bragg reflection at the Brillouin zone boundary, rather due to domain oscillation as in the Gunn device. I really think we should all recognize that many new ideas were not cleverly conceived, but instead were discovered by accident or for the wrong reason as with the discovery of the type-III superlattices. By judging the rapidity of laser developments, injection lasers as well as gas lasers, the complexity of the QCL is the main reason for the slow development. Recently, University of Wisconsin-Madison team, Botez et al. (2010) and Shin et al. (2009), demonstrated the improved efficiency by using higher barrier to prevent leakage from the top of the confining barriers with Al up to 75% instead of 50%.

Before I close with this section, I would like to point out why we do not put the component active regions in parallel instead of in series. Since the wavelengths of the photons are so much greater than the de Broglie wavelength of electrons, in order to have cavity feedback, the QCL must be at least half wavelength thick. And that calls for many active components in series. In the last chapter of this book, I shall discuss systems that components should be placed in parallel.

10.7 Graphene Adventure

My fascination with graphite, the precursor of graphene, started when I was taking a sabbatical in Brazil at Instituto de Fisica Gleb Wataghin, Universidade Estadual de Campinas, during 1977—1978 working on graphite. I discovered after I arrived that there were lots of activities in energy-related research. My colleague Carlos Luengo, Director of Energy Research, convinced me to take part in the coal project. After spending a year at the Max Planck Stuttgart doing Raman scattering, naturally I got involved with Raman scattering with coal, good coal with lots of carbon, and bad coal with lots of Kaolin, combination of silicates and aluminates. To my surprise, the Raman spectra consist of 1370 cm^{-1} from amorphous diamond and 1606 cm^{-1} from graphite. We found that the ratio matched the carbon-to-mineral ratio used for the classification of the quality of coal. Furthermore, using a sample of single crystalline graphite purchased from Union Carbide, we discovered that the E_{2g} is split into 1590 cm^{-1} and 1627 cm^{-1} (Tsu, 1978). Subsequently we discovered that this splitting even occurs in benzene; therefore, the assignment cannot be D_{6h}. It was suggested that Raman is fast so that momentary electronic configuration is actually D_{3h}. I showed this result to Peter Sorokin, who thought that we might be able to use resonant Raman to slow down the process in order to reveal D_{6h}. We did not do the experiment because we were unable to find a UV laser line. Although the 1371 cm^{-1} line only showed up in polycrystaline graphite, the second order at 2742 cm^{-1} always showed up. The ratio between the E_{2g} and second order served to classify the degree of polycrystallinity quite well.

What was most surprising is the 13 orders of magnitude change of conductivity of coal on annealing between 300°C and 2000°C under vacuum. In fact, the greatest rate of change is between 600°C and 700°C with 12 orders of magnitude change of conductivity, due to conductivity percolation (Hernandez et al., 1982). That was the largest range of change of anything I have ever experienced in my life. It is interesting to reveal that I hired Jesus Hernandez in 1980 working for Stan Ovshinsky at ECD. Within a year, he has obtained 8 orders of magnitude change in conductivity by annealing amorphous silicon using various gas adsorption, ranging from H_2 to O_2, even with N_2 using a hot filament (Tsu et al., 1987). It seems that all of a sudden I have developed interest in graphene. Actually, back in my mind, I have been thinking of ways to develop graphite-related superlattices for quite some time.

Recently, Yong Zhang and I published our DFT computation on Gr/Si superlattice (Zhang and Tsu, 2010), shown in Figure 10.15 is the DFT computed structure. The silicon is located in the rings of the carbon as one would have guessed intuitively. Figure 10.16 shows the band structures of graphene (A) and the Si layer (B). Note that the Fermi level in graphene is at the k point with a zero energy gap, commonly known as the Dirac point in the π-bonds of the p_z orbitals. However the Fermi level in Si passes through the p-orbital, the conduction band, therefore the silicon sheet is metallic.

The band structure of Gr/Si superlattice is shown in Figure 10.17, indicating the coupling is relatively weak. The linear dispersion of graphene, a property of symmetry, is basically preserved (Slonczewski, 1958). E_F moves to 234 meV above the k-point of graphene due to electron transferred from the Si to graphene, while there is a downward shift relative to Si. Thus, this type of superlattice, lacking a miniband, is quite similar to the Si/O superlattice treated in Chapter 6: this is very different from the conventional superlattices, which are dominated by NDC and Bloch oscillation involving electrons in the minibands.

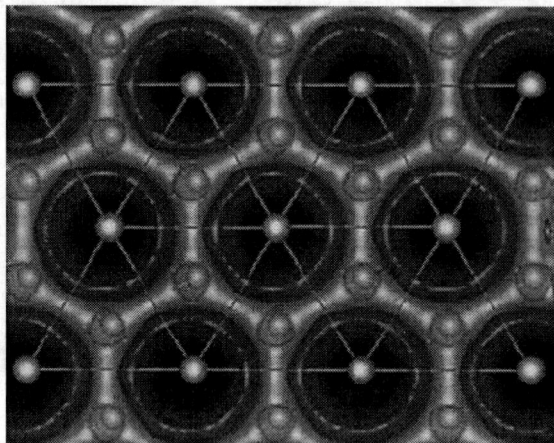

Figure 10.15 Atomic structure of Gr/Si superlattice computed with DFT, with Si atoms on top of C atoms, with the charge density of Si, circles shown in dark gray (blue in the web version) and that of C, rings shown in light gray (red in the web version).
Source: After Zhang and Tsu (2010), with permission.

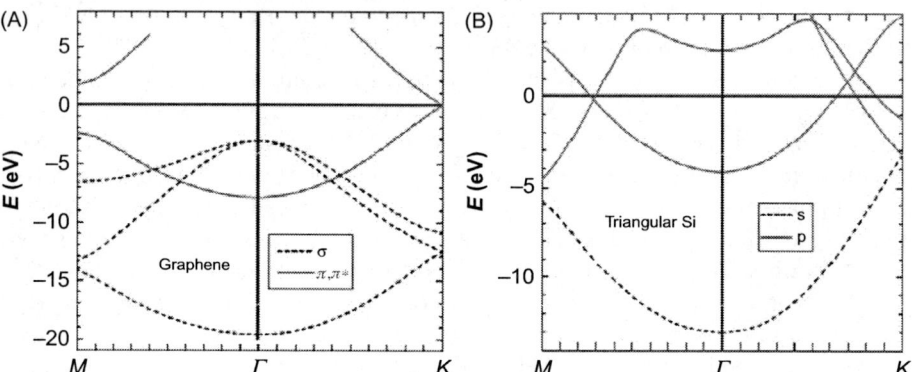

Figure 10.16 Band structure of graphene (A), and Si layer (B), with the Fermi level set to zero.
Source: After Zhang and Tsu (2010), with permission.

Figure 10.18 shows the DFT computed adsorbed Cu on graphene taken from Khomyakov et al. (2009). This is an important work giving many adsorbed metals from Al to Pt. The importance of this work is because metal is not only used for contacts to graphene, but also as catalysts such as Cu, Ni, Ta, etc. (Li et al., 2009; Cao et al., 2010), essential for nucleation of graphene in almost all epitaxial growth. It was shown that more than a single layer of graphene are involved in up to 5% of the deposited area, covering a dimension of several centimeters, has more than a single layer of graphene. This is a very significant development toward implementation of graphene-based materials for electronic and optoelectronic applications. Personally, I consider most of the devices fabricated with flakes obtained by exfoliation such as Li et al. (2009) no more than a bench-top demonstration. And it was the fabrication facilities at major laboratories such as IBM, Lin

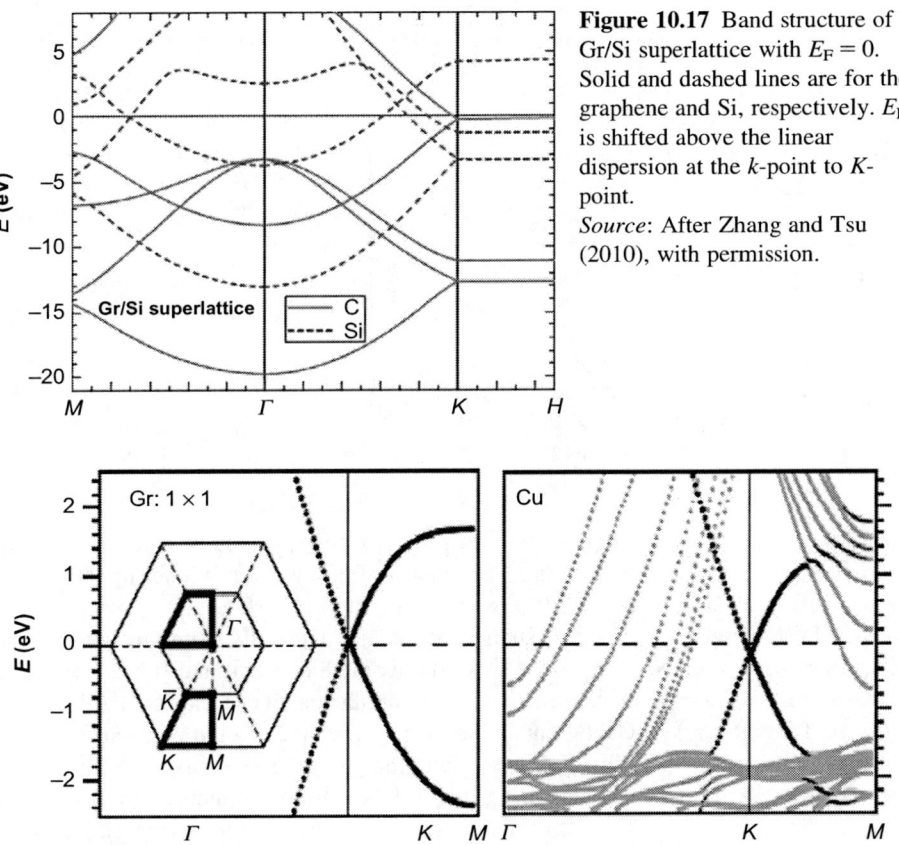

Figure 10.17 Band structure of Gr/Si superlattice with $E_F = 0$. Solid and dashed lines are for the graphene and Si, respectively. E_F is shifted above the linear dispersion at the k-point to K-point.
Source: After Zhang and Tsu (2010), with permission.

Figure 10.18 Band structure of graphene adsorbed on Cu. The darkness indicates the amount of p_z character. Note that Cu adsorption moves up the Fermi level slightly.
Source: After Khomyakov et al. (2009), with permission.

et al. (2009), using specially designed mask and device configuration that allowed some degree of success. I consider this type as typical benchtop demonstration hardly in the mainstream of device research and development.

I want to close this section by showing the tool needed to distinguish 1 ML from 3 ML shown in Figure 10.19. I think the point of significant development in graphene-related research using metal catalysts for large area deposition and Raman for distinguishing single monolayer has arrived, particularly when the concept of superlattice is adopted.

10.8 Summary

The cold cathode is a device for electron field emission replacing the hot filament in vacuum electronics. High-power high-frequency devices such as the TWT are

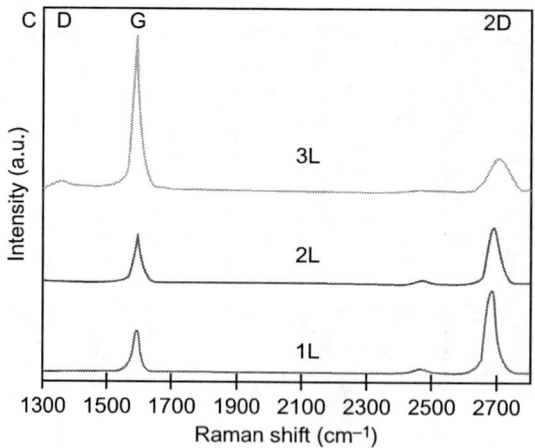

Figure 10.19 Raman spectra of 1, 2, and 3 ML of CVD graphene on Cu, with 532 nm laser. There shows some shifts, however, it is the intensity ratio that separates the layers of different ML.
Source: After Li et al. (2009), with permission.

still the backbone of the high-power amplifier. Recently, a new type of cold cathode has come on the scene: field emission with resonant tunneling (FERT), a scheme that involves placing a QW structure at the surface of a semiconductor with a fairly low work function such as the GaN. The QW structure consists of a GaAlN barrier, a GaN well, and a second barrier, the vacuum. With a field at the surface, the vacuum is transformed into a triangular barrier at the surface. Electrons emerge from the $n^+ - GaN$, tunnel resonantly into the vacuum providing a high current, as well as lowering the work function, because the quantum state is flat across the potential profile allowing the effective work function to be lowered by $\sim E_1/2$; where E_1 is the quantum state involved. In fact, the device outperformed the design by a huge margin, giving a factor of 5 lower electric field, because it was found that the space charge in the well gives rise to additional lowering of the work function, by eV_{SC}, where V_{SC} is the potential from the space charge ρ_{SC}. This is another typical example of research. One should not hesitate about doing new research or finding complete understanding, because this is hardly ever achieved even after a given device has been in use for a long time. I spent several years at Energy Conversion Devices, Inc. when we were working closely with Japanese scientists and engineers from Sharp. I discovered that the Japanese would go ahead and launch a given device before we in the USA would consider it fully field-tested. To my mind, there is no such thing as complete understanding. I remember that PbS was the backbone of a near-infrared detector, and that HgCdTe was used further into the infrared. Putting PbS in a sol-gel provided an efficient luminescence material embedded in a highly porous medium that allowed decoupling of phonons as well as serving as a vacuum barrier for PbS in quantum confinement. We found several orders of magnitude reduction in absorption saturation when the particle size of PbS below 10 nm becomes a QD. This work opened the door to a whole class of schemes, such as embedding CdSe in porous silicon for nonlinear optics at low power! What happens results from each QD having only one quantum state. This is not achievable with QWs because quantum confinement in one dimension leaves a large 2D density of states in the transverse degree of freedom. Optical transition still involves nearly the same number of oscillators, while in QD, being a zero-dimensional system, optical transition involves only one state! I asked an organic chemist why ethylene-glycol is so good in preserving QDs such as CdSe. The answer is that these organic liquids are covalent, which prevents oxygen from getting in. This gives us something to think about when looking for a matrix to serve as a passivation agent.

Let us again emphasize that multipole potentials are short-ranged. If you have any doubts just remember that the radiation field in the near field is short-ranged and consists of multipoles. In fact, many people, even without having a clear understanding of what a multipole field is, are capable of offering designs for electrodes, basically creating higher-order multipole fields! Coupling a hybrid structure with heterojunction and multipole electrodes seems to present the future of quantum devices. The best part is the fact that there is no problem with I/O, because the contact at high current can still be planar across a planar heterojunction with multipole electrodes on the planar surface. In fact what I am suggesting is nothing more than reconfiguration of the FET into a structure with a double-gate in the planar surface instead of one on the top surface and a second one below the planar surface. Actually there can be two pairs, with one pair on the top and bottom surface of the QS, and a second pair on the top surface of the QS. Figure 10.9 illustrates a possible design of a hybrid heterostructure of 3D QD-FET. Note that of all the benchtop success stories in nanoelectronics with QDs, the connections to the active region are invariably *via* contacts lying in a plane that includes this structure. I predict that the forerunners of quantum nanodevices will all be connected with contacts in a plane. There seems no other way of implementing the I/O. I have added two subsections: one on the recent perfection of coherent THz sound from superlattice with phonon-assisted tunneling when neighboring wells are localized; and a long-overdue section on the QCL. Five years ago I thought that the subject was too complex in engineering design without realizing that using the quantum structure in component fashion offers the new direction in applications involving the basic components, the DBQW. I shall elaborate on this point in the closing chapter of this second edition. A section on graphene adventure has been added. I stumbled onto the discovery that coal is basically graphite intercalated with Kaolin molecule consisting of aluminates and silicates. Depending on the fraction of Kaolin, coal can be hard like rock to something easily crumbled by rubbing with one's feet. I knew that intercalation of graphite cannot work because it is not different from stuffing things into a book already bound. When the bandwagon for graphene started to roll, I was tempted to enter the race. It was during the APS March Meeting in Pittsburgh when I realized the magnitude of the graphene craze. Upon returning, I started a few depositions of carbon without silicon. I secured funding on developing ways to deposit large area SiC. The method we have adopted involves using C(60) as source for carbon with Si from e-beam on AlN substrate or plain Si wafer. I asked my students to try using C(60) alone without Si using various substrate varying from sapphire to Si wafer. Well, the first try resulted in no deposition at all, particularly on graphite substrate. I called my son David, who worked as a postdoctoral fellow under Bob Chang at Northwestern, who worked extensively on diamond. David told me that the only way one can deposit carbon onto graphite substrate is either using e-beam damage, or simply rubbing pencil over the graphite surface. Then I remembered Dick Rutz, who succeeded depositing AlN on tungsten, more than 30 years ago. He told me that Ni or Ta also worked. All of a sudden the picture cleared for me after I got hold of a paper on the deposition of graphene on Cu. I want to stop here by repeating what I have been saying. There is really no such thing as a stroke of genius. Ideas come and go, but big success is really very rare. We need to remind ourselves the old saying that progress is made by climbing on each other's shoulders.

References

Alferov, Z.I., Kazarinov, R.F., 1965. USSR Patent 181,737.

Alivisato, A.P., 1996. J. Chem. Phys. 100, 13226.

Beardsley, R.P., Akimov, A.V., Henini, M., Kent, A.J., 2010. Phys. Rev. Lett. 104, 085501.

Binh, V.T., Adessi, C., 2000. Phys. Rev. Lett. 85, 864.

Binh, V.T., Semet, V., Dupin, J.P., Guillot, D., 2001. J. Vacuum Sci. Tech. B19 (3), 1044.

Botez, D., Helin, C., Qingkai, Yu., Jauregui, L.A., Xuesong, Li., Weiwei, C., et al., 2010. Proc. SPIE 7616, 76160N.

Brus, L., 1994. J. Chem. Phys. 98, 3573.

Brus, L., 1996. Phys. Rev. B53, 4649.

Brus, L.E., Efros, Al.L., Itoh, T., 1996. J. Lumin. 70, 1.

Cao, H., Yu, Q., Jauregui, L.A, Tian, J., Wu, W., Liu, Z., et al., 2010. Appl. Phys. Lett. 96, 122106.

Deutsch, D., 1985. Proc. R. Soc. Lond. A400, 97.

Esaki, L., Tsu, R., 1977. US Patent 4.163.238.

Faist, J., Capasso, F., Sivco, D.L., Sirtori, C., Hutchinson, A.L., Cho, A.Y., 1994. Science 264, 533.

Glavin, B.A., Kochelap, V.A., Linnik, T.L., Kim, K.W., Stroscio, M.A., 2002. Phys. Rev. B 65, 085303.

Hayashi, I., 1984. Heterostructure lasers. IEEE Trans. Electron Dev. ED-31, 1630.

Hernandez, J., Calderon, I., Luengo, C., Tsu, R., 1982. Carbon 20, 201.

Kang, K., Daneshvar, K., Tsu, R., 2004. Microelectron. J. 35, 629.

Kazarinov, R.F., Suris, R.A., 1971. Sov. Phys. Semicond. 5, 163.

Khomyakov, P.A., et al., 2009. Phys. Rev. B79, 195425.

Kini, R.N., Kent, A.L., Stanton, N.M., Henini, M., 2005. J. Appl Phys. 98, 033514.

Korotkov, A.N., Likharev, K.K., 1999. Appl. Phys. Lett. 75, 2491.

Kroemer, H., 1963. Proc. IEEE 51, 1782.

Li, X., Cai, W., An, J., Kim, S., Nah, J., Yang, D., et al., 2009. Science 324, 1312.

Lin, Y.M., et al., 2009. Nano Letters, 9, 422.

Meirav, U., Kastner, M.A., Wind, S.J., 1990. Phys. Rev. Lett. 65, 771.

Mullins, J., 2001. IEEE Spectrum, 38, 42−49.

Reed, M.A., Randall, J.N., Aggarwel, R.J., Matyi, R.J., Moore, T.M., Wetsel, A.E., 1988. Phys. Rev. Lett 60, 535.

Semet, V., Binh, V.T., Tsu, R., 2008. Microelectronic J. 39, 607.

Semet, V., Binh, V.T., Zhang, J.P., Yang, J., Khan, M.A., Tsu, R., 2004. Appl. Phys. Lett. 84, 1937.

Shin, J.C., et al., 2009. Electron. Lett. 45, 741.

Slonczewski, J.C., Weiss, P.R., 1958. Phys. Rev. 109, 272.

Sollner, T.C.L.G., Goodhue, W.D., Tannenwald, P.E., Parker, C.D., Peck, D.D., 1983. Appl. Phys. Lett. 43, 588−590.

Song, A.M., Lorke, A., Kriele, A., Kotthaus, J.P., Wegscheider, W., Richler, M., 1998. Phys. Rev. Lett. 80, 383−385.

Stern, F., Howard, W.E., 1967. Phys. Rev. 163, 816.

Trigo, M., Eckhause, T.A., Reason, M., Goldman, R.S., Merlin, R., 2006. Phys Rev. Lett. 97, 124301.

Tsu, R., Hernandez, J., Calderon, I., 1978. Solid State Comm. 27, 507.

Tsu, R., 1994. Nature 369, 442.

Tsu, R., 2001a. ECS Proc. 2000−2028, 167.

Tsu, R., 2001b. Nanotechnology 12, 625.

Tsu, R., 2002. In: Cahay, M. (Ed.), Proc. ECS Int. Symp. Adv. Lum. Material & Quantum Confinement, Philadelphia, May 13−14.

Tsu, R., 2003. In: Dutta, M., Stroscio, M. (Eds.), Adv. Semiconductor Heterostructures. World Scientific, Singapore, pp. 1159−1171.

Tsu, R., Datta, T., 2002. ICPS Edinburgh, UK.

Tsu, R., Döhler, G., 1975. Phys. Rev. B. 12, 680.

Tsu, R., Martin, D., Hernandez, J., Ovshinsky, S.R., 1987. Phys. Rev. B. 35, 2385.

Tsu, R., Greene, R.F., 1999. ECS Solid-State Lett. 2, 645.

von Klitzing, K., Dorda, G., Pepper, M., 1980. Phys. Rev. Lett. 45, 494.

Washburn, S., Webb, R.A., 1992. Rep. Prog. Phys. 55, 1311.

Wong, H.-S., 2002. IBM J. Res. Dev. 46, 133−168.

Wong, H.-S., Chan, K., Taur, Y., 1997. IEDM Tech. Digest 427.

Yu, Y., Greene, R.F., Tsu, R., 2002. IJHSES 12, 1083−1100.

Zhang, Y., Tsu, R., 2010. Nano Res. Lett. 5, 805−808.

This page intentionally left blank

11 Quantum Impedance of Electrons

In circuits, particularly linear systems, impedance is a universal concept that characterizes input/output (I/O). For waves, an additional parameter, the propagation constant, together with the impedance, characterizes the operation of a system. Since electrons are waves, they should also be represented by these two parameters, propagation constant and impedance. Nevertheless, electromagnetic waves are considered to be "true waves" because the energy—momentum relation is linear, allowing a wave packet to be maintained at all times, whereas for electrons, a Gaussian packet spreads because of the dispersion in nonlinear $E-k$. (See, e.g., any book on quantum mechanics.) I have had several discussions on this point, particularly with those working on the popular subject of quantum computing. The prevailing view is that the electron, perhaps, is not really a wave. With this possible point of fundamental controversy, nonetheless I present my derivation of the impedance of electrons.

11.1 Landauer Conductance Formula

Sometime ago, almost as long ago as I first tackled the problem of resonant tunneling *via* a finite superlattice (Tsu and Esaki, 1973), I noticed that the conductance consists of discrete components that depend on the number of longitudinal modes, the quantum states in a quantum well (Chapter 2), as well as the transverse degree of freedom. In the Tsu—Esaki expression for the resonant tunneling, integration was performed over the transverse degree of freedom first, noting that the two-dimensional density of states (2D-DOS) for an unbounded case is simply $m*/\pi\hbar^2$. Let us instead integrate over the longitudinal direction first, dk_1 or dE_1. Defining the function $F(E) \equiv 2\sum_t[1 + \exp(E + E_t)/k_B T]$, then the net tunneling current between two contacts becomes (Mitin et al., 1999)

$$I = \frac{2e}{L}\sum_t\sum_{k_1}\frac{1}{\hbar}\frac{dE_1}{dk_1}\{F(E + eV - E_F) - F(E - E_F)\}. \tag{11.1}$$

With $T \to 0$ and $V \to 0$, $F(E + eV - E_F) - F(E - E_F) \to eV\, \partial(E_F - E)$, the conductance $G = \partial I/\partial V$ from Eq. (11.1) becomes the Landauer's conductance formula

$$G = 2G_0 \sum_t |T|^2(E_F,\, E_t), \tag{11.2}$$

Superlattice to Nanoelectronics. DOI: 10.1016/B978-0-08-096813-1.00011-4

where the sum over the transverse degree of freedom without confinement should have an extra factor of $m*/\pi\hbar^2$, as in Tsu and Esaki (1973) in which the conductance per spin, $G_0 = e^2/h = 38.6\ \mu S$, with its inverse $Z_0 = 25.9\ k\Omega$. The main conceptual difference between the conductance and the resonant tunneling formulas lie in the term $|T|^2$, which is broad compared with the sharp Fermi function in the former and is almost a δ function for the latter. There are issues raised by Landauer (1957, 1970) concerning the assumption of zero reflectance and by Datta (1995) concerning the potential drop at the contact interface. Both will be discussed more in detail after the section on the wave impedance of an electron.

11.2   Electron Quantum Waveguide

Facing the serious I/O problem of quantum devices (Chapter 7), I thought that the electron quantum waveguide (EQW) from an application point of view has at least no serious I/O problem. But before we discuss in more detail the issues involved, a more in-depth account of EQW (Tsu, 2003) is presented because there are fundamental issues regarding contact conductance and fundamental wave impedance. Since the subject is so intimately connected with the wave impedance and characteristic impedance of a photonic waveguide, some conventional waveguide aspects will first be reviewed. Stratton (1948) introduced the concept of the ratio defining a wave impedance,

$$\eta = |\mathbf{E}|/|\mathbf{H}| = |\mathbf{E}|^2/2|\mathbf{S}|, \tag{11.3}$$

with $\mathbf{S}$ being the Poynting vector. Stratton, referring to Schelkunoff (1938), stated: "Impedance offered by a given medium to a wave is closely related to energy flow."

The electromagnetic wave impedance in an unbounded region is given by $\eta_0 = \sqrt{\mu/\varepsilon}$. In free space with $\mu = \mu_0$ and $\varepsilon = \varepsilon_0$, $\eta_0 = 377\ \Omega$ and the characteristic impedance in free space $Z_0 = \eta_0$. This is not true for a waveguide. The wave impedances for transverse electric (TE) and transverse magnetic (TM) are η_{TE} and η_{TM}, having the specific forms $\eta_{TE} = \eta_0/\kappa$ and $\eta_{TM} = \eta_0\kappa$, with $\kappa = \sqrt{1 - k_c^2/k_0^2}$, in which the cutoff wave vector $k_c^2 = (m\pi/a)^2 + (n\pi/b)^2$ and $k_0 = 2\pi/\lambda_0$, where λ_0 is the free space wavelength. Note that $\eta_{TE}\,\eta_{TM} = \eta_0^2$. At $k_c \sim k_0$, $\eta_{TE} \to \infty$, but $\eta_{TM} \to 0$. Thus the wave impedance depends on the geometrical boundary conditions as well as the field configurations. However, this wave impedance is not the same as the characteristic impedance in circuits (Marcuvitz, 1951). The problem arises because at a fixed location along the length of the waveguide, the fields in the transverse direction depend on the location in the transverse plane. Using an appropriate averaging of the fields over the transverse plane in terms of the crest vales, the voltage and current can now be definitively defined and for both TE and TM modes, the characteristic impedance Z (TE or TM) $= (a/b)\ \eta$ (TE or TM),

following Reich et al. (1953). For $a = b$, the definition of the characteristic impedance Z and the wave impedance η are the same but not the same as η_0. Since the
propagation constant along the guide $k_z = \sqrt{k_0^2 - k_c^2}$, when $k_0 < k_c$ the propagation
k_z is imaginary and the guide is cut off. This is why k_c is called the cutoff wave
vector.

The I/O of conventional microwave waveguides is not child's play. Many great
minds were involved in the excitation, coupling, and matching of microwaves using
waveguides during World War II, mainly conducted at the MIT Lincoln Laboratory
and Bell Telephone Laboratories. Let us discuss how we launch a wave into a
waveguide. For simple visualization we use a laser source and fiber optics as
a guide. The laser consists of a cavity, with a set of Bragg reflectors, usually with a
reflectivity R close to 1, forming a high Q Fabry–Perot resonator. The reason we
use a high Q cavity is because stimulated transition is proportional to the photon
intensity. A higher Q leads to a higher intensity and thus higher efficiency. Since
reflectivity $R = (Z_i - Z_0)/(Z_i + Z_0) \sim 1$, and because the impedance of the laser cavity $Z_i \gg Z_0$, the transmission coefficient $T \sim 0$. If $T \sim 0$ at the source–guide interface, how does the excitation enter the guide? The answer is provided by the
definition of Q, being the number of cycles the excitation lasts if the source of the
excitation is turned off. Since the field inside the laser cavity is so high, even with
$T \ll 1$, the "leak-out" rate is sufficient to ensure that a significant amount of power
flows into the waveguide. The reason we go through all this to explain the excitation of optical waveguide is to lay the foundation for discussion of the excitation of
the electron waveguide. In a typical microwave waveguide, a matching section,
$E–H$ tuner, is located near the source–guide interface to tune out any mismatch. A
matching network like the $E–H$ tuner is usually nothing more than a section of a
waveguide with adjustable length, or simply put, a resonating section. This, inci-
dentally, is similar to the electron case, where a section consists of at least two barriers with the length of the section being adjustable. We shall explain this point in
more detail later. We want to mention briefly that in EQW it is not necessary to
adjust the length, because an applied voltage can change the de Broglie wavelength, and therefore, effectively change the length. *I want to stress that the "contact" is far from reflectionless for electromagnetic waveguides; rather, reflectivity
is near unity.*

Next, let us take the electron waveguide as shown in Figure 11.1. Following
Tsu (2003), the propagating wave vector k_z including the potential energy eV is
given by

$$k_z^2 = \frac{2m_e}{\hbar^2}(E + eV) - k_{t,nm}^2. \tag{11.4}$$

The transverse momentum vector $k_{t,mn}$ at the mode (m, n) is given by

$$k_{t,nm}^2 = \left(\frac{m\pi}{a}\right)^2 + \left(\frac{n\pi}{b}\right)^2 \equiv k_c^2, \tag{11.5}$$

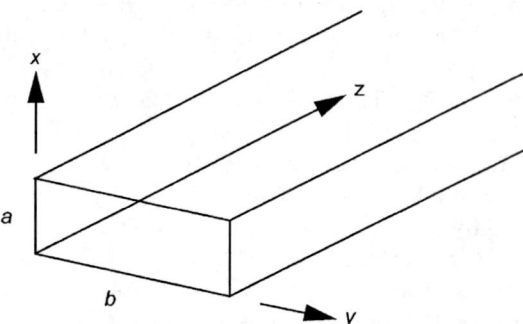

Figure 11.1 Section of an electron waveguide embedded in a potential barrier on all sides with coordinates shown. The applied potential V is such that it is 0 for $z < 0$ and V for $z > 0$.

and $E_{t,mn} = (\hbar^2 k_c^2)/2m$, we see that the transverse energy is nothing other than the energy at the cutoff propagation vector, k_c in the usual waveguide case. The DOS for a cross-sectional area A is

$$\text{DOS} = \frac{1}{A} \sum_{n,m} \int_0^{k_{zM}} dk_z, \tag{11.6}$$

where k_{zM} is the maximum value for a given set (m, n), E, and eV. At $T = 0$, $E = E_F$ for

$$\frac{2m_e}{\hbar^2}(E_F + eV) < k_c^2, \tag{11.7}$$

k_z is purely imaginary and propagation is not possible. Counting only the propagating modes, including two spins, the current density is

$$j = \frac{2e}{2\pi A} \sum_{n,m} \int_0^{k_{zM}} \frac{1}{\hbar} \frac{\partial E}{\partial k_z} dk_z,$$

with $E = E_F$, the current,

$$I = \frac{2e}{h} \sum_{n,m} (E_F + eV) - E_{t,mn}. \tag{11.8}$$

At $m = n = 0$, $I_{00} = 2(e/h)(E_F + eV)$, and for $eV_{00} + E_F > 0$, $G_{00} = (\partial I_{00}/\partial V) = 2(e^2/h) \equiv 2G_0$; at $m = 1$, $n = 0$, $I_{10} = 2(e/h)(E_F + eV) - E_{t,10}$, and for $eV_{10} + E_F > E_{t,10}$, $G_{10} = (\partial I_{10}/\partial V) = 2(e^2/h) \equiv 2G_0$, etc., resulting in $G = G_{00} + G_{10}$, continuing to the general case of (m, n), $G = \partial I/\partial V$ is

$$G = \sum_{n,m} 2G_0 \theta(E_F + eV - \hbar^2 k_{t,nm}^2/2m), \tag{11.9}$$

in which θ is the unit step function, having a series of steps depending on how many modes (m, n) are included. With a negative sign for e, $+eV$ becomes $-eV$ in Eq. (11.9). The factor of 2 in front of G_0 is for the two spins. Thus for the spin-polarized case, there should be $G_0(+1/2)$ and $G_0(-1/2)$ without the factor 2. It is important to recognize that for single mode operations, only one step in G appears, depending on the condition given by Eq. (11.7). The origin of these extra modes is due to the inclusion of modes (m, n) coming from the incident electrons that have transverse energy. In free space, we simply take $a = b = \infty$, then $G = G_0 \theta \ (E_F - eV)$, which ensures that the potential of the waveguide is below that of E_F. Otherwise no transmission is possible from the source of electrons if the energy range from $E = 0$ to E_F. It is important to recognize that an incident electron with transverse energy can enter the waveguide, but without transverse—longitudinal scattering, only the longitudinal energy contributes to conductance. What is this conductance? From the derivation, clearly it is an input conductance, which Datta (1995) referred to as contact conductance. What happens to the output impedance? The setup of the problem implicitly assumes the output end is terminated by its own characteristic impedance. What happens to the sending end? Since all transverse modes forming the allowed modes entering the EQW are assumed to be uncoupled, the reflection coefficient of each mode is zero for the planar boundary conditions. More on this point will be presented later.

From Heisenberg's uncertainty principle, the manifestation of the wave nature of an electron for G is simple starting from $I = e/\Delta t$, with $\Delta t \ \Delta E \geq \hbar$, and $\Delta E = eV$, $I = e^2 V/\hbar$ giving rise to $G = e^2/\hbar$, which is a factor of 2π greater than G_0 per spin. Nevertheless, experimental results from Van Wees et al. (1988), as well as the results for Si quantum dots (QDs) presented in Chapter 7, clearly give $G_0 \sim 40 \ \mu S$ per spin, confirming $G_0 = e^2/h$, rather than $G_0 = e^2/\hbar$. I think the difference between the two expressions is due to the fact that Heisenberg's relation, $\Delta t \ \Delta E \geq \hbar$ in most quantum mechanics books, refers to a minimum packet.

The quantized conductance steps in units of $2e^2/h$ for the transport of electrons in constricted geometry were first pointed out by Landauer (1957, 1970), and often mistaken for ballistic behavior or even some mechanism of energy loss in these systems. However, these conductance steps are entirely a wave phenomenon: the dependence of the longitudinal component of the wave vector k_z on the potential without dissipation. In EQW, the above derivation shows that, unlike photons, electrons from a contact on a Fermi surface are involved with electrons in all directions. For example, a cone of electrons at a solid angle $\Delta\Omega$ emerging from a spherical Fermi surface involves electrons both in a given direction such as the z-direction as well as in the transverse directions. In general, multimode operation dominates over a single mode operation. With the application of a potential V, many transverse modes are included, resulting in propagation that has a sum of modes given by the expression for G. *It should be apparent that this sum of modes is totally controllable by the potential V, thus providing a very useful electronic device, not only as a filter, but also for selectable tunneling.*

As pointed out in Figure 7.4, when the applied voltage sweeps the quantum state into the forbidden gap of the n-doped contact, the disappearance of the tunneling current leads to negative differential conductance (NDC). However, with a metal contact, only one step appears. There is an important point. Going back to Figure 7.4B, the current after V_1 is shown as a constant slope that needs more explanation, but it was not explained in Chapter 7. This is because of the role of the electron wave impedance in terms of the conductance per channel that is presented in this chapter. Before we go into detail, I want to briefly touch on this point. In a quantum wire or EQW, the 2D-DOS is a constant, being $m^*/\pi\hbar^2$. Thus each channel contributes to a constant factor, the conductance $G_0 = e^2/h$. Now, for a QD, the state is denoted by three quantum numbers, therefore each channel is enumerated by three quantum numbers. Whenever the applied voltage is swept through a state, the conductance is increased by G_0. In addition to this conductance, Figure 7.4B also shows conductance peaks originating from the sharp vertical rise of the current. Experimentally, this peak has a linewidth very close to $k_B T$ as shown in Figure 7.7.

11.3   Wave Impedance of Electrons

Several years ago, shortly after I published my version of the electron waveguide (Tsu, 2003), I asked Datta whether he liked controversial topics. If he did, I suggested that he should join with me in introducing the concept of the electron wave impedance in the same way as the wave impedance of photons, or for that matter any wave, including classical systems like sound waves and wave propagation over a one-dimensional string.

There is a conceptual problem as to why this conductance has no imaginary part and yet represents a lossless case. As in any wave phenomenon such as in electromagnetics, the wave impedance in free space, $\eta = (\mu/\varepsilon)^{1/2}$, has a value of 376.6 Ω. In reality, the concept of wave impedance manifests itself in the transport of energy either to a load with reflection and dissipative losses, or simply flowing on. It is entirely a wave phenomenon. Although electrons are not pure waves, with entanglement due to Pauli's exclusion principle, the differential equation, at least represented by the Schrödinger equation, is a typical wave equation and as such should have similar wave impedance. As long as we are dealing with a lossless case, the conductance is real. Reflection and transmission, which are the consequence of boundary conditions in sections where the differential equation is piecewise analytical, are handled through the use of impedance matching at these boundaries, an identical manner to the role of impedances in circuit theory.

11.3.1   Wave Impedance in a Solid with a Plane Wave in One Direction

Following Datta and Tsu (2003), the wave impedance is defined *via* a plane wave normalized in a volume AL, with A transverse to the direction z, and length L

sufficiently large for normalization. However, in solids, L must be less than the mean free path, otherwise waves have no meaning. Then

$$E = \int_0^L dz \int_0^b dy \int_0^a dx\psi^*H\psi = \frac{\hbar^2 k_0^2}{2m}, \tag{11.10}$$

and

$$I = e \int_0^b dy \int_0^a dx \frac{\hbar}{i2m}(\psi^*\psi_z - \psi\psi_z^*) = e\frac{\hbar k_z}{mL}, \tag{11.11}$$

where H is the kinetic energy operator. To get an expression for impedance in units of ohms, we need to divide E in Eq. (11.10) by the charge to obtain the potential V. The impedance $Z = V/I$, so that

$$Z = \frac{\hbar}{2e^2}\frac{k_0^2 L}{k_z}, \tag{11.12}$$

and in terms of $k_z = \sqrt{k_0^2 - k_c^2}$, as in Section 11.2, Eq. (11.12) may be recast into

$$Z = \frac{\hbar}{2e^2}\frac{k_0 L}{\sqrt{1 - k_c^2/k_0^2}}, \tag{11.13}$$

where $k_c^2 \equiv (m\pi/a)^2 + (n\pi/b)^2$, which looks similar to but is not quite the same as the waveguide case for photons. Furthermore, along the direction of propagation, unlike the sine dependences in the x and y directions, the wave function dependence on z is $\exp(ik_z z)$, a propagating function, therefore, periodic boundary conditions must be applied, i.e., $k_z L = 2\ell\pi$, with ℓ being any integer. Then Eq. (11.12) becomes

$$Z = Z_0\ell[1 - k_c^2/k_0^2]^{-1}, \tag{11.14}$$

where

$$Z_0 = h/2e^2. \tag{11.15}$$

The factor $[1 - k_c^2/k_0^2]^{-1}$ leads to a different expression from G for the EQW in the last chapter. However, in deriving the conductance of the EQW, we allow electrons with transverse energy to enter the waveguide, although only the longitudinal energy contributes to the conductance. Here, an incident electron that has a transverse degree of freedom applies to an electron incident at an angle to the impedance along the z-axis. Since our formulation does not allow the transverse energy to be channeled into the z-direction for comparison with the derivation for EQW, we should not have included the transverse energy in the present derivation. In other words, we should have taken an electron that has energy only in the

z-direction. Then $k_z = k_0$, and the factor $[1 - k_c^2/k_0^2]^{-1}$ does not appear. Then $Z_0 = h/2e^2$ and $Z_0^{-1} = 2G_0$, which is a factor of 2 larger than the derivation for EQW. Eq. (11.9) applies to contact conductance, as elaborated by Datta (1995), while $Z_0^{-1} = 2G_0$, applies to the wave conductance of the electrons. This is because in the derivation of G_0, a contact represented by a tunneling barrier is present, which is quite different from wave impedance even without an applied potential. There is really no reason that the two expressions should be the same, because whenever contacts are present, we are talking about a closed system, while without contacts, an open system is considered, which has exactly the same ratio for a capacitor with contact and without contact and whose stored electrostatic energy is CV^2 and $0.5\ CV^2$, respectively. We shall delve into this point in detail later to be sure that the factor of 2 increase in the conductance obtained from the wave impedance compared to G_0 is not a trivial mistake in algebra.

11.3.2   Quantum Wave Impedance of Open and Closed Systems

For a section of the EQW with cross-sectional area $A = a \times b$ and length L, the energy density

$$\varepsilon = (\hbar^2 k_0^2/2m)/abL, \tag{11.16}$$

power per unit area

$$P/ab = \varepsilon(\hbar^2 k_0^2/2m)(\hbar k_z/m)/L, \tag{11.17}$$

and current

$$i = e(\hbar k_z/m)/L. \tag{11.18}$$

With $P = i^2 Z$, Eqs. (11.16)–(11.18) gives

$$Z = \left(\frac{\hbar}{2e^2}\right) \frac{k_0^2 L}{k_z} = Z_0 \ell/(1 - k_c^2/k_0^2), \tag{11.19}$$

where the periodic boundary condition for a traveling wave, $k_z L = 2\pi\ell$, is used to arrive at the right-hand side of Eq. (11.19). Note that Eq. (11.19) is identical to Eqs. (11.12) and (11.14). The inverse of Z_0, $G_{QW} \equiv Z_0^{-1} = 2G_0$. At this point I was certain that a satisfactory explanation must be found for the factor of 2 difference. To start, we shall go back to the case of our definition of quantum capacitance C_Q or C_{eff}. The main physics involves the difference between an open system compared with a closed system. A closed system consists of two contacts such as the resonant tunneling diode (RTD) and an open system consists of a QD or quantum well not connected to leads.

The capacitance of a closed system involves a capacitor with two leads or a quantum well connected to two reservoirs as the double-barrier resonant tunnel diode (DBRTD). Like in Chapter 8, the difference between two-electron occupation of a QD E_2 and 1-electron occupation E_1 defines the capacitance, i.e.,

$$C_c = q^2/2(E_2 - E_1). \tag{11.20}$$

Because $(E_2 - E_1) = q\Delta V/2$, as for a symmetrical DBRTD, $C_c = q/\Delta V$. However, for an open system, we still have $C_0 = q/\Delta V$, but $(E_2 - E_1) = q\Delta V$, so that

$$C_0 = q^2/(E_2 - E_1). \tag{11.21}$$

Therefore

$$C_0 = 2C_c. \tag{11.22}$$

Here is the factor of 2 for the capacitance: the open system is twice the value for the closed system using the symmetric model. In fact, I knew even if the DBRTD were not symmetrical, the conductance is still represented by the closed system, with a proper treatment of the potential, although in a computer calculation, the potential function varies from barrier to barrier anyway. Only when a simple model is used to estimate resonant tunneling, the potential seen by the quantum well is half the applied voltage across the two contacts.

The EQW without an applied voltage is an open system with G_{QW} as derived resulting in Eq. (11.14). However, in the derivation of G_0, a voltage is applied at the contact, representing a closed system, and thus

$$G_{QW} = 2G_0. \tag{11.23}$$

In attributing the conductance G_0 to a contact conductance and including discussion of the need to smear out the voltage at the interface to a screening length (Datta, 1995), we take a step forward from other hypotheses. The explanation in terms of the open system compared with a closed system circumvents the question of where the voltage is dropped. Rather, the voltage is dropped between two contacts as in any semiconductor devices. No one is concerned with the detail of the screening going on inside the contact. As I have shown, the appearance of G_0 is because of the wave nature. Contact certainly plays an important role in bringing a voltage to the EQW, however, the existence of impedance is purely a wave phenomenon. Throughout this section, EQW is used. There are other nomenclatures, such as quantum wire, Qwire or nanowire, for the same system, although strictly speaking, there are differences. An electromagnetic wire guides the energy mostly outside of the wire. On the other hand, in microwave waveguides, energy is mostly inside the structure. In optical fibers, energy is guided along the fiber optics both inside and outside the fiber. Electron waveguides, by virtue of the carrier of energy being electrons, the induced charge travels in the energy barriers confining the electrons allows energy to flow outside in addition to the energy inside the structures. Another difference between the microwave waveguides and EQW is the fact that in photonic guides, the lowest-order mode capable of propagating is the TE_{01} mode, which shows variation in the fields only along one of the two transverse directions. In EQW, as in the energy states of QD (Chapter 7), the lowest mode is (1,1,1). Not only is there always a cutoff for propagation determined by the conditions in Eq. (11.7), but the lowest mode already consists of charge variation in all three directions.

The potential energy term in the derivation of the conductance for EQW is taken as eV. In all the resonant tunneling calculations I came across, even without symmetry, the potential seen by the structure is $eV/2$, and then $G = \partial I/\partial V$, and has this factor of 2 increase. What is going on? Placing two contacts, the voltage drop across the structure is V and the voltage seen by the structure is $V/2$. The voltage V is used for the open system, while $V/2$ is used for the closed system. Thus, we have clearly established that for an open system the conductance is twice G_0 per spin. Datta (1995) is correct to call it a contact conductance, but he should have gone farther by calling it an input conductance.

There is another point that needs clarification. This conductance G_0 as a sum of unit step functions is because the incident electrons from a Fermi sphere have a transverse degree of freedom, i.e., the excitation consists of transverse modes in parallel, leading to adding the contributions forming a ladder for G_0. For the derivation of wave impedance, apart from the factor of 2 which we have attributed to an open system, the wave impedance with propagation along the z-axis only is $Z = Z_0\ell$, thus there is still this factor ℓ that needs further consideration. The question lies in the length we take for the normalization along the direction of propagation of the wave. The length L must be less than the coherence length Λ, otherwise the wave picture loses its meaning. In a solid, we can set $L = na_0$, with a_0 being the size of the unit cell, then $k_0 = 2\pi\ell/na_0 = 2\pi p/a_0$. If we take a_0 for the normalization, then $Z = Z_0\, p$. At high energy and high k_0, the impedance Z goes up with p. There is an intrinsic difference between electron waves and photons, where Z is a constant. Why then is this the basis for comparing Z with G? The lowest allowed Z and G appear to be from the same origin, which forms the basis of my statement that the contact conductance is only the excitation of the quantum structure consistent with the wave conductance of electron, and each transverse degree of freedom is perfectly matched to the wave conductance of electron in a quantum structure, whether a QD or a Qwire.

To summarize, uniqueness is established by taking $p = 1$, the lowest allowed state, thereby fixing the length L for normalization. The wave impedance is the current due to wave propagation for a given kinetic energy. (I have explained in detail that the so-called contact conductance is really an input wave conductance in a closed system; however, for historical reasons, I shall continue to use the term contact conductance or universal conductance here.) As in Eq. (11.2), conductance, involving a sum of the transverse degree of freedom increases in steps, but in wave impedance, adding inverse impedances, the conductance increase is not in equal steps. In fact, it has another entirely different origin. Referring to research conducted by Nicollian and myself for tunneling *via* quantum states of silicon QDs several nanometers in size, the conductance jumps are not equally spaced for the following reasons. There are degeneracies in the quantum states (as discussed in

Chapter 7; see also Van Houten and Beenakker, 1996). Filling 1s-like state takes two electrons, but the p-states are not threefold degenerate because of the nonisotropic effective mass as well as various valleys as in the $\langle 111 \rangle$, $\langle 110 \rangle$, and $\langle 100 \rangle$ directions. The real picture is much more complicated. But why do we keep seeing "fairly" regular steps in published measurements? These measurements come from the charging of the capacitor caused by some arrangements that allow tunneling to take place. They represent the discreteness of electronic charge together with a region of constriction being expressible by a constant capacitance. These data do not appear in Si QDs. I am certain of this because I wasted or enjoyed 5 years of my life trying to sort out the complicated $I-V$ in tunneling via Si QDs.

What happens for a very small energy with very small k_0 such that the length of normalization exceeds the mean free path of coherence length Λ? The concept of Z is only definable for an energy greater than this minimum k fixed by L as the greatest length allowed for the definition of Z. Only the ground state has a wave impedance of Z_0. Suppose another solid has an electron coherence length twice that of the other, then the lowest ground state has Z_0, but the energy of this lowest state is 4 times lower. The wave impedance loses its meaning altogether. In fact even for contact conductance to be meaningful, the de Broglie wavelength must be shorter than Λ. We see that the wave impedance $Z \approx \hbar k L / 2e^2$ before we put in the periodic boundary condition for k. In photons, the energy momentum involves a linear relationship, but for electrons, the energy momentum involves a square relationship, resulting in $Z \propto k$. The consequence is that as energy increases, Z also increases.

The lack of uniqueness in Z is more troublesome. What we need to do is to pick a lowest energy, so that k_0 is fixed by this energy. This in turn defines L for $\ell = 1$. In other words $L = 2\pi/k_0$. For this lowest energy, $Z = Z_0$. At all energies greater than this lowest energy, Z is increased, given by $\ell = 2$, 3, etc. This procedure of normalizing $Z = Z_0$ for the lowest energy requires further thought. In reality, because of Coulomb interaction for electrons, the electron wave has a finite coherent length; therefore, limiting L is same as limiting this lowest energy.

Now we want to discuss the meaning of this wave impedance. Suppose this wave is incident onto a region described by the differential equation with piecewise analyticity, then the boundary conditions lead to the reflection coefficient

$$\Gamma = \frac{|k - k'|}{|k + k'|} = \frac{|Z_0 - Z_0'|}{|Z_0 + Z_0'|}.$$

What we stress here is that once the impedances are calculated or expressed in terms of simple parameters, as in circuit theory, there is no need to solve the wave equation every time a new situation arises. This is just the beginning of our goal of developing circuit theory applicable to the designs and analyses of nanoelectronic systems. In our view, we are not far away from developing a general circuit theory based on all that is used in microwave circuits, eventually including dissipation and complicated geometrical shapes.

11.3.3   Wave Impedance in Unbounded Space

In an arbitrary direction of propagation and using periodic boundary conditions for all three directions, the wave impedance in an unbounded free space is given by

$$Z_{\ell,m,n} = Z_0 \Xi_{\ell,m,n},\tag{11.24}$$

where

$$\Xi_{\ell,m,n} = \left[\frac{\frac{\ell^2}{L^2} + \frac{m^2}{a^2} + \frac{n^2}{b^2}}{\frac{\ell}{L^2} + \frac{m}{a^2} + \frac{n}{b^2}}\right].\tag{11.25}$$

Even with $L = a = b$, the function $\Xi_{\ell,m,n}$ listed as follows consists of fractions except in the one-dimensional case, reminding the fractional quantum numbers in the fractional Hall effects (Chakraborty and Pietilainen, 1988). Note that there are degeneracies in the three-dimensional case. Suppose there is only one electron traversing the space, one can always pick, in this case, one of the axes of the cube to align with the direction of propagation, so that the wave impedance will be given by the fourth column in Table 11.1 marked 1D. Now, a second electron is propagating in a direction not collinear with the first. Since we cannot align the coordinates with both,

Table 11.1 Quantum Number Dependence of $\Xi_{\ell,m,n}$ in 1, 2, and 3 Dimensions

ℓ	m	n	1D	2D	3D
1	1	1	1	1	1
2	1	1	–	5/3	3/2
2	2	1	–	–	9/5
2	2	2	2	2	2
3	1	1	–	5/2	11/5
3	2	1	–	–	7/3
3	2	2	–	13/5	17/7
3	3	1	–	–	19/7
3	3	2	–	–	11/4
3	3	3	3	3	3
4	1	1	–	17/5	3
4	2	1	–	–	3
4	2	2	–	10/3	3
4	3	1	–	–	13/4
4	3	2	–	–	29/9
4	3	3	–	25/7	17/5
4	4	1	–	–	11/3
4	4	2	–	–	18/5
4	4	3	–	–	41/11
4	4	4	4	4	4

complicated impedances will appear at the detector. Therefore in principle these fractional terms will play a role. Moreover, our derivation is for noninteracting electrons. If there is more than one interacting electron, the subject becomes complicated and controversial—discussed in the next section.

11.3.4 Some Fundamental Issues in Quantum Systems

Because photons are bosons, lasers and microwave sources like the magnetron involve excited states with many photons occupying the same state or nearly the same state. For fermions, such as electrons, each state can only be occupied by one with a particular set of quantum numbers. Simply put there will be one electron per state. Then how do we make an amplifier? The usual explanation is that e—e interactions split the state into a band. The totality of responses by all the electrons in a band constitutes amplification. Then the spectrum will be sufficiently broad.

For many years I have thought about how a many-electron system wipes out the coherent effects of a wave behaving almost classically. Semiconductor oscillators such as gunn and avalanche diodes certainly fall into this category. RTDs like Esaki tunnel diodes are NDC devices that belong to a class where the transmitted electrons maintain their phase relationship with the incident electrons. However, the transmitted electrons lose their phase coherence after cascading down to the Fermi level of the collector, as in the emitter, retaining no phase coherency. In short, the signal from thousands or even millions of electrons behaves classically. In this respect, the situation is quite similar to my example of the policeman detecting the speeding motorist and recording a traffic violation. During detection a phase-sensitive scheme is involved, not unlike the tunneling process involving constructive and destructive interference during the tunneling phase, and forgetting these during the collecting phase. In QD devices, things are quite different.

First, unlike RTDs with all the transverse degrees of freedom, a given longitudinal discrete state can accommodate very large numbers of electrons with different transverse energies. The transverse energy, together with their longitudinal energy, "reassembles" the nearly spherical Fermi energy surface in the collector! Without the planar interface, the transverse and longitudinal energies do not separately satisfy the boundary conditions. Let us be reminded that in RTDs, the longitudinal energy and the transverse momentum are conserved. However, in tunneling through a QD, the total energy and momentum are conserved separately. Therefore, in a small QD, strictly speaking only one electron per state is allowed. Suppose there are 10 electrons, as presented in Chapter 8, they are coupled by a number of processes: *via* their induced charges due to the differences in dielectric constants from the matrix; direct Coulomb and exchange interactions; *via* phonons, vibronic states, and even defect states; above all, coupling *via* geometrical shapes, because geometrical boundaries do mix up the state functions—as, for example, the eigenstates of spherical harmonics will form new linear combinations on the surface of a cube! In other words, there are plenty of strong couplings of the 10 electrons that result in a band, in a practically identical way to the cases described for RTDs.

The end results are similar, but the dynamics leading to the collector are quite different in detail. For example, with a handful of electrons, how do we assign occupation and how do we apply the equilibrium distribution for a few electrons? If we do, we are assuming, for example, that strong interaction with phonons justifies the use of the equilibrium distribution function. In fact this is not a bad assumption because, as mentioned previously, a QD may be isolated in terms of potential energy barriers, but barely isolated in terms of phonons. Simply because elastic constants are not all that different between the QD and the matrix—unless we are talking about a sol—gel as a matrix—then of course we cannot talk about conduction, although we can talk about the optical properties of the QD. Although I have not championed magnetic devices yet, a magnetic QD with a handful of atoms much smaller in extent than the magnetic domain is indeed a very interesting subject. In this regard, quantum Hall effects, Arharonov—Bohm effects, and magnetic superlattices indeed represent good physics; however, I still have doubts that any of these will ever become mainstream devices. On the other hand, I can see that devices like charge couple devices certainly would acquire new dimension with QDs. We can state with certainty that the overall knowledge and techniques needed to analyze and to engineer QD devices are becoming increasingly complex, which remind me of something attributed to Wigner. Each generation needs to rediscover the accumulated knowledge and acquire new working skills.

11.3.5   Electron–Photon Coupled Systems

Circuit representations are very useful in analyzing complicated I/O systems. The relationship between input and output should be expressed in terms of the operator $\mathscr{H}$, generally an integration involving Green's Function,

$$O(t') = \mathscr{H}(t, t')\, I(t), \tag{11.26}$$

Or expressed in transformed space, as

$$O(\omega') = \mathscr{H}(\omega, \omega')\, I(\omega'). \tag{11.27}$$

If spatial variation is included, we may generalize this relationship in terms of

$$O(t', \mathbf{r}') = \mathscr{H}(t, t'; \mathbf{r}, \mathbf{r}')\, I(t, \mathbf{r}), \tag{11.28}$$

and

$$O(\omega', \mathbf{q}') = \mathscr{H}(\omega, \omega'; \mathbf{q}, \mathbf{q}')\, I(\omega, \mathbf{q}). \tag{11.29}$$

We can further generalize it into multiple inputs and outputs with matrices

$$O_n(\omega', \mathbf{q}') = \mathscr{H}_{nm}(\omega, \omega'; \mathbf{q}, \mathbf{q}')\, I_m(\omega, \mathbf{q}). \tag{11.30}$$

This is how it is done with generalized circuits. We should be reminded the comment after Eq. (11.30), about the need to come up with some averaging procedure when transverse variation is present at a given longitudinal position discussed by Marcuvitz in the case of circuit parameters for waveguides. Let us discuss something more complex, generalization to coupled systems, e.g., a particle with both electrical charge and mass, or ultimately to a coupled system of electrons and photons. For example, what would we find if we replaced the momentum operator P with $\left(P - \frac{e}{-c}A\right)$ in the Hamiltonion operator when both electron and photon are present, and proceeded to derive an expression for the wave impedance as we did in Section 11.31s. There should not be any real difficulties in computing the expectation value of the Hamiltonian and the current as in Eqs. (11.10) and (11.11) to arrive at an expression for the generalized impedance for the coupled system, even with both electric and magnetic fields including static parts, for example, with an applied dc potential as well as a dc magnetic field. The question is what would be the meaning of such an impedance. My view is that we probe as if the system is uncoupled. For example, we measure the weight of a charged particle created by another one. There are interactions from both masses and charges. Because the electrical force is so much more than the gravitational force, we ignore the gravitational force while measuring the electrical force. Certainly if the coupling is significant, the value of the impedance used for designing a matching circuit is certainly affected by the coupling! In closing, what we discussed here is certainly more than a concept, something should be taken into account. In short, what Marcuvitz discussed was certainly important to the RADAR system. Therefore, I suggest that the definition of generalized impedance for coupled waves and systems in general should be considered. When that is done, I predict that even the quantum Hall effect would look quite different.

11.4 Summary

In the electromagnetic case, the wave nature leads to a wave impedance of free space that is purely real. The wave impedance of waveguides is different, depending on the propagation constants. The characteristic impedance of a waveguide contains further geometrical factors, as discussed by Marcuvitz. These three are all different, yet share the same origin, the wave nature of photons. We have derived the wave impedances for electrons in free space, in a quantum wire or in EQW, for both open and closed systems. The prefactor for the conductance per spin $G_0 = e^2/h$ for a closed system and is double this value, $2G_0$ for an open system. Similar to electromagnetic waves, the wave impedance or wave conductance for various cases are different in detail, although all of them share the same origin, the wave nature of the electron. We have clearly identified the so-called universal conductance as the input conductance from a contact to a structure, whether a section of a Qwire, a QW or a QD. What led Landauer (1970) to assume that the contact is reflectionless? Generally contacts are not reflectionless, but the effects of reflection, as in the case of resonant tunneling, are accounted for by the transmission term in addition to the "prefactor" G_0. If the transmission is very small, reflection is very large, so that the input impedance will be very large and the conductance will be very small. All this fits into the description of input impedance. However, the question is why experimental data give unity for the transmission. At low temperatures, different modes from different transverse degree of freedoms are truly independent. As soon as mixing of the longitudinal and transverse modes is present, longitudinal and transverse momenta are mixed, and these equal steps of conductance are smeared. But why, in the case of Si QD that Nicollian and I worked on, are conductance jumps clearly in units of G_0 even at room temperatures? I think the answer lies in the fact that, for a size of a few nanometers, the quantized energies are so far apart that they are almost unaffected by phonons, a primary contributor to mixing of modes. And last, I discussed the need to consider a coupled system of photons and electrons, and general coupled waves.

References

Chakraborty, T., Pietilainen, P., 1988. The Fractional Hall Effects, Springer, Berlin.

Datta, S., 1995. Electronic Transport in Mesoscopic Systems, Cambridge University Press, Cambridge.

Datta, T., Tsu, R., 2003. QWI_LANE2. 19. Nov. 2003, <http://arXiv.org/cond-mat/0311479>.

Landauer, R., 1957. IBM J. Res. Dev. 1, 223.

Landauer, R., 1970. Phil. Mag. 21, 863.

Marcuvitz, N., 1951. Waveguide Handbook, Rad. Lab. Series, vol. 10. McGraw-Hill, New York.

Mitin, V.V., Kochelap, V.A., Stroscio, M.A., 1999. Quantum Heterostructures, Cambridge University Press, Cambridge.

Reich, H.J., Ordung, P.F., Krauss, H.L., Skalnik, J.G., 1953. Microwave Theory and Techniques, D. van Nostrand Co., Inc., New York.

Schelkunoff, S.A., 1938. Bell Syst. Tech. J. 17, 17.

Stratton, J., 1948. Electromagnetics, McGraw Hill, New York,

Tsu, R., Datta, T., 2002. Proceedings of ICPS, Edinburgh, UK IOP Bristol, 2003, in CD form.

Tsu, R., Esaki, L., 1973. Appl. Phys. Lett. 22, 562.

Tsu, R., 2003. In: Dutta, M., Stroscio, M. (Eds.), Advanced Semiconductor Heterostructures. World Scientific, Singapore, p. 221.

Van Houten, H., Beenakker, C., 1996. Phys. Today 22 July.

Van Wees, B.J., van Houton, H., Beenakker, C.W.J., Williamson, J.G., Kouwenhoven, L.P., van der Marel, D., et al., 1988. Phys. Rev. Lett. 60, 848.

12 Why Super and Why Nano?

Before I discuss the subject of superlattice on the 40th anniversary of its inception in this new edition, I want to recall the diagram depicted in Figure 12.1 which I organized 16 years ago (Tsu, 1988), based on a report by Dickinson (1970) that I received when I was at IBM, which in turn was based on an unpublished work by J.C. Slater at the Lincoln Laboratory. The concept of covalent, metallic, and ionic radii has played such an important role in my career that I cannot part with it in this 2nd edition.

12.1 Finite Solid, Giant Molecule, and Composite

Some of us think in terms of discreteness and continuum. Continuum has no beginning and end, no inside and outside, no boundaries, and no surfaces. Thus, a giant molecule or a polymer is essentially a solid with a finite size. Let us differentiate between several systems:

1. Finite solids terminate in a surface with every portion of the interior region identical accept within a transition region from the interior to the exterior.
2. A giant molecule has a transition region that embodies the entire structure.
3. A composite consists of various entities. If the entire entity is a repetition of a single unit, then the composite is nothing but a superlattice.

However, we need to distinguish between two cases:

(a) A crystalline solid has all the individual parts phase-related to one another. Obviously there are degrees of phase coherence such as within the first nearest neighbor, or extending to the third nearest neighbor, etc.
(b) A random phase defines an amorphous substance as in liquids.

12.2 Generalization of Superlattices into Components

What has been discussed in reality has been a part of research in solid-state physics and material science, as well as polymer science and even composite materials in general. However, superlattices consist of repetition of a single layer of semiconductor such as GaAs walled in by a double-barrier such as GaAlAs on each side forming a single unit called double-barrier quantum well, DBQW. The repetition of such unit forms a superlattice having new electronic and optical properties for

Superlattice to Nanoelectronics. DOI: 10.1016/B978-0-08-096813-1.00012-6

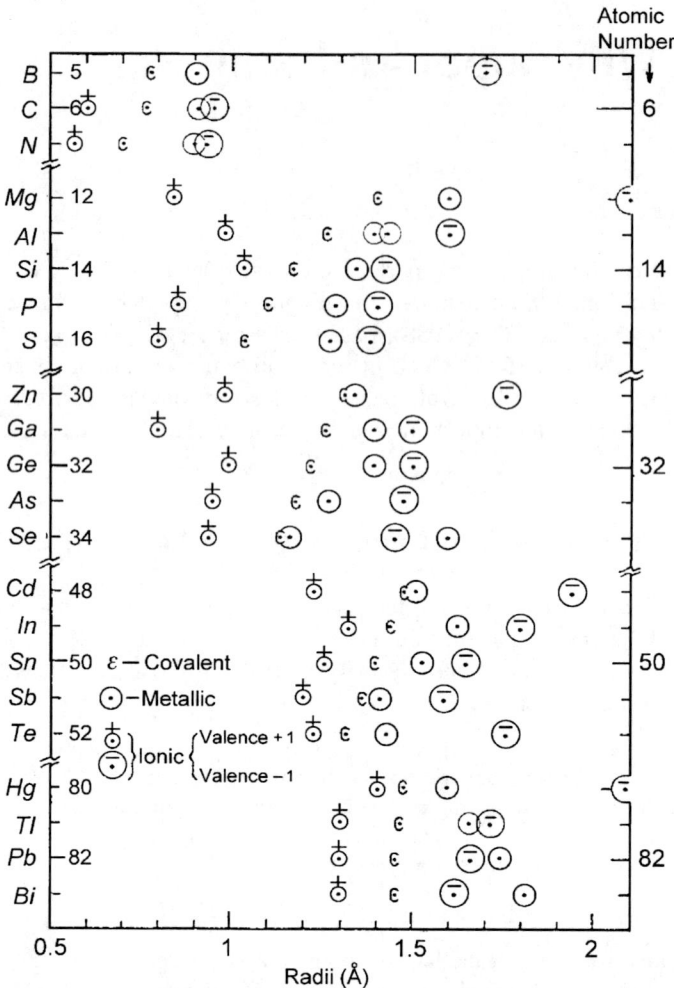

Figure 12.1 Covalent, metallic, and ionic radii of selected elements taken from Dickinson 1970 and organized and interpolated by Tsu (1988).

device applications. Each individual unit may be further arranged into something more than a simple periodic structure. Before we touch on a variety of ways for the arrangement, let us first discuss some variations of the single unit.

(i) Single unit is basically one-dimensional layered structure with its repetition similar to the Fabry–Perrot interferometer, or simply parallel plate resonator. In such case, all we need is phase coherence length exceeding the period of the layers. Optically it is not difficult to have the phase coherence length in excess of the basic period. However, electronically, the size and periods must be shorter than the mean free path, for most materials at room temperature, approximately a few nanometers in length. That require-ment limits the size of the individual entity, the size of a single unit, a single period, or the size of the quantum dot, QD,

(ii) It is really not necessary, neither desirable, to arrange the entire device in terms of a single individual unit, such as a single DBQW for a superlattice in one dimension, or in general, a single QD repeated in three dimension, 3D. As discussed before, QCL consists of three parts, one for injection into an upper level, another one for downward transition to a lower level with emission of a photon, and a third one allowing the collection of the electron prepared to be launched into the next stage. Even in a simple He-Ne laser, the excited state in He is transferred to the Ne *via* collision.

(iii) An ultimate multicomponent parametric system may involve QWs or QDs as individual components designed as the pump, signal, and idler frequencies. The parametric amplifier is considered as the ultimate sophisticated optoelectronic device with low noise.

12.3   QDs as Individual Components

There is no question that DBQW can be integrated into such a system. During the past 10 years, however, so much expectation has been focused onto QDs. And yet I have not seen any real system using QDs as individual components. The reason is simple: even if individual QDs are ready for applications, integrating them or package them into a part of a device is still not possible at this stage. However, the real problem is the fact that we do not have readymade QDs to be assembled into a device. The closest attempts involve the package of CdSe QDs for protection against oxidation by ZnO coating. Even so, we are talking about packaging into photonic crystals (Liu et al., 2008), at least a factor of 10 greater in working dimensions than a truly electronic device involving the de Broglie wavelength instead of the optical wavelength.

12.4   Size Requirements

Ultimately it is the power handling requirement that determines the size. A given MOSFET only needs to drive the next one; certainly much more power is needed to drive the display. Therefore, we need to increase the size of the system to handle a higher power. Let us summarize using a very important point. Larger size does not necessarily mean that the whole active region need to be larger, i.e., nanoscale components are sufficient for the process stage, and a larger size for the interface with display. What I am emphasizing here is that we should go much beyond the concept of superlattices as a means to mimic nature in building a large entity by repetition. In fact superlattices were not introduced for such purposes; rather, it was to take advantage of the phase coherence of nanoscale materials in much the same spirit as to take advantage of the strong bonding of the molecule and by repeating, forming a large entity for practical applications. If one needs a clear example of why we need nano, we take the modern computer. Being small, individual process of logic may be made very fast. To couple to the user *via* a screen, we need integration for display.

12.5 Superlattice and the World of Nano

There is another aspect of nano somewhat different from the considerations discussed: symmetry. Are we to say that the role of symmetry induced by the interaction of two electrons in a left—right relationship and 120° symmetry with three electrons are more fundamental for those working on catalysis than those working on the bond strength between two substances studied by a mechanical engineer? The common role of nanoscience has created a need for better integration of different disciplines. I think it is time to totally re-examine the division of human endeavor into arts and science, and science and engineering, etc. Nanoscience has brought us together into one endeavor for all of us!

Now I would like to point out the topics chosen for this book. Part of the materials provides a working knowledge for the subject. In particular, being involved with something quite new, I had to acquire the working tools needed for a somewhat sufficient degree of understanding of the physics involved, the sort of materials covered in graduate level solid-state physics. However, I want to emphasize that ultimately it is the experiment with measurements that counts. In the broader sense, I do like to emphasize the old saying that if you cannot explain something in 5 min, you may not understand as you thought you did. Discussion in terms of open and closed systems is rather fundamental. The wave impedance of electrons, like photons, will play an important role some day when quantum devices become routine components. Following this, I want to mention a few things about the direction taken by nanoelectronics. It took almost 20 years before superlattices and quantum wells moved from benchtop demonstrations to actual devices. After 30 years, the best application for the superlattice is perhaps high-mobility devices with modulation doping, THz oscillators including QCL as well as detectors, particularly using type-III schemes. Porous silicon did not make the grade in view of its fragile structure. Nevertheless, control of porosity is developing into a field of its own. QDs and quantum interference devices have been demonstrated often enough for more than 15 years, and yet we have not seen the development of any real devices. With all the problems of input/output to a QD, it appears that only contacts lying entirely in a plane, used by all those benchtop demonstrations, can work. Integrated circuits are all in a plane, i.e., contacts and active elements lie in a plane. Metallic regions allow us to define a voltage; perhaps we should make QD devices from metals! After all, a small metallic dot is an "insulator." I challenge the reader to come up with totally new ways of defining what a nanodevice is, perhaps in areas connected with biotechnology. How should we judge the impact of superlattices and resonant tunneling through quantum wells? It is clear that in comparison with the field effect transistors, the impact of these quantum devices is minuscule. However, their contribution to electronic devices lies in the fact that they brought many researchers to the field of quantum phenomena in man-made structures leading to the enormous breadth of nanotechnology today. Finally, I want to tell a story. Many years ago Brian Schwartz showed me a copy of his notes on superconductivity that were to be published as a book. I told him that I had a better way of understanding superconductivity. I made an impression on him when I referred to phonons as the "devil in my world of science," but that Cooper pairs are formed in partnership with the devil. When I repeated this to a first class biologist recently, he said that oxygen is the devil in his world of science. However, millions of years of natural selection allow the formation of partnership between cells and oxygen in the energy production process of ATP (adenosine triphosphate). He asked me whether I have found my devil to hold hands within nanoelectronics. My reply is that I am still looking for it. I predict that progress in applying nanoscale physics in devices will come from those working in metal oxide silicon field effect transistor (MOSFET), step-by-step, slowly but surely inching ahead. I asked Quiyi Ye, who played an important role in 30 nm Souce–Drain MOSFET at IBM, to give me the smallest number of active electrons in a modern MOSFET. Her best estimate is ~400 electrons. The researchers continue their slow but sure progress in MOSFET. However, when the number is reduced by another factor of 10, I predict the game is altogether different, as discussed in Chapter 8. On the other hand, great leaps in nanoelectronics will come from those who have to identify a function and look for a particular system to deliver the mission, rather than those who already have the invention and are looking for applications. But most of all, success belongs to those who are willing to take a risk. Stability issues are very real in quantum nanodevices. Switching and hysteresis, apart from nonlinear effects, are all caused by strong coupling between the intended active parts with the unintended defects forming deep traps. This is particularly prominent in structures in a few nanometer regimes where the structure and defects have similar wave functions allowing them to couple strongly. Nearly 10 of my students were involved with the strange data presented in Chapter 7. Many of those results were never even written up for publication. Ten years have gone by, and now we know that these weird data may be lumped together as telegraph-like noise or more generally, $1/f$ noise. There is a message: Scientific and technical communities should have found an outlet for strange and inexplicable results, so that these students did not have to abandon their work in order to find a more acceptable path to glory.

12.6   Some New Opportunities

12.6.1   Beyond Chemistry

Unlike chemistry which mainly concerns with group symmetry in the formation of molecules, QDs, together with the boundary and shape, provide extra-symmetry relationships. I cannot help but imagine how a wide range of possibility opens up with nano-size QDs offering *new shapes* and *boundaries* to the wave functions. Remember that the bulk of the concepts such as sp^3 or dp-hybridization may not be a good basis set for general shapes and size. I am extremely interested in experiments enriching the understanding of the symmetry role in these QDs. *For example, we can use e-beam lithography to produce arrangements of dots representing various symmetries to study catalysis and nucleation in material growth.*

12.6.2   Beyond RPA

As we know that RPA (random phase approximation), introduced by Bohm and Pines as a catch phrase, with the approximation by adding square-moduli to avoid cancellations, is no more than the recognition of not being able to take into account of phase relationship in totaling an interacting system. We do that in most constitutive relationships such as dielectric function and elastic constants. We should be seriously considering the alternatives to using RPA. We know that the most powerful amplifier is the parametric amplifier, where we cannot simply add oscillator strengths. Ed Stern once told me that EXAFS is so powerful because, with a giant computer, one can account for multiple scatterings without resorting to the use of RPA. As we pursue the nanoscience with ever-increasing vigor using modern instruments such as AFM and STM having piezoelectric control of distances measured in nanometers, I think we should be seriously considering *"beyond RPA."*

12.6.3   Beyond Solid-State Physics

When we are dealing with a macroscopic entity, nature shows us how to assemble atoms and molecules into a solid using translational symmetry. The subject of applying translational symmetry constitutes a discipline commonly referred to as solid state physics. In simple terms, it is not solid-state physics without translational symmetry, but perhaps we should use the term *nanosolid*. As we know that nothing is perfect, traditionally we resort to statistics to arrive at an average such as current and flow for the description of cause–effect as voltage–current, so useful for the description as well as the design of devices. As the size shrinks to dimensions in nanometers, the defects may be down to a few even zero in such way that statistical average becomes meaningless. However, we can still use statistics in an *assemble average*, but not summing and averaging the individual scatterings! Moreover, if the size is still represented by several unit cell distances, superlattice definitely is the only definable entity. In fact, even in the very first paper (Esaki and Tsu, 1970), we pointed out that all one needs are three periods in defining a superlattice or a QD in 3D.

12.6.4   Beyond Composites

We shall go beyond electronics and optoelectronics to include the consideration of mechanical composites without glue or hooks. I envision a new kind of composite material consisting of components such as nanoscale entities dispersed in a matrix forming a composite instead of using nanorivets or glue, bonded together chemically as in superlattices, e.g., amorphous carbon as matrix, with embedded QDs of silicates. This recipe is not far from coal put together by nature.

12.7   A Word of Caution

Superlattices have already broken into organic substances. It is only time to get involved with living organisms such as chlorophyll. Basically, now we have the tool to do it. I conclude here with one thought: Survival of the fittest for biological evolution should not be impeded by "intelligent human technological advances" in nanoscience. However we must be super-vigilant to avoid possible disasters to mankind.

References

Dickinson, S.K., 1970. Airforce Cambridge Res. Lab. Report 70-0727.
Liu, Ke., Tsu, R., 2008. Microelectronic J. 38, 700–705.
Tsu, R., 1988. Mater. Res. Soc. Proc. 102, 219–222.